MW00532979

Studies in the History of Mathematics and Physical Sciences

16

Studies in the History of Mathematics and Physical Sciences

Vol. 1: O. Neugebauer
A History of Ancient Mathematical Astronomy

Vol. 2: H.H. Goldstine
A History of Numerical Analysis from the 16th through the 19th Century

Vol. 3: C.C. Heyde/E. Seneta
I.J. Bienaymé: Statistical Theory Anticipated

Vol. 4: C. Truesdell
The Tragicomical History of Thermodynamics, 1822-1854

Vol. 5: H.H. Goldstine
A History of the Calculus of Variations from the 17th through the 19th Century

Vol. 6: J. Cannon/S. Dostrovsky
The Evolution of Dynamics: Vibration Theory from 1687 to 1742

Vol. 7: J. Lützen
The Prehistory of the Theory of Distributions

Vol. 8: G.H. Moore
Zermelo's Axiom of Choice

Vol. 9: B. Chandler/W. Magnus
The History of Combinatorial Group Theory

Vol. 10: N.M. Swerdlow/O. Neugebauer
Mathematical Astronomy in Copernicus's De Revolutionibus

Vol. 11: B.R. Goldstein
The Astronomy of Levi ben Gerson (1288-1344)

Vol. 12: B.A. Rosenfeld
A History of Non-Euclidean Geometry

Vol. 13: B. Stephenson
Kepler's Physical Astronomy

Vol. 14: G. Graßhoff
The History of Ptolemy's Star Catalogue

Vol. 15: J. Lützen
Joseph Liouville 1809-1882: Master of Pure and Applied Mathematics

Vol. 16: A.I. Dale
A History of Inverse Probability: From Thomas Bayes to Karl Pearson

Andrew I. Dale

A History of Inverse Probability

From Thomas Bayes to Karl Pearson

Springer-Verlag
New York Berlin Heidelberg London Paris
Tokyo Hong Kong Barcelona Budapest

Andrew I. Dale
Department of Mathematical Statistics
University of Natal
King George V Avenue
Durban, Natal 4001
South Africa

With seven illustrations.

Library of Congress Cataloging-in-Publication Data
Dale, Andrew I.
A history of inverse probability: from Thomas Bayes to Karl Pearson / Andrew I. Dale.
p. cm. – (Studies in the history of mathematics and the physical sciences ; 16)
Includes bibliographical references and index.
ISBN 0-387-97620-5
1. Bayesian statistical decision theory – History.
2. Probabilities – history. I. Title. II. Series.
QA279.5.D35 1991
519.5′42 – dc20 91-17794

Printed on acid-free paper.

Photocomposed copy prepared from the author's LaTeX file.
Printed and bound by Edwards Brothers, Inc., Ann Arbor, MI.
Printed in the United States of America.

9 8 7 6 5 4 3 2 (corrected second printing, 1995)

ISBN 0-387-97620-5 Springer-Verlag New York Berlin Heidelberg
ISBN 3-540-97620-5 Springer-Verlag Berlin Heidelberg New York

To F. J. H.

PROLOCUTION

It will be no strange thing at all for some to dislike the matter of this work, and others to be displeased with the manner and method of it. Easily can I foresee that my account will be too long and tedious for some, while others, perhaps, may be apt to complain of its being too short and concise.

Edmund Calamy

Preface

It is thought as necessary to write a Preface before a Book, as it is judged civil, when you invite a Friend to Dinner, to proffer him a Glass of Hock beforehand for a Whet.

John Arbuthnot, from the preface to his translation of Huygens's "De Ratiociniis in Ludo Aleae".

Prompted by an awareness of the importance of Bayesian ideas in modern statistical theory and practice, I decided some years ago to undertake a study of the development and growth of such ideas. At the time it seemed appropriate to begin such an investigation with an examination of Bayes's *Essay towards solving a problem in the doctrine of chances* and Laplace's *Théorie analytique des probabilités*, and then to pass swiftly on to a brief consideration of other nineteenth century works before turning to what would be the main topic of the treatise, *videlicet* the rise of Bayesian statistics from the 1950's to the present day.

It soon became apparent, however, that the amount of Bayesian work published was such that a thorough investigation of the topic up to the 1980's would require several volumes — and also run the risk of incurring the wrath of extant authors whose writings would no doubt be misrepresented, or at least be so described. It seemed wise, therefore, to restrict the period and the subject under study in some way, and I decided to concentrate my attention on inverse probability from Thomas Bayes to Karl Pearson.

Pearson was born in 1857 and died in 1936, and in a sense a watershed in statistics was reached during his lifetime. The somewhat cavalier approach to inverse probability that one finds in many writings in the century following the publication of Bayes's *Essay* was succeeded in the fullness of time (even if destined only by Tyche) by the logical and personal approach to probability grounded on the works of Jeffreys, Johnson, Keynes, Ramsey and Wrinch in the first third of this century (and Jeffreys in fact gained his inspiration from Pearson's *Grammar of Science*). At roughly the same time Fisher was making himself a statistical force — indeed, one can perhaps view the rigorous development of Bayes's work into a statistical tool to be reckoned with as a reaction to Fisher's evolution of sampling theory. The thirties also saw the birth of the Neyman-Pearson (and later Wald) decision-theoretic school, and subsequent work of this school was later incorporated into the Bayesian set-up, to the distinct advantage of both.

One must also note the rise of the biometric school, in which Pearson of course played a considerable rôle, and which owed its growth to the appearance of Francis Galton's *Natural Inheritance* of 1889 and his work on correlation. This work also awoke Walter Frank Raphael Weldon's interest in correlation, and he in turn did much to turn Pearson's thoughts to evolution. W.S. Gosset's work c.1908 foreshadowed an attenuation in inverse probability, a tendency which was to be reversed only in the mid-twentieth century.

It would not be too great a violation of the truth to say that, after roughly the beginning of this century, inverse probability took a back seat to the biometric, Fisherian and logical schools, from which it would only rise around 1950 with the work of Good and Savage and the recognition of the relevance of de Finetti's earlier studies. Pearson, whose writings cover both inverse probability and what would today be grouped under "classical" methods, seems then to be a suitable person with whom to end this study.

Todhunter's classic *History of the Mathematical Theory of Probability* was published in 1865. For reasons as to which it would be futile to speculate here, nothing in similar vein, and of such depth, appeared for almost a century (I except books nominally on other topics but containing passages or chapters on the history of statistics or probability, anthologies of papers on this topic, and works on the history of social or political statistics and assurances) until David's little gem of 1962. Several works in similar vein followed, the sequence culminating in Stigler's *History of Statistics* of 1986 and Hald's *History of Probability and Statistics*, the latter appearing in 1990 as the writing of this book nears completion (for trying to write a preface before the actual text is complete is surely as awkward as trying to "squeeze a right-hand foot into a left-hand shoe").

Before I am carelessly castigated or maliciously maligned let me say what will *not* be found here. Firstly, there will be little biographical detail, apart from that in the first chapter on Thomas Bayes. Secondly, little will be found in the way of attempt at putting the various matters discussed in the "correct" historical and sociological context. To interpret early results from a modern perspective is at best misguided, and I lack the historian's ability, or artifice, to place myself in the period in which these results were first presented. Those interested in these aspects will find abundant satisfaction in the *Dictionary of National Biography*, the *Dictionary of Scientific Biography*, and the books by Hald and Stigler cited above. Daston's *Classical Probability in the Enlightenment* of 1988 may also be useful: like the work by Hald it appeared too late to be consulted in the writing of this text.

Our aim is more modest — and the captious critic will no doubt opine with Winston Churchill that there is much to be modest about! It is to present a record of work on inverse probability (that is, crudely speaking, the arguing from observed events to the probability of causes) over some 150 years from its generally recognized inception to the rise of its sample-theoretic and logical competitors. Since this is a record, it has been thought advisable to preserve the original notations and the languages used — at least almost everywhere. For while translations may well help the thoughtful reader, the serious scholar will need the original text to avoid being misled by the translator's inability to render precise any nuances taxing his linguistic capabilities.

We have not considered only the most important works published during the period under consideration: minor writings, sometimes by seldom cited authors, have also been examined, in order that the effect of the greater works on the wider community of scholars be also noted. It is to be hoped, though, that this consideration has not led to a book of which it can be said, as M.G. Kendall [1963] said of Todhunter's magnum opus, that "it is just about as dull as any book on probability could be."

It is not claimed that this is *the* history of inverse probability: rather, it is one man's view of the topic, a view, it is hoped, in which any peculiarities observed will be ascribed to innocent illusion rather than deliberate delusion, and in which the seeds of future research may be nurtured.

Is there not something essentially diabolical in keeping the impatient reader, even for one moment, from the joys that await him?

D. N. Brereton, introduction to Charles Dickens's "Christmas Books", British Books edition.

Acknowledgments

Many there are who have had a hand in the completing of this work, even if only in some small degree. Particular gratitude is due to the following (in random order): H.W. Johnson, of the Equitable Life Assurance Society, for providing a copy of Bayes's notebook; C. Carter, of the reference section of the Library of Congress, and W.J. Bell, Jr., of the American Philosophical Society, for their search for information on a possible American publication of Bayes's Essay; J. McLintock, of the archives of the University of Glasgow, for verification of the award of Price's D.D. by Marischal College; J. Currie, of the special collections department of the library of Edinburgh University, for her discovery of documents relating to Bayes's attendance at the College of James the Sixth; and D.V. Lindley, for his providing a copy of a hitherto unpublished note by L.J. Savage. This note is printed, by permission of I.R. Savage, as the Appendix to the present work. It has been edited by D.V. Lindley.

Many too are the librarians who have helped by providing photo-copies or microfilms of rare items. Their assistance is greatly appreciated.

Then there are the authors who generously provided copies of their papers. Without the benefit of their historical insights I would have found my task much more difficult.

Financial support during the preparation of this work was provided by the University of Natal and the Council for Scientific and Industrial Research. The particularly generous grants by the latter facilitated many overseas trips for the consultation of rare documents, and thus contributed to the accuracy of the matters reported here.

The department of Philosophy of Cambridge University (and less directly the department of Mathematical Statistics) and the department of Statistics of the University of Chicago were gracious enough to have me as a visiting scholar during two sabbaticals: access to their excellent libraries was a great incentive in pursuing this work.

I am grateful to the following for granting permission for quotation from the works mentioned: Almqvist & Wiksell, from the paper published in the

Scandinavian Journal of Statistics by A.W.F. Edwards in 1978; the American Philosophical Society, from the paper published in the *Proceedings* of that body by C.C. Gillispie (1972); Edward Arnold, from R.W. Dale's *A History of English Congregationalism* (1907); Basic Books, Inc., from M. Kac's *Enigmas of Chance: an autobiography* (1985); the Bibliothèque de l'Institut de France—Paris, from MS 875, ff. 84–99; the Bibliothèque Nationale, from the manuscript FF 22515, f 96 v/r (m.a.), ff. 94–95 (copy); the Biometrika Trustees, from the papers published in *Biometrika* by K. Pearson (1920, 1924, 1925, and 1928), W. Burnside (1924), J.B.S. Haldane (1957), G.A. Barnard (1958), E.S. Pearson (1967) and S.M. Stigler (1975); Albert Blanchard, from P. Crepel's paper published in *Sciences a l'èpoque de la revolution française*, ed. R. Rashed (1988); Cambridge University Press, from E.S. Pearson's *Karl Pearson: An Appreciation of Some Aspects of His Life and Work* (1938), and from I. Hacking's *Logic of Statistical Inference* (1965); Dover Publications, Inc., from C.C. Davis's translation of C.F. Gauss's *Theoria Motus Corporum Coelestium* (1963); Edinburgh University Library, from their manuscripts from which details of Bayes's education have been taken; Edinburgh University Press, from D.A. MacKenzie's *Statistics in Britain 1865–1930. The Social Construction of Scientific Knowledge* (1981); A.W.F. Edwards, from his paper in the *Proceedings of Conference on Foundational Questions in Statistical Inference*, ed. O. Barndorff-Nielsen et al. (1974); Encyclopædia Britannica, from F.Y. Edgeworth's article on Probability in the 11th edition; the Faculty of Actuaries, from the papers published in the *Transactions* of that body by J. Govan (1920) and E.T. Whittaker (1920); I. Hacking, from his 1971 paper published in the *British Journal for the Philosophy of Science*; Hodder & Stoughton Ltd, from (i) M. Boldrini's *Scientific Truth and Statistical Method* (1972), and (ii) K. Pearson's *The History of Statistics in the 17th & 18th Centuries* (1978); the Institute of Actuaries, from the papers published in the *Journal* of that body by W.M. Makeham (1891), E.L. Stabler (1892) and W. Perks (1947), and from T.G. Ackland & G.F. Hardy's *Graduated Exercises and Examples for the Use of Students of the Institute of Actuaries Textbook*; the Institute of Mathematical Statistics, from (i) Q.F. Stout & B. Warren's paper in the *Annals of Probability* (1984), (ii) I.J. Good's paper in *Statistical Science* (1986), (iii) L. Le Cam's paper in *Statistical Science* (1986), (iv) G. Shafer's paper in the *Annals of Statistics* (1979) and (v) D. Hinkley's paper in the *Annals of Statistics* (1979); Macmillan Publishers Inc., from *Life and Letters of James David Forbes, F.R.S.* by J.C. Shairp, P.G. Tait & A. Adams-Reilly (1873), and from J.M. Keynes's *A Treatise on Probability* (1921); Manchester University Press, from H. McLachlan's *English Education under the Test Acts: being the history of non-conformist academies, 1662–1820* (1931); *The Mathematical Gazette*, from G.J. Lidstone's 1941 paper; J.C.B. Mohr (Paul Siebeck), from the second edition of J. von Kries's *Die Principien der Wahrscheinlichkeitsrechnung* (1927); Oxford University Press, from (i) R.A. Fisher's *Statistical Methods and Scientific Inference* (1956) (re-issued by Oxford University Press in 1990), (ii) *The Dictionary of National Biography*, (iii) A.G. Matthews's *Calamy Revised. Being a revision of Edmund Calamy's Account of the ministers ejected and silenced, 1660–1662* (1934) and (iv) F.Y. Edgeworth's papers published in *Mind* in 1884 and 1920; Peter Smith Publishers Inc., from K. Pearson's *The Grammar of Science* (1969 reprint); Princeton University Press, from T.M. Porter's *The Rise of Statistical Thinking* (1986); Springer–Verlag, from my papers of 1982 and 1986, published in the *Archive for History of Exact Sciences*; the Royal Society, from R.A. Fisher's paper published in the *Philosophical Transactions* in 1922; the Royal Statistical Society, from the papers published in the *Journal* of that body by F.Y. Edgeworth (1921), J.D. Holland (1962) and S.M. Stigler (1982); Taylor & Francis, Ltd., from the papers published in *The London, Edinburgh, and Dublin Philosophical Magazine and Journal of Science* by F.Y. Edgeworth (1883, 1884) and K. Pearson (1907); John Wiley & Sons, Ltd., from B. de Finetti's *Probability, Induction and Statistics* (1972). Excerpta from Thomas Bayes's election certificate and G. Boole's "Sketch of a theory and method of probabilities founded upon the calculus of logic" are reproduced by kind permission of the President and Council of the Royal Society of London. Extracts are reprinted from "Note on a Scholium of Bayes", by F.H. Murray, *Bulletin of the American Mathematical Society*, vol. 36, number 2 (February 1930), pp. 129–132, and from "The Theory of Probability: Some Comments on Laplace's Théorie Analytique", by E.C. Molina, *Bulletin of the American Mathematical Society*, vol. 36, number 6 (June 1930), pp. 369–392, by permission of the American Mathematical Society. The material quoted from (i) W.L. Harper & C.A. Hooker's *Foundations of Probability Theory, Statistical Inference, and Statistical Theories of Science*, vol. 2 (1976), (ii) J. Hintikka, D. Gruender & E. Agazzi's *Pisa Conference Proceedings*, vol. 2 (1980) and (iii) B. Skyrms & W.L. Harper's *Causation, Chance, and Credence* (1988), is reprinted by permission of Kluwer Academic Publishers.

The quotations from the manuscripts in Dr Williams's Library, 14 Gordon Square, London, reprinted by permission of the Librarian, Mr J. Creasey, are made subject to the following declaration: "(a) that the Trustees have allowed access to the manuscript but are not responsible for the selection made, and (b) that the author, both for himself and his publisher, waives whatever copyright he may possess in the extracts made, as far as the exercise of that right might debar other scholars from using and publishing the same material and from working for that purpose on the same manuscripts."

The original will of Thomas Bayes is in the custody of the Public Record Office, Chancery Lane, London (ref. PROB 11/865).

This tribute would be incomplete without mention of my indebtedness to Linda Hauptfleisch and Jackie de Gaye, for their typing of the manuscript, and to the editorial staff of Springer-Verlag, New York, for their assistance.

A. I. DALE
Durban, Natal
January, 1991

Contents

Chapter 1

Thomas Bayes: a biographical sketch

If those whose names we rescue from oblivion could be consulted they might tell us they would prefer to remain unknown.

Matthew Whiteford.

Most authors of papers or articles devoted to biographical comments on Thomas Bayes preface their remarks with an *Apologia* for the paucity of pertinent particulars. In 1860 we find de Morgan publishing a request in *Notes and Queries* for more information on Bayes, listing, in no more than a few paragraphs, all that he knows. In 1974 Maistrov, in what was probably to that date the most complete and authoritative[1] history of probability theory since Todhunter's classic of 1865, bemoans the fact that

> biographical data concerning Bayes is scarce and often misleading... Even in the "Great Soviet Encyclopedia" (BSE) there is no mention of his birthdate and the date of his death is given incorrectly as 1763. [pp.87-88]

But no national shame need be felt by the Soviets on this account: the *Dictionary of National Biography* (ed. L. Stephen), though devoting space to Thomas's father, is stubbornly silent on the perhaps more illustrious son[2], while the *Encyclopædia Britannica* has apparently[3] no entry under "Bayes" until the fourteenth edition, post 1958, where a brief biographical note may be found. The only earlier work of general reference to contain a biographical note on Thomas Bayes, as far as has been ascertained, is J.F. Waller's edition of the *Imperial Dictionary of Universal Biography*[4] of 1865.

The information conveyed in the present work is, unfortunately, almost as exiguous: indeed, for one whose work has come to play such an important rôle in modern statistical theory and practice (and hence in modern science in general), Thomas Bayes has been singularly successful in preserving a large measure of personal (and public) privacy.

Thomas, the eldest child of Joshua and Ann Bayes, was born in 1701 or 1702 (the latter date seems generally favoured, but the present epitaph in the Bunhill Fields Burial Ground, by Moorgate, merely gives his age at death[5], in April 1761, as 59). The place of his birth is subject to similar uncertainty: the received preference seems to be for London[6], but Holland surmises that "his birthplace was in Hertfordshire" [1962, p.451]. As luck would have it, however, the parish registers of Bovingdon, Hemel Hempstead, Herts. (where Joshua is supposed to have ministered at Box Lane[7]) for 1700-1706 have apparently gone astray.

Of Thomas Bayes's early childhood little is known. While some sources[8] assert that he was "privately educated", others[9] believe he "received a liberal education for the ministry": the two views are perhaps not altogether incompatible. Some light can perhaps be shed on the question of Thomas's schooling from the existence of a Latin letter to him from John Ward, a letter dated 10. kal. Maii 1720 and distinctly schoolmasterish in its advocation of the importance of the cultivation of style in writing. John Ward (1679?-1758), the son of the dissenting minister John Ward, was, according to the *Dictionary of National Biography*, a clerk in the navy office until leaving it in 1710 to open a school in Tenter Alley, Moorfields. *The Imperial Dictionary of Universal Biography* is perhaps more careful in stating merely that Ward, in 1710, "exchanged his clerkship for the post of a schoolmaster in Tenter Alley". Ward was elected a fellow of the Royal Society on 30th November 1723 and, on his death, was interred in Bunhill Fields.

John Eames was assistant tutor in classics and science at the Fund Academy[10] in Tenter Alley, succeeding Thomas Ridgeley as theological tutor on the latter's death in 1734. It is indeed tempting to suppose that Thomas Bayes was a pupil at the school at which both Eames and Ward taught, but this is mere conjecture (see Appendix 1.2 for further discussion of this matter). In fact, Bayes's name does not appear in a still extant list of Ward's students.

What Thomas could have studied at the Fund Academy is uncertain, the Latin letter referred to above merely indicating the importance Ward attached to the classics and the mathematical sciences ("mathesi")[11]. Where he could have picked up his knowledge of probability is unknown: there is, to our mind, little evidence supporting Barnard's theory that he might have had some contact with "poor de Moivre"[12], at that time eking out a precarious existence by teaching mathematics at Slaughter's Coffee House in St Martin's Lane[13], or, according to Pearson [1978]

> sitting daily in Slaughter's Coffee House in Long Acre, at the beck and call of gamblers, who paid him a small sum for calculating odds, and of underwriters and annuity brokers who wished their values reckoned. [p.143]

There is, however, more evidence for Holland's [1962, p.453] tentative suggestion that he might, after all, have been educated further afield, as recent research has disclosed[14]. For in a catalogue of manuscripts in the Edinburgh University Library the following entry may be found:

> Edinburgi Decimo-nono Februarij Admissi sunt hi duo Juvenes praes. D. Jacobo. Gregorio Math. P. Thomas Bayes. Anglus. John Horsley. Anglus.

The year of admission is 1719. The entries in this manuscript bear the signatures of those admitted: that of Bayes is markedly similar to the one found in the records of the Royal Society.

Bayes's name also appears in the Matriculation Album of Edinburgh University under the heading

> Discipuli Domini Colini Drummond qui vigesimo-septimo die Februarii, MDCCXIX subscripserunt

and further evidence of his presence may be found in the *List of Theologues in the College of Edinburgh since October 1711* (the date is obscure), in which Thomas's entry to both the College and the profession is given as 1720. He is stated as being recommended by "Mr Bayes", presumably his father Joshua. What are possibly class lists give Thomas's name in the fifth section in both 1720 and 1721. In a further list, this time of the prescribed theological exercises to be delivered, we find Bayes mentioned twice: on 14th January 1721 he was to deliver the homily on Matthew 7, vs 24-27, and on 20th January 1722 he was to take the same rôle, the text in this case being Matthew 11, vs 29-30. Finally, he is mentioned in the list of theological students in the University of Edinburgh, from November 1709 onwards, as having been licensed, but not ordained. A full list of references to Bayes in the records of that University is given in Appendix 1.4.

It is perhaps hardly surprising that Thomas, coming as he did from a family strong in nonconformity, should have sought ordination as a nonconformist minister. When this ordination took place we do not know:

the only thing we know with some degree of certainty is that it must have been during or before 1727; for in Dr John Evans's (1767-1827) list of "Approved Ministers of the Presbyterian Denomination" for that year we find Thomas's name[15]. We suspect also that Thomas had assisted his father at Leather Lane for some years[16] from 1728 before succeeding[17] the Rev. John Archer as minister at the meeting-house, Little Mount Sion[18], in Tunbridge Wells[19]. Whiston [1749, Pt.II] describes Bayes as "a successor, tho' not immediate to Mr. *Humphrey Ditton*"[20] [p.390]. James [1867], in his second appendix, entitled "Particular account of Presbyterian chapels, and list of Baptist chapels in England, 1718-1729", has the following entry:

> Tunbridge Wells, John Archer [Presbyterian congregation extinct, chapel reopened by Independents]. [p.664]

This reopening must have occurred after the death of Bayes, who was a presbyterian.

The 1730's saw a virulent attack on Sir Isaac Newton's work on fluxions[21]. The metaphysical side of this work was attacked by Bishop Berkeley in 1734 in his *The Analyst; or, a Discourse addressed to an Infidel Mathematician*, London[22]. This prompted replies from Dr Jurin[23] and J.A. Walton, followed by further rebuttal from Berkeley in 1735[24]. A strong defence of Newton appeared in a tract[25] entitled *An Introduction to the Doctrine of Fluxions, and Defence of the Mathematicians against the Objections of the Author of the Analyst, so far as they are designed to affect their general Methods of Reasoning*, John Noon, London, 1736. In his question in *Notes and Queries*, de Morgan writes "This very acute tract is anonymous, but it was always attributed to Bayes by the contemporaries who *write in* the names of authors; as I have seen in various copies: and it bears his name in other places" [1860, p.9].

It appears, on the face of it, that this latter work was the sufficient cause[26] of Bayes's election as a Fellow of the Royal Society in 1742, for it was not until about 1743 that a resolution was taken by the Society[27] "not to receive any person as a member who had not first distinguished himself by something curious"[28]. The certificate (dated London April 8, 1742) proposing Bayes for election reads as follows[29]

> The Rev^{d}. M^{r}. Thomas Bays [sic] of Tunbridge Wells, Desiring the honour of being Elected into this Society; We propose and recommend him as a Gentleman of known merit, well skilled in Geometry and all parts of Mathematical and Philosophical Learning, and every way qualified to be a valuable member of the same.

It is signed: Stanhope James Burrow
Martin Folkes Cromwell Mortimer
John Eames.

In the *New General Biographical Dictionary* Rose writes: "He [i.e. Thomas Bayes] was distinguished for his mathematical attainments, which led to his being elected a fellow of the Royal Society" [1848]. From those of Bayes's writings that have come down to us, we can only assume, as already stated, that his fellowship came about as a result of his contribution to the Berkleian dispute[30].

While no other scientific or mathematical work published by Bayes before his election (and in the light of which the latter might prove more explicable) has come to light, a notebook[31] of his is preserved in the muniment room of the Equitable Life Assurance Society, through the careful offices of Richard Price and his nephew William Morgan[32]. Here, among other curiosities, are details of an electrifying machine, lists of English weights and measures, notes on topics in mathematics, natural philosophy and celestial mechanics, the complete key to a system of shorthand[33], and, most important for our purposes, a proof of one of the rules in the Essay, to which proof we shall return in Chapter 4.

Two further works by Thomas Bayes appeared after his death. In 1764, a "Letter from the late Reverend Mr. Thomas Bayes, F.R.S. to John Canton, M.A. & F.R.S." was published in the *Philosophical Transactions* (read 24th November 1763). This short note (a scant two pages) deals with divergent series, in particular the Stirling-de Moivre Theorem[34], viz.

$$\log x! = \log\sqrt{2\pi} + (x + \frac{1}{2})\log x - S ,$$

where

$$S = \left[x - \frac{1}{12x} + \frac{1}{360x^3} - \frac{1}{1260x^5} + \frac{1}{1680x^7} - \frac{1}{1188x^9} + \cdots\right] .$$

The same volume (LIII) of the *Philosophical Transactions* contains, as article LII, "An Essay towards solving a Problem in the Doctrine of Chances. By the late Rev. Mr. Bayes, F.R.S. communicated by Mr. Price, in a Letter to John Canton, A.M. F.R.S", and it is to this essay that we now turn our attention[35]. (This essay was followed by Bayes's (and Price's) "A Demonstration of the Second Rule in the Essay towards the Solution of a Problem in the Doctrine of Chances, published in the Philosophical Transactions, Vol. LIII". This memoir occupies pp.296-325 of Volume LIV of the *Philosophical Transactions*.)

1.1 Appendix 1.1

While almost all that is known about Thomas Bayes has been mentioned above, there are some facts about other members of his family which might be of some interest to the reader.

Thomas's paternal grandfather was Joshua Bayes, who was baptised on the 6th May 1638 and was buried on the 28th August 1703. Like his father Richard, Joshua was a cutler in Sheffield, and in 1679, like his father before him, he was Master of the Company of Cutlers of Hallamshire. In 1683-1684 he was Town Collector, and he also served a spell as Trustee for the town[36].

According to the Reverend A.B. Grosart, writing in the *Dictionary of National Biography*[37], Joshua's elder brother Samuel was "ejected by the Act of Uniformity of 1662 from a living in Derbyshire, and after 1662 lived at Manchester until his death". (This act, passed by the anti-puritan parliament after the restoration of Charles II, provided that "all ministers not episcopally ordained or refusing to conform should be deprived on St. Bartholomew's Day, the 14th of August[38] following".) It is possible that Samuel did not in fact leave his parish until 1665, when[39] "ejected ministers were forbidden to come within five miles of their former cures".

Grosart is substantially correct, apart from the fact that he refers to Samuel rather than Joshua as Thomas's grandfather, for in Turner [1911] we find the following records[40]:

> Licence to Sam: Buze to be a Pr[eacher] Teacher in his howse in Manchester

and

> Licence to Sam̃: Bayes of Sankey in Lancash̃: to be a Pr[eacher]: Teachr. Sep^{t} 5th [1672]

(Turner [1911, vol. 1, pp.518,556]), while in volume 2 [p.677] of the same work we find

> Sankey.(1) Samuel Bayes (t^{r}) (cal. iii, 35), ej. from Grendon, Northants. (2) New Meeting House (m[eeting] pl[ace]).

The most complete, and most accurate, biographical sketch of Samuel Bayes is to be found in Matthews [1934]. It runs in full as follows:

> Bayes, Samuel. Vicar of Grendon, Northants. 1660. Adm. 16 Dec. 1657. Successor paid cler. subsidy 1661. Son of Richard, of Sheffield, cutler, by 2nd wife, Alice Chapman. Bap. there 31 Jan. 1635-6. Trinity, Camb. mc. 1652: Scholar 1655: BA.

> 1656. Minister at Beauchief Abbey, Derbs. Licensed (P.), as of Sankey, Lancs., 5 Sep. 1672; also, as Buze, at his house, Manchester. Mentioned in father's will 15 March 1675-6: p.13 July 1677. Died c.1681, when Joshua Bayes, of Sheffield, was found his brother and heir. Joshua Bayes (1671-1746), minister in London, his nephew, not his son. [p.40]

Even Joshua Bayes (Thomas's father) is not immune from biographical confusion. Holland [1962] states (correctly) that "Joshua was the nephew of Samuel Bayes of Trinity College, Cambridge, ejected minister of Grendon in Northamptonshire" [p.452], a view which is supported by Rose [1848] who asserts further that Joshua was "the son of Joshua Bayes of that town [viz. Sheffield], and nephew to Samuel Bayes". Wilson writes that Samuel Bayes (father of Joshua), a native of Yorkshire and educated at Trinity College, Cambridge,

> enjoyed the living of Grendon in Northamptonshire, which he lost at the Restoration; and he seems afterwards to have had another living in Derbyshire, but was obliged to quit that also upon the passing of the Bartholomew Act, in 1662. Upon being silenced, he retired to Manchester, where he lived privately until his death. [1814, vol. 4, p.396]

On the 15th November 1686, Joshua was entrusted to the tender care of the "reverend and learned Mr." Richard Frankland[41] of Attercliffe, Yorkshire, the founder of the first academy for nonconformists[42] and one who, subjected to the buffeting of the winds of orthodox persecution, moved his academy, together with his pupils, from place to place[43].

There Joshua pursued his studies "with singular advantage"[44], and at their conclusion proceeded to London, where, on the 22nd of June 1694, he was one of the first seven candidates[45] (not *the* first, as stated by Pearson[46]) to be publicly ordained "according to the practice of the times"[47]. This ordination, the first public ceremony of such nature among dissenters in the city after the Act of Uniformity, took place at the meeting-house of Dr Annesley, Bishops-gate Within, near Little St Helens[48].

Having been ordained "preacher of the gospel and minister" [Stephen 1885], Joshua seems to have become a peripatetic preacher, serving churches around London[49] before settling down at St Thomas's Meeting-house, Southwark, as assistant[50] to John Sheffield ("one of the most original of the later puritan writers")[51] in 1706 or thereabouts. Since this calling required his attendance on Sunday mornings only, Joshua also acted as assistant to Christopher Taylor[52] of Leather Lane in Hatton Garden, London. While engaged in this two-fold assistantship, Joshua was one of a panel of presbyterian[53] divines engaged to complete Matthew Henry's (1662-1714) "Commentary on the Bible", his special charge being the Epistle to the Galatians[54].

On succeeding to Taylor's pastorate on the latter's death[55] in 1723, Joshua resigned his morning service duties at St Thomas's. Feeling the weight of advancing years, he "confined his labours chiefly to one part of the day" [Wilson 1814], being assisted on the other part firstly by John Cornish[56] (d.1727) and then by his own son Thomas[57] (appointed in 1728). When Dr Calamy died in 1732, the Merchants' lectureship at Salters' Hall[58] fell vacant, and Joshua was chosen to fill the vacancy. In a special course of lectures delivered by a company of divines at Salters' Hall in 1735, directed against Popery, Joshua expounded[59] on "The Church of Rome's Doctrine and Practice with relation to the Worship of God in an unknown tongue."

As far as can be ascertained, Joshua's only other published writings were some sermons. These are listed by Nicholson and Axon [1915] as, in addition to the above, (1) A funeral sermon occasioned by the death of Mr. J. Cornish, preached Dec.10, 1727, [1728]; (2) A funeral sermon occasioned by the death of the Rev. C. Taylor, [1723]; and (3) A sermon preach'd to the Societies for the Reformation of manners, at Salters' Hall, July 1, 1723 [1723]. There is no evidence of any mathematical or scientific discourse, and we may (must?) therefore view with some measure of suspicion the statement that he was a Fellow of the Royal Society[60]. Joshua died[61] on 24th April, 1746, (in his 76th year and the 53rd of his ministry[62]) being buried in Bunhill Fields[63], in a grave later to be shared by other members of his family.

Before taking leave of Joshua Bayes, let us see what Wilson had to say:

> Mr. Bayes was a man of good learning and abilities; a judicious, serious and exact preacher; and his composures for the pulpit exhibited marks of great labour. In his religious sentiments he was a moderate Calvinist; but possessed an enlarged charity towards those who differed from him. His temper was mild and amiable; his carriage free and unassuming; and he was much esteemed by his brethren of different denominations. Though his congregation was not large, it consisted chiefly of persons of substance[64], who contributed largely to his support, and collected a considerable sum annually for the Presbyterian fund. [1814, p.399]

Thomas was the eldest son of Joshua Bayes (1671-1746) and Ann Carpenter (1676-1733). He had six siblings:[65] Mary (1704-1780), John (1705-1743), Ann (1706-1788), Samuel (1712-1789), Rebecca (1717-1799) and Nathaniel (1722-1764). The only references to any of the children, apart from Thomas, we have managed to find are (a) the mention of John, and his father, in the list of subscribers to Ward's *Lives of the Professors of Gresham College*, and (b) the following obituary from *The Gentleman's Magazine and Historical Chronicle* for 1789:

> Oct. 11. At Clapham, Sam. Bayes, esq. formerly an eminent linen-draper in London, son of the Rev. Mr. Sam [sic] Bayes, an eminent dissenting minister. His lady died[66] a few weeks before him. [vol. 59, p.961]

In the 1730's vitilitigation arose on the following matter: God was not compelled to create the universe; why, then, did He do so? The Anglican divine Dr John Balguy (1686-1748) started the (published) debate with his pamphlet *Divine Rectitude, or a Brief Inquiry concerning the Moral Perfections of the Deity; Particularly in respect of Creation and Providence*, London, 1730. This was followed by a rebuttal[67], attributed to Thomas Bayes, entitled *Divine Benevolence, or an attempt to prove that the Principal End of the Divine Providence and Government is the Happiness of his Creatures. Being an answer to a Pamphlet entitled: "Divine Rectitude: or an Inquiry concerning the Moral Perfections of the Deity". With a Regulation of the Notions therein advanced concerning Beauty and Order, the Reason of Punishment, and the Necessity of a State of Trial antecedent to perfect Happiness*, London, printed by John Noon at the White Hart in Cheapside, near Mercers Chapel, 1731. Not satisfied with either "Rectitude" or "Benevolence" as the motive for creation, Henry Grove[68] (1684-1738) found the answer in "Wisdom", and expounded this in his tract of 1734: *Wisdom, the first Spring of Action in the Deity; a discourse in which, Among other Things, the Absurdity of God's being actuated by Natural Inclinations and of an unbounded Liberty, is shewn. The Moral attributes of God are explained. The Origin of Evil is considered. The Fundamental Duties of Natural Religion are shewn to be reasonable; and several things advanced by some late authors, relating to these subjects, are freely examined.*

The first two of the above-mentioned pamphlets were published anonymously, but there seems little doubt that the authorships have been correctly attributed[69]. Remarking on the polemic in general, Pearson [1978] writes

> On the whole Balguy and Grove may be held to have had the better of the controversy because they considered in opposition to Bayes that God may have ends in view, distinct from and sometimes interfering with the happiness of his creatures. This controversy rather shows Bayes as a man desiring a loving and paternal deity than as a good logician or a fluent writer. [p.359]

At the time, however, Bayes's tract was apparently well received[70], for we read in Walter Wilson's *The History and Antiquities of Dissenting Churches and Meeting Houses*[71] that it "attracted notice and was held in high esteem", and that, compared to those of Balguy and Grove, "Mr. Bayes's scheme was more simple and intelligible" [Wilson 1814, p.402].

The next recorded reference to Thomas Bayes that we have is due to William Whiston[72] (Newton's successor in the Lucasian Chair at Cambridge[73]), in whose *Memoirs of his Life* [1749, part II], we find the following[74]

> *Memorandum.* That on *August* the 24th this Year 1746, being *Lord's* Day, and St. *Bartholomew's* Day, I breakfasted at Mr. *Bay's* [sic], a dissenting Minister at *Tunbridge Wells*, and a successor, tho' not immediate to Mr. *Humphrey Ditton*, and like him a very good Mathematician also. [p.390]

In his authoritative biographical note to his 1958 edition of Bayes's Essay in *Biometrika*, Barnard states that "Whiston goes on to relate what he said to Bayes, but he gives no indication that Bayes made reply" [p.294]. That this is a slip is evidenced by the continuation of the preceding quotation from Whiston's *Memoirs*, viz.[75]

> I told him that I had just then come to a resolution to go out always from the public worship of the Church of England, whenever the Reader of Common Prayer read the Athanasian Creed; which I esteemed a publick cursing [of] the Christians: As I expected it might be read at the Chapel that very Day, it being one of the 13 Days in the Year, when the Rubrick appoints it to be read. Accordingly I told him that I had fully resolved to go out of the Chapel that very Day, if the Minister of the Place began to read it. He told me, that Dr. Dowding the Minister, who was then a perfect Stranger to me, had omitted it on a Christmas-Day, and so he imagined he did not use to read it. This proved to be true, so I had no Opportunity afforded me then to shew my Detestation of that Monstrous Creed; Yet have I since put in Practice that Resolution, and did so the first Time at Lincolns Inn Chapel on St. Simon and St. Jude's Day October 28, 1746, when Mr. Rawlins began to read it, and I then went out and came in again when it was over, as I always resolved to do afterwards.

In April 1746, as already mentioned, Joshua Bayes died, leaving £2,000 and his library to Thomas, with similar bequests to his other children and his siblings amounting to some £10,000 in all [76]. A little over a month after drawing up his will Joshua added a codicil in which the bequest of £1,400 to his daughter Rebecca was revoked, so that she might not be subject to the debts of her husband, Thomas Cotton. She was, however, left £40 for mourning, and the original amount was left in trust, with her brothers Thomas and Samuel as trustees, for her son, Joshua Cotton.

In 1749 Thomas Bayes became desirous of retiring from his cure, and to this end he opened his pulpit to various Independent ministers from London[77]. This arrangement was suddenly terminated on Easter Sunday in 1750, when, disliking the Independents' doctrine, Bayes resumed his pulpit[78]. (This point is reported rather differently by Barnard [1958], who states that Bayes "allowed a group of Independents to bring ministers from London to take services in his chapel week by week, except for Easter, 1750, when he refused his pulpit[79] to one of these preachers" [p.294].) There is something strange about all this; why, after the successful implementation of this system in 1749 ("All that summer of 1749 we had supplies from London, Sabbath after Sabbath; 'twas indeed a summer to be remembered")[80], did Bayes suddenly put a stop to it? We shall probably never know. However, he seems to have left his cure in about 1750 (though he remained in Tunbridge Wells until his death), his successor at Little Mount Sion being the Rev. William Johnson[81] (or Johnstone or Johnston).

On the 7th April 1761 Thomas Bayes died[82], and he was interred in the family vault[83] in Bunhill Fields. Most of Thomas's inheritance from his father was left to his (Thomas's) family and friends, including £200 to be divided between John Hoyle and Richard Price. Also named were "my Aunt Wildman...my cousin Elias Wordsworth and my cousin Samuel Wildman". A substantial bequest of "five hundred pounds and my watch made by Ellicot and all my linnen and wearing apparell and household stuff" was made to Sarah Jeffery, "daughter of John Jeffery living with her father at the corner of Jourdains lane at or near Tonbridge Wells".

1.2 Appendix 1.2

Holland [1962, p.452] has somewhat hesitantly put forward the suggestion that Thomas Bayes might have been educated at Coward's Academy[84]. The discussion in this appendix will, I trust, set this suggestion at nought.

In 1695 the Congregational Fund Board, originally supported by both Presbyterians and Independents, established an academy in Tenter Alley, Moorfields. Thomas Godwin was appointed Tutor to the Board in 1696 or 1697 (Dale [1907, p.506]), and was succeeded in the principal charge of the students by Isaac Chauncey[85] (or Chauncy), who had initially been appointed in 1699. Chauncey died in 1712, and Thomas Ridgeley[86] followed him as theological tutor, being succeeded in turn by John Eames[87] (F.R.S. 1724), who had previously "held the chair of Philosophy and Languages" (Dale [1907, p.501]). In 1744 the Fund Academy was united with the Academy of the King's Head Society, the union being represented by Homerton College until 1850.

Philip Doddridge (1702-1751) opened an academy[88] at the beginning of July 1729 at Market Harborough. In December of that year the academy was moved to Northampton, Doddridge having been called by an Independent congregation at Castle Hill. In 1733 "an ecclesiastical prosecution was commenced against Doddridge for keeping an Academy in Northampton" (Dale [1907, p.518]), a case speedily quashed by the Crown, King George II refusing to allow persecution for conscience' sake. After Doddridge's death the Academy was moved to Daventry, its deceased head being succeeded in turn by Caleb Ashworth, Thomas Robins and Thomas Belsham. The latter resigned on finding that he could not conscientiously teach the doctrines required by the Coward Trustees, who maintained the Academy and had subsidized it from 1738. The latter was moved back to Northampton, with John Horsey as theological tutor: he, being suspected of unorthodoxy, was removed in 1798 by the Trustees and the Academy was dissolved. It was restarted the next year in Wymondley, Hertfordshire, where it remained until 1832 when it was established as Coward College in Torrington Square, London. Here the theological teaching was carried out by Thomas Morell, the former Tutor of the Academy, while other subjects were taught by University College, London.

In 1778 the *Societas Evangelica* (founded 1776) established an "Academy" for the training of evangelists. In the next few years a more liberal course of education was adopted, and in 1791 the Evangelical Academy moved to Hoxton Square as the Hoxton Academy. In 1825 it was moved to Highbury Park and became Highbury College.

In 1850 the three colleges — Homerton, Coward and Highbury (or Hoxton) — were united to form New College.

William Coward, a London merchant noted for what the *Dictionary of National Biography* calls "his liberality to dissent", continued, while alive, "to assist the poorer ministers and to aid in the teaching of their children." On his death, at age 90, at Walthamstow on 28th April 1738, his property was valued at £150,000, the bulk of which was left in charity. As we have mentioned, it was Coward's Trustees who later took over Doddridge's Academy.

From the preceding discussion it seems quite clear that anything known as Coward's Academy would have been formed far too late to have been attended by Bayes. Since, however, Holland cites as evidence for Bayes's possible attendance at Coward's the fact that John Eames was one of his sponsors for election to the Royal Society on 4th November 1742, it is possible that he was in fact referring to the Fund Academy.

1.3 Appendix 1.3

There exists an anecdote concerning Bayes which is reported by Bellhouse [1988b]. The passage, from Phippen [1840], runs as follows:

> During the life of Mr. Bayes, an occurrence took place which is worthy of record. Three natives of the East Indies, persons of rank and distinction, came to England for the purpose of obtaining instruction in English literature. Amongst other places, they visited Tunbridge Wells, and were introduced to Mr. Bayes, who felt great pleasure in furnishing them with much useful and valuable information. In the course of his instructions, he endeavoured to explain to them the severity of our winters, the falls of snow, and the intensity of the frosts, which they did not appear to comprehend. To illustrate in part what he had stated, Mr. Bayes procured a piece of ice from an ice-house, and shewed them into what a solid mass water could be condensed by the frost — adding that such was the intense cold of some winters, that carriages might pass over ponds and even rivers of water thus frozen, without danger. To substantiate his assertion, he melted a piece of the ice by the fire, proving that it was only water congealed. 'No', said the eldest of them, 'It is the work of Art! – we cannot believe it to be anything else, but we will write it down, and name it when we get home'. [p.97]

It is not known who these travellers were, or when their visit took place. Similar tales are recounted in David Hume's essay *Of Miracles* and in John Locke's *Essay concerning Human Understanding.*

1.4 Appendix 1.4

The complete list (as far as has been ascertained) of references to Bayes in the archives of Edinburgh University, in no particular order, runs as follows (the references in square brackets are the shelf-marks of the university's special collections department):

1. [Da]. *Matriculation Roll of the University of Edinburgh. Arts-Law-Divinity. Vol. 1, 1623-1774. Transcribed by Dr. Alexander Morgan, 1933-1934.* Here, under the heading "Discipuli Domini Col-

ini Drummond qui vigesimo-septimo die Februarii, MDCCXIX subscripserunt", we find the signature of Thomas Bayes. This list contains the names of 48 students of Logic.

2. [Da.1.38] *Library Accounts 1697-1765.* Here, on the 27th February 1719, we find an amount of £3-0-0 standing to Bayes's name — and the same amount to John Horsley, Isaac Maddox and Skinner Smith. All of these are listed under the heading "supervenientes", i.e. "such as entered after the first year, either coming from other universities, or found upon examination qualified for being admitted at an advanced period of the course" (Dalzel, [1862, vol.II, p.184]).

3. *Leges Bibliothecae Universitatis Edinensis. Names of Persons admitted to the Use of the Library.* The pertinent entry here runs as follows:

   > Edinburgi Decimo-nono Februarij Admissi sunt hi duo Juvenes praes. D. Jacobo Gregorio Math. P. Thomas Bayes. Anglus. John Horsley. Anglus.

   Unfortunately no further record has been traced linking Bayes to this eminent mathematician.

4. [Dc.5.24^2]. In the *Commonplace Book of Professor Charles Mackie,* we find, on pp.203-222, an *Alphabetical List of those who attended the Prelections on History and Roman Antiquitys from 1719 to 1744 Inclusive. Collected 1 July, 1746.* Here we have the entry

   Bayes (   ), Anglus. 1720,H.   21,H.   3

   The import of the final "3" is uncertain.

5. *Lists of Students who attended the Divinity Hall in the University of Edinburgh, from 1709 to 1727. Copied from the MSS of the Revd. Mr. Hamilton, then Professor of Divinity, etc.* Bayes's name appears in the list for 1720, followed by the letter "*l*", indicating that he was licensed (though not ordained).

6. *List of Theologues in the College of Edin[burgh] since Oct:1711. the 1st. columne contains their names, the 2d the year of their quūmvention, the 3d their entry to the profession, the 4th the names of those who recommend them to the professor, the 5th the bursaries any of them obtain, the 6th their countrey and the 7th the exegeses they had in the Hall.* Here we have

Tho.Bayes|1720|1720|Mr Bayes| — |London|E. Feb. 1721. E. Mar. 1722.

In a further entry in the same volume, in a list headed "Societies", we find Bayes's name in group 5 in both 1720 and 1721. (These were perhaps classes or tutorial groups.) In the list of "Prescribed Exegeses to be delivered" we have

1721. Jan. 14. Mr. Tho: Bayes. the Homily. Matth. 7.24, 25, 26, 27.

and

1722. Ja. 20. Mr Tho: Bayes. a homily. Matth. 11. 29, 30.

The final entry in this volume occurs in a list entitled "*The names of such as were students of Theology in the university of Edinburgh and have been licensed and ordained since Nov. 1709. Those with the letter .o. after their names are ordained, others licensed only.* Here we find Bayes's name, but without an "o" after it.

There is thus no doubt now that Bayes was educated at Edinburgh University. There is unfortunately no record, at least in those records currently accessible, of any mathematical studies, though he does appear to have pursued logic (under Colin Drummond) and theology.

That Bayes did not take a degree at Edinburgh is in fact not surprising. Grant [1884, vol.I] notes that "after 1708 it was not the interest or concern of any Professor in the Arts Faculty ⋯ to promote graduation ⋯ the degree [of Master of Arts] rapidly fell into disregard" [p.265]. Bayes was, however, licensed as a preacher, though not ordained.

The manuscript volume in the library of Edinburgh University that contains the list of theologues also contains a list of books. The range of topics covered seems too narrow for this to be a listing of books in the University library, and it is possible that the works listed were for the particular use of the theologues. But be that as it may: only two of these books are recognizable as being distinctly mathematical: they are

(i) *Keckermanni systema mathem:* and,

(ii) *Speedwells geometrical problems.*

At least that is what appears to be written. The first is probably a book by Bartholomaeus Keckermann, who published other "systema" during the early part of the seventeenth century. The second work is most probably John Speidell's *A geometrical extraction, or a collection of problemes out of the best writers*, first published in 1616 with a second edition appearing in 1657.

Chapter 2

Bayes's Essay

Et his principiis, via ad majora sternitur.

Isaac Newton.

2.1 Introduction

As we have already mentioned, Bayes's books and papers were demised — or so one is sometimes given to believe — to the Reverend William Johnson, his successor at the Pantile Shop[1] at Little Mount Sion. Timerding [1908] concludes that

> nach seinem Ableben betrauten seine Angehörigen *Price* mit der Durchsicht seiner hinterlassenen Papiere, in denen verschiedene Gegenstände behandelt waren, deren Veröffentlichung ihm aber seine Bescheidenheit verboten hatte [p.44]

but it is difficult to see, on the basis of Bayes's posthumous publications, why he should have papers on "sundry matters" ascribed to him, and why his not publishing should be attributed (or even attributable) to a modesty[2] Miranda might well have envied.

Whether some, or all, of the papers were passed on to Richard Price, or whether he was merely called in by Johnson or Bayes's executors to examine them, is unknown. However, on the 10th November 1763 Price sent a letter to John Canton[3] which opens with the words

> Dear Sir, I now send you an essay which I have found among the papers of our deceased friend Mr. Bayes, and which, in my opinion, has great merit, and well deserves to be preserved.

It seems probable, therefore, that, apart from the Essay and a letter[4] on asymptotic series (published in 1764 in the *Philosophical Transactions* 53 (1763), pp.269-271), Bayes left behind no other significant unpublished mathematical work[5].

The Essay has undergone a number of reprintings[6] since it was first published. In view of this fact, I shall content myself with giving, in this chapter, a fairly detailed discussion, in modern style and *more geometrico*, of the Essay. The latter, divided into two sections[7], is preceded by Price's covering letter, and it is to this that we first turn our attention.

2.2 Price's introduction

Price clearly states [p.370] that Bayes had himself written an introduction to the Essay. For reasons best known to himself, Price omitted forwarding this proem to Canton, contenting himself with giving, in his accompanying letter, a report of Bayes's prefatory remarks. Here we find clearly stated the problem that Bayes posed himself, viz.

> to find out a method by which we might judge concerning the probability that an event has to happen, in given circumstances, upon supposition that we know nothing concerning it but that, under the same circumstances, it has happened a certain number of times, and failed a certain other number of times [pp.370-371].

Several points should be noted in this quotation: firstly, the event of current concern is supposed to take place *under the same circumstances* as it has in the past. This phrase is missing both from Bayes's own statement of the problem [p.376] and from his scholium [pp.372 et seqq.]. Whether it is in fact implicit in his Essay will be examined later in this work. Secondly, what does the phrase "judge concerning the probability" mean? Are we to understand by it that a specific value should be attached to the probability of the happening of the event, or merely that a (possibly vague) inference about the probability should be made? In Bayes's statement of his problem, Edwards [1974, p.44] finds the latter interpretation meant: we shall return to this point later.

Continuing his reporting of Bayes's introduction, Price points out that Bayes noted that the problem could be solved (and that not with difficulty — p.371)

> provided some rule could be found according to which we ought to estimate the chance that the probability for the happening of

> an event perfectly unknown, should lie between any two named degrees of probability, antecedently to any experiments made about it. [p.371]

Three points come to mind from this passage: firstly, we are required to *estimate* the *chance* of a *probability.* The difficulty that the word "judge" in an earlier quotation occasioned (as discussed in the preceding paragraph) presents itself again in the phrase "estimate the chance": does this denote a point or an interval estimate? And is this estimate to be used for prediction? From the previous quotation this certainly seems to be the case, but, as we shall see later, the problem as posed by Bayes at the start of his Essay is silent on this point, and the matter of prediction is only taken up in the Appendix, which is by Price. One can indeed but regret the latter's suppression of Bayes's own introduction.

Secondly, note that the statement of the problem refers only to inference about "degrees of probability": inference about an *arbitrary* parameter is not mentioned. And thirdly, the estimation is to be undertaken prior to any experimental investigation.

We read further, in Price's introduction, that Bayes's first thought was that, for the solution to be effected,

> the rule must be to suppose the chance the same that it [i.e. the probability p of the unknown event] should lie between any two equidifferent degrees [of probability] [p.371]

(i.e. $p_2 - p_1 = q_2 - q_1 \Rightarrow \Pr[p_1 \leq p \leq p_2] = \Pr[q_1 \leq p \leq q_2]$) — the rest, he believed, would then follow easily from "the common method of proceeding in the doctrine of chances" [p.371]. (It seems, then, that a certain generally received corpus of probability rules was already in use by this time.) In this quotation we see the origin of the notorious "Bayes's postulate", an hypothesis whose tentative advocation (let alone definite adoption) has engendered more heat than light in numerous statistical and philosophical papers and proceedings.

Proceeding on this assumption, Bayes proposed[8] "a very ingenious solution of this problem". Second thoughts, however, persuaded him that "the *postulate* on which he had argued might not perhaps be looked upon by all as reasonable". Fisher [1956, pp.9-10] was persuaded[9] that it was the realization of these doubts that prevented Bayes from publishing his essay during his lifetime (doubts apparently not shared by Price), though this is not suggested in Price's covering letter. Indeed, the latter informs us that Bayes laid down "in *another form* * the proposition in which he thought

---

*Emphasis added.

the solution of the problem is contained" [p.371], defending his reasons in a *scholium*. In §3.5 of the present work it is argued that Bayes's original solution is given in his tenth proposition, the ninth, which is followed by the scholium, containing the alternative form. Karl Pearson, writing of Bayes's initial postulate, says that, according to Price, "he [i.e. Bayes] rejected it and proceeded on another assumption" [Pearson 1978, p.364]: but as I have already suggested, such a conclusion seems unwarranted.

The importance of this problem was not lost on Price[10], and a long paragraph [pp.371-372] is devoted to a discussion of this matter. Price notes here that the discussion of the present problem is necessary to determine "in what degree repeated experiments confirm a conclusion" [p.372], and mentions further that the problem

> is necessary to be considered by any one who would give a clear account of the strength of *analogical* or *inductive reasoning*. [p.372]

Price concludes his comments on this point by saying

> These observations prove that the problem enquired after in this essay is no less important than it is curious. [p.372]

The problem that Bayes considered was new[11], or at least it had not been solved before [p.372]. Price mentions de Moivre's improvement of Bernoulli's Law of Large Numbers[12], and sees in Bayes's problem a converse to this[13]. Clearly, to de Moivre at least, Bayes's problem was not as difficult as the Law of Large Numbers [p.373], yet it has undoubtedly been more eristic. De Moivre's theorem was thought applicable to "the argument taken from final causes for the existence of the Deity" [Bayes 1763a, p.374]: Price claims that the problem of the Essay is more suited to that purpose,

> for it shows us, with distinctness and precision, in every case of any particular order or recurrency of events, what reason there is to think that such recurrency or order is derived from stable causes or regulations in nature, and not from any of the irregularities of chance. [p.374]

The last two rules of the Essay were presented without their proofs, such deductions being, in Price's view, too long: moreover the rules, Price claims, "do not answer the purpose for which they are given as perfectly as could be wished" [p.374]. Price later published (in 1765) a transcription[14] of Bayes's proof of the second rule, together with some of his own improvements. In connexion with the first rule he writes, in a covering letter to Canton,

> Perhaps, there is no reason about being very anxious about proceeding to further improvements. It would, however, be very agreeable to me to see a yet easier and nearer approximation to the value of the two series's in the first rule: but this I must leave abler persons to seek, chusing now entirely to drop this subject.

The improvements were in the main limited to a narrowing of the limits obtained by Bayes[15].

Price also added short notes where he considered them necessary, and appended

> an application of the rules in the essay to some particular cases, in order to convey a clearer idea of the nature of the problem, and to show how far the solution of it has been carried [p.374]

any errors being his.

Thus far Price's introduction.

2.3 The first section

Bayes's Essay opens with a clear statement of the problem he proposes to solve[16]:

> *Given* the number of times in which an unknown event has happened and failed: *Required* the chance that the probability of its happening in a single trial lies somewhere between any two degrees of probability that can be named. [p.376]

This problem, says Savage in an unpublished note[17],

> is of the kind we now associate with Bayes's name, but it is confined from the outset to the special problem of drawing the Bayesian inference, not about an arbitrary sort of parameter, but about a "degree of probability" only. [1960]

In modern notation, the solution to this problem (given as Proposition 10 in the Essay) can be expressed thus:

$$\Pr\left[x_1 \le x \le x_2 \mid p \text{ happenings and } q \text{ failures of the unknown event}\right]$$
$$= \int_{x_1}^{x_2} x^p(1-x)^q\,dx \Big/ \int_0^1 x^p(1-x)^q\,dx.$$

Bayes, of course, gives the solution in terms of the ratio of areas of rectangles, as Todhunter [1865, art.547] notes. In his edition of Bayes's Essay, Timerding [1908] explains this avoidance of the integral notation in the interesting (albeit faintly chauvinistic) sentence

> Um *Bayes*' Darstellung zu verstehen, muß man sich erinnern, daß in England die Integralbezeichnung verpönt war, weil ihr Urheber *Leibniz* als Plagiator *Newtons* galt. [p.50]

But before an attempt at solution is essayed, however, Bayes devotes some pages to various definitions, propositions and corollaries in elementary probability[18]. Price relates that Bayes

> thought fit to begin his work with a brief demonstration of the general laws of chance. His reason for doing this, as he says in his introduction, was not merely that his reader might not have the trouble of searching elsewhere for the principles on which he has argued, but because he did not know whither to refer him for a clear demonstration of them. [p.375]

Now this is a somewhat curious statement. It is difficult to believe that Bayes was completely ignorant of de Moivre's *The Doctrine of Chances*, of which three editions were published (in 1718, 1738 and 1756) during Bayes's lifetime[19]. De Moivre was, moreover, elected to a fellowship of the Royal Society in 1697, and since he did not die until 1754, it seems unlikely that Bayes did not know of his work. The third edition of *The Doctrine of Chances* contained a 33 page Introduction explaining and illustrating the main rules of the subject. However, Bayes's definition of probability differs from that of de Moivre[20], and this might well be the reason for the detailed first section of the former's Essay.

The definition of probability given by Bayes, viz.

> the *probability of any event* is the ratio between the value at which an expectation depending on the happening of the event ought to be computed, and the value of the thing expected upon it's happening [p.376]

is slightly unusual[21], as Bayes apparently realized himself since he chose to give a definition of that sense of the word "which all will allow to be its proper measure in every case where the word is used" [p.375].

We have already mentioned (§2.2) the possible ambiguity in Price's use of the phrase "judge concerning the probability" in his statement of Bayes's problem. Notice that Bayes, by using "chance" as synonymous[22] with "probability" [p.376], failed to resolve the difficulty[23].

The rest of this first section of the Essay, following the definitions, is devoted to seven routine (at least by today's standards) propositions and a number of corollaries, including a lucid definition of the binomial distribution. One might note, however, that Bayes regarded the failure of an event as the same thing as the happening of its contrary [1763a, pp.376, 383, 386], a view which has bearing on the question of additivity of degrees of belief[24]. Notice too that Bayes takes pains to point out that the happening or failure of the same event, in different trials (i.e. as a result of certain repeated data), is in fact the same thing as the happening or failure of as many distinct independent events, all similar[25] [1763a, p.383].

2.4 The second section

Before we undertake any critical exegesis of this section, it might perhaps be advisable to reformulate certain parts of it in modern notation. Similar accounts have been given by Fisher, Barnard and Edwards[26], but it will be useful to have a "translation" here also.

This Section opens with two postulates[27]. In the first of these it is suggested that a level square table[28] be so made that a ball W thrown upon it will have the same probability of coming to rest at any point as at any other point[29]. The second postulate is that this throw of the first ball is followed by $p+q$ or n throws of a second ball, each of these latter throws resulting in the occurrence or failure of an event M according as to whether the throw results in the second ball's being nearer to or further from a specified side of the table than is the first ball. Examination of Bayes's proof of the results of this Section shows that we may, without loss of generality, express these postulates in the following form [30]:

(i) a single value x is drawn from a uniform distribution concentrated on [0,1], and

(ii) a sequence of Bernoulli trials, with probability x of success, is generated.

These postulates are followed by two lemmata which essentially provide their geometrization.

Let us suppose, without loss of generality, that the square table is of unit area, and let A have co-ordinates (0,0). Let x be the abscissa of the point on the table at which the first ball comes to rest.

Lemma 1. For any x_1, x_2 such that $0 \leq x_1 < x < x_2 \leq 1$,

$$\Pr[x_1 < x < x_2] = x_2 - x_1.$$

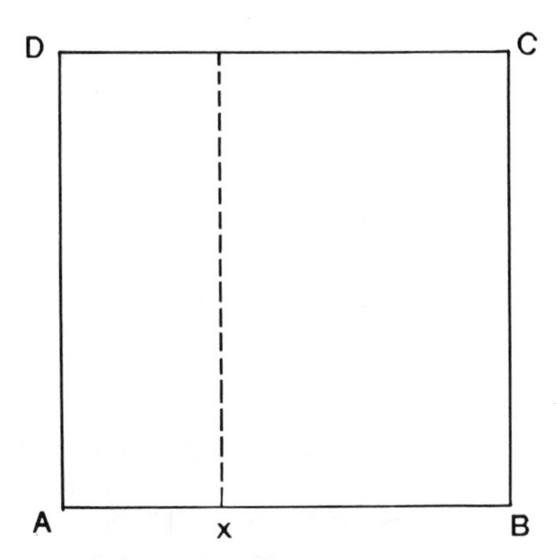

Figure 2.1: Bayes's square table, showing the abscissa x of the point at which the first ball thrown comes to rest.

Lemma 2. Suppose that the second ball is thrown once on the table. Then

$$\Pr[\text{success}] = x.$$

Proposition 8. For any x_1, x_2 such that $0 \le x_1 < x_2 \le 1$,

$$\Pr[x_1 < x < x_2 \text{ \& } p \text{ successes and } q \text{ failures in } p+q=n \text{ trials}]$$
$$= \int_{x_1}^{x_2} \binom{p+q}{p} x^p(1-x)^q\, dx \ .$$

It is not clear whether Bayes interpreted "x lies between A and B" in the sense of *included* or *excluded* end points: I (like Edwards [1978]) have used "$0 < x < 1$" rather than "$0 \le x \le 1$", and similar statements, throughout (the distinction is a fine one, of course, and of little significance here).

Corollary. $\Pr[0 < x < 1 \text{ \& } p \text{ successes and } q \text{ failures in } p+q \text{ trials}]$

$$= \int_0^1 \binom{p+q}{p} x^p(1-x)^q\, dx \qquad \left(= \frac{1}{p+q+1}\right) \ .$$

Proposition 9. For any x_1, x_2 such that $0 \le x_1 < x_2 \le 1$,

$$\Pr[x_1 < x < x_2 \mid p \text{ successes and } q \text{ failures in } p+q \text{ trials}]$$

$$= \int_{x_1}^{x_2} \binom{p+q}{p} x^p(1-x)^q\,dx \Big/ \int_0^1 \binom{p+q}{p} x^p(1-x)^q\,dx$$

$$\left(= \frac{(p+q+1)!}{p!\,q!} \int_{x_1}^{x_2} x^p(1-x)^q\,dx\right) .$$

Corollary. $\Pr[x < x_2 \mid p \text{ successes and } q \text{ failures}]$

$$= \frac{(p+q+1)!}{p!\,q!} \int_0^{x_2} x^p(1-x)^q\,dx.$$

Scholium[31]: suppose one knows how often a success has occurred (and how often it has not occurred) in n trials. One may then "give a guess whereabouts it's probability is", and hence (by the preceding proposition) find "the chance that the guess is right" [Bayes 1763a, p.392]. Bayes now asserts that the same rule is to be used when considering an event whose probability, antecedent to any trial, is unknown. In support of this assertion he adduces the following argument (paraphrased here): let us suppose that to know nothing of the (antecedent) probability is equivalent to being indifferent between the possible number of successes in n trials (i.e. each possible number of successes is as probable as any other)[32]. Writing of "an event concerning the probability of which we absolutely know nothing antecedently to any trials made concerning it" [pp.392-393], Bayes in fact goes on to say

> that concerning such an event I have no reason to think that, in a certain number of trials, it should rather happen any one possible number of times than another. [p.393]

But, by the Corollary to Proposition 8, this is precisely the situation of the proposed model.

> In what follows therefore I shall take for granted that the rule given concerning the event M [i.e. success] in prop. 9. is also the rule to be used in relation to any event concerning the probability of which nothing at all is known antecedently to any trials made or observed concerning it. And such an event I shall call an unknown event. [pp.393-394]

Then, following a corollary in which, in essence, the table is assumed to be of unit area, one finds Proposition 10, which provides the solution to the problem initially posed[33]:

Proposition 10. Let x be the (prior) probability of an unknown event A. Then

$\Pr[x_1 < x < x_2 \mid A$ has happened p times and failed q times in $p+q$ trials$]$

$$= \int_{x_1}^{x_2} \binom{p+q}{p} x^p (1-x)^q \, dx \bigg/ \int_0^1 \binom{p+q}{p} x^p (1-x)^q \, dx.$$

It should be noted that Proposition 9, framed as it is in terms of "table and balls thrown", does *not* furnish the desired solution[34]: the preceding quotation provides the link between this result and that for the "unknown event" in Proposition 10.

Having stated this proposition, in which the solution to the problem posed at the outset of his paper lies, Bayes finds its proof "evident from prop. 9. and the remarks made in the foregoing scholium and corollary" [p.394]. He then turns his attention to the evaluation of the incomplete beta-integral[35] appearing in this proposition (or, for that matter, in the ninth). The details of five Articles [pp.395-399] are summarized in Rule 1 as follows:

Rule 1. $\Pr[x_1 < x < x_2 \mid p$ successes and q failures$]$

$$= (n+1)\binom{p+q}{p}\left[\frac{x_2^{p+1}}{p+1} - \binom{q}{1}\frac{x_2^{p+2}}{p+2} + \binom{q}{2}\frac{x_2^{p+3}}{p+3} - \text{\&c.}\right.$$

$$\left. - \left\{\frac{x_1^{p+1}}{p+1} - \binom{q}{1}\frac{x_1^{p+2}}{p+2} + \binom{q}{2}\frac{x_1^{p+3}}{p+3} - \text{\&c.}\right\}\right]$$

This essentially completes Bayes's contribution: the next few pages (up to p.403) contain (in two further rules) particular methods of approximating the solution given in Rule 1, and are in the main due to Price.

Noting that the formula of Rule 1 is impractical for large values of p and q, Price states that Bayes deduced another expression, summarized in Rule 2 (which in turn was deduced "by an investigation which it would be too tedious to give here" [p.400]) as follows:

Rule 2. If nothing is known concerning an event but that it has happened p times and failed q in $p+q$ or n trials, and from hence I guess that the probability of its happening in a single trial lies between $p/n + z$ and $p/n - z$; if $m^2 = n^3/pq$, $a = p/n$, $b = q/n$, E the coefficient of the term in which occurs $a^p b^q$ when $(a+b)^n$ is expanded, and $\Sigma = \frac{n+1}{n} \times \frac{\sqrt{2pq}}{\sqrt{n}} \times E a^p b^q \times^d$ by the series $mz - \frac{m^3 z^3}{3} + \frac{n-2}{2n} \times \frac{m^5 z^5}{5} - \frac{(n-2)(n-4)}{(2n)(3n)} \times \frac{m^7 z^7}{7} + \frac{(n-2)(n-4)(n-6)}{(2n)(3n)(4n)} \times \frac{m^9 z^9}{9}$ &c. my chance to be in the right is greater than

$$\frac{2\Sigma}{1 + 2Ea^pb^q + 2Ea^pb^q/n}$$

and less than

$$\frac{2\Sigma}{1 - 2Ea^pb^q - 2Ea^pb^q/n} \cdot$$

And if $p = q$ my chance is 2Σ exactly.

[p.400; notation slightly modernized.] The term $2Ea^pb^q/n$ occurring in the denominator of each of the two last expressions was apparently omitted by Bayes "evidently owing to a small oversight in the deduction of this rule", which oversight Price goes on to say, "I have reason to think Mr. Bayes had himself discovered" [p.400]. A further *culpa levis* occurs in the definition of m^2: it should be taken equal to $n^3/2pq$: this was pointed out by Price in the paper of 1764 in the twenty-eighth article.

The third rule, "which is the rule to be used when mz is of some considerable magnitude" [p.403], may, I suspect, be due to Price as it is stated: for whereas the latter is most punctilious in referring to Bayes in his (i.e. Price's) discussion of the second rule, there is no direct mention of Bayes in the immediate preamble to the third rule. However Bayes did give a theorem for use when mz is large (see p.402), a theorem whose application effects the desired modification of the second rule.

In the Supplement to the Essay Price went into more detail. As he wrote in the accompanying letter to John Canton,

> I have first given the deduction of Mr. Bayes's second rule chiefly in his own words; and then added, as briefly as possible, the demonstrations of several propositions, which seem to improve considerably the solution of the problem, and to throw light on the nature of the curve by the quadrature of which this solution is obtained. [Bayes, 1764, p.296]

Strictly speaking this brings us to the end of this section. However, Price's remarks at the start of the Appendix are pertinent, and we accordingly adduce them here. He begins by saying

> The first rule gives a direct and perfect solution in all cases; and the two following rules are only particular methods of approximating to the solution given in the first rule, when the labour of applying it becomes too great. [p.404]

Then follows a paragraph setting out more succinctly than before the cases (depending on the magnitudes of p, q and mz) in which the various rules may be used.

2.5 The Appendix[36]

The last fifteen pages contain some applications of the preceding rules.

The first of these applications runs as follows: let M be an event concerning whose probability (antecedently to any trials) nothing is known. Denoting by S_i the occurrence of M on the i-th trial, we have

(i) $\Pr\left[\frac{1}{2} < x < 1 \mid S_1\right] = \frac{3}{4}$;

(ii) $\Pr\left[\frac{1}{2} < x < 1 \mid S_1, S_2\right] = \frac{7}{8}$;

(iii) $\Pr\left[\frac{1}{2} < x < 1 \mid S_1, S_2, S_3\right] = \frac{15}{16}$;

(iv) $\Pr\left[\frac{1}{2} < x < 1 \mid p \text{ successes}\right] = \left(2^{p+1} - 1\right)/2^{p+1}$;

(v)[37] $\Pr\left[\frac{2}{3} < x < \frac{16}{17} \mid 10 \text{ successes and no failures}\right] = 0.5013.$

Price next goes on to consider a particularly noteworthy example: suppose we have a die of unknown number of faces and unknown constitution (it will not, I suppose, do any harm to suppose the faces numbered $n_1, n_2, ..., n_k$ — not necessarily distinct). The die is thrown once, the face n_i (say) resulting (which shows only that the die has this face). It is only at this stage, i.e. *after* the first throw, that the situation of the Essay obtains; the occurrence of n_i in any subsequent trial being an event of whose probability we are completely ignorant. If, at the second trial, n_i appears again, then by the first application, the odds will be three to one on that n_i is favoured (either through being more numerous, or (equivalently) because of the die's constitution). We shall return to this matter in the next chapter.

Price then emphasizes that improbability is not the same thing as impossibility, and goes on to discuss applications to "the events and appearances of nature" [p.408]. Once again he takes pains to point out that the first experiment merely shows that some particular occurrence is possible: no notion of uniformity of nature is suggested, though further observations of the same occurrence may tend to support that view. As an illustration[38] Price cites the well-known example of the Rising of the Sun, emphasizing once again that a "previous total ignorance of nature" [p.410] is required for the validity of his arguments.

Having considered the case where only "successes" have occurred, Price now turns his attention to the case in which either "success" or "failure" may arise. As a particular illustration of the procedure, he considers a lottery of unknown scheme in which the proportion of blanks to prizes is unknown. Price in fact evaluates by Rules 1 to 3, $\Pr\left[\frac{9}{10} < x < \frac{11}{12} \mid p \text{ blanks}\right.$

and q prizes] for various values of p and q, where x denotes the proportion of blanks to prizes.

He concludes this Appendix by noting that

> what most of all recommends the solution in this *Essay* is, that it is compleat in those cases where information is most wanted, and where Mr. De Moivre's solution of the inverse problem can give little or no direction; I mean, in all cases where either p or q are of no considerable magnitude [p.418]

and he emphasizes that, while it is fairly easy to see that

$$\Pr[\text{success}] : \Pr[\text{failure}] :: p : q$$

(for large values of p and q), the Essay demonstrates the folly of such a judgement when either p or q is small.

In 1764 Price forwarded a supplement to the Essay to John Canton. In this paper, published in the volume of the *Philosophical Transactions* for 1764, may be found proofs, and some development, of the Rules given in the Essay. This supplement will be considered in Chapter 4.

2.6 Summary

Before we pass on to a closer examination of the Essay, it might be useful to provide a recapitulation of its main results. From Price's introductory remarks and Bayes's own work one sees that the scheme of the Essay, and the thought prompting it, can be summarized as follows:

Problem 1.[39] An event M has occurred (under the same circumstances) p times and failed to occur q times. How can we estimate the probability of this event's happening?

The solution can be effected if one can solve

Problem 2. Let $P(M)$ denote the probability of the (perfectly unknown) event M. For any α and β, with $\alpha < \beta$, what is $\Pr[\alpha < P(M) < \beta]$? (This is to be determined before any experimentation.)

This in turn can be solved by using

Rule 1. If $\beta_1 - \alpha_1 = \beta_2 - \alpha_2$ then

$$\Pr[\alpha_1 < P(M) < \beta_1] = \Pr[\alpha_2 < P(M) < \beta_2]$$

— i.e. a uniform distribution for $P(M)$.

Being unhappy with this procedure, Bayes next considers

Problem 3. M has happened p times and failed to happen q times. For any α and β, what is $\Pr[\alpha < P(M) < \beta \mid p, q]$?

Turning to the "table and balls" example, we see that θ, the position of the first ball on the horizontal axis, is distributed $U((0,1))$. If X denotes the number of "successes" obtained in n throws of the second ball, then, for a given $\theta, X \sim b(n,\theta)$. It then follows that (unconditionally) X has a discrete uniform distribution on $\{0,1,2,...,n\}$ — i.e.

$$\Pr[X = k] = 1/(n+1), \qquad k \in \{0,1,2,...,n\}.$$

Assuming that this holds for *all* k and n, we have in fact

$$\theta \sim U((0,1)) \Leftrightarrow X \sim U(\{0,1,...,n\}).$$

Bayes proposes in his *Scholium* that the number of occurrences of the unknown event should be taken to have a discrete uniform distribution.

Chapter 3

Commentary on Bayes's Essay

The labours of others have raised for us an immense reservoir of important facts.

Charles Dickens, Pickwick Papers.

3.1 Introduction

In the preceding chapter several points, arising from Bayes's Essay, were either glossed over or omitted altogether. It is now time to fill in these lacunae, though certain of the topics to be discussed here will in fact undergo further development later in this tractate (in particular, we shall not consider here any *elaboration* of the main results of the Essay, and the Supplement to the Essay will be dealt with in Chapter 4).

3.2 Price's introduction

In his statement of Bayes's problem, Price says (see §2.2) that the event whose probability is sought should be known to take place "under the same circumstances" [pp.370-371] as it occurred under in the past. According to Price, this phrase was in fact used by Bayes in his own (suppressed) introduction to the Essay: it is, however, missing from the statement of the problem on p.376, although its implicit assumption is made clear, I believe, from the following observations.

In his postulate at the end of Section II, Bayes refers to "the happening of the event M in a single trial" [p.385], and the word "trials" appears in each of the propositions of that section. But not for Bayes any escape from the precise meaning called for of this word: he grasps the nettle firmly, and in the first part of his Essay we find the following passage:

> Definition. If in consequence of certain data there arises a probability that a certain event should happen, its happening or failing, in consequence of these data, I call its happening or failing in the 1st trial. And if the same data be again repeated, the happening or failing of the event in consequence of them I call its happening or failing in the 2d trial; and so on as often as the same data are repeated. [p.383]

It is, I think, quite clear from this quotation that the conditions under which the event of current concern takes place are supposed to be the same as those under which it happened in the past. (Such an assumption is of course frequently tacit in this sort of work: that Bayes bothers to state it — and that most carefully — is surely a tribute to the rigour of his thinking, if not indeed to his mathematical ability.)

A more difficult matter, also stemming from Price's statement of the problem, arises in connexion with the phrase "judge concerning the probability" (see §2.2), a phrase which it is expedient to consider in conjunction with his later one "to estimate the chance that the probability..." Two interpretations of the first phrase are possible, as Edwards [1974] notes in the following words:

> Does 'judge concerning the probability' mean 'attach a specific value to the probability of the next event' or does it mean 'make an inference – possibly vague – about the probability'? [p.44]

If the phrase "of the next event" may be assumed to qualify the last word in this quotation, then there can, I think, be little doubt that Price intended the latter interpretation (the justification for this assertion may become more apparent when, in a later section in this chapter, we consider Price's applications of the results of the Essay).

Moreover, in view of Bayes's own statement of the problem he proposed to solve (see §2.3) and his words "by *chance* I mean the same as probability" [p.376], it seems to me, as it indeed did to Edwards, that Bayes was in fact only interested in an inference (possibly vague) about the probability: the second of Price's introductory phrases quoted above also supports this view, I suggest.

It is perhaps significant, though I do not wish to urge the point, that, according to the *order* in which the comments are reported by Price [pp.370, 371], the *first* idea was to find out a method by which we might "judge concerning the probability" [p.370] (i.e. a possibly vague inference), and *then* that this could be done by estimating the chance of the probability's being between any two degrees of probability.

3.3 The first section

Remarks on this part of the Essay are wide-ranging. Todhunter [1865, art.544] describes it as "excessively obscure", and he comments further[1] that it "contrasts most unfavourably with the treatment of the same subject by De Moivre." Savage [1960] finds in this section "a whole short course on probability", and he provides a paraphrase of it on pp.2-3 of his unpublished note. Stigler [1982a, p.250] sees here "an intriguing development of rules of probability, most of which we would now regard as elementary". A useful summary is given in Dinges [1983, pp.75-80], in which work it is also suggested that "Wir möchten Th. Bayes gerne als ersten Zeugen für einen theoretischen Wahrscheinlichkeitsbegriff in Anspruch nehmen" [p.94].

The section opens with seven definitions. While these are unexceptionable (although some may perhaps be slightly unusual), two points are worth noting. The first of these concerns Bayes's definition of the probability of an event in terms of expectation (see §2.3). This is certainly different to the (more usual) definition given by de Moivre[2], and the fact that Bayes's problem required such an approach for its solution might well be the cause of his giving his own "probability primer" and Price's statement that Bayes "did not know whither to refer him [i.e. the reader] for a clear demonstration of them [i.e. the principles on which Bayes argued]" [p.375]. The second point to be noted concerns Bayes's definition of independence. The seventh definition reads as follows:

> Events are independent when the happening of any one of them does neither increase nor abate the probability of the rest. [p.376]

It seems that Bayes saw no distinction[3] between "independence" and "pairwise independence".

The rest of this section is devoted to seven propositions[4], five corollaries and a further definition. The first proposition states (in modern terminology) that, if $\{E_i\}$ is a sequence of mutually exclusive events, then $\Pr[\cup E_i] = \sum \Pr[E_i]$. Bayes was apparently the first to state this fact[5].

Propositions 3 (and its corollary) and 5 require comment, and are accordingly given here:

> Proposition 3. The probability that two subsequent events will both happen is a ratio compounded of the probability of the 1st, and the probability of the 2d on supposition the 1st happens. [p.378]
> Corollary. Hence if of two subsequent events the probability of the 1st be a/N, and the probability of both together be P/N, then the probability of the 2d on supposition the 1st happens is P/a. [p.379]
> Proposition 5. If there be two subsequent events, the probability of the 2d b/N and the probability of both together P/N, and it being 1st discovered that the 2d event has happened, from hence I guess that the 1st event has also happened, the probability I am in the right is P/b. [p.381]

At first sight the Corollary to Proposition 3 and Proposition 5 appear to be saying the same thing. Thus if E_1 and E_2 are two events (E_1 preceding E_2 in time), one might be tempted to phrase these two results in modern notation as

$$\Pr[E_2 \mid E_1] = \Pr[E_1 \cap E_2] / \Pr[E_1]$$

$$\Pr[E_1 \mid E_2] = \Pr[E_1 \cap E_2] / \Pr[E_2] .$$

But since it is "well-known" and "universally accepted" that "the timing of events is irrelevant to the concept of conditional probability" (Shafer [1982, p.1076]), one might well be perplexed at Bayes's deliberateness. Shafer (op. cit.) has forcefully argued that while an argument using rooted trees can establish the validity of Bayes's Corollary to Proposition 3, such an argument fails to establish Proposition 5. Since the latter in turn is crucial in the proof of Proposition 9, Shafer's thrust is to the very heart of the Essay.

However, if we view Bayes's fifth definition in terms of *subjectively* determined values of expectations[6], on Shafer's admission "the fifth proposition would then become merely a subjective version of the third" [op. cit. p.1086].

In the fifth proposition Bayes introduces as a new factor the order in which we *learn* about the happening of the events. Shafer [1982] concludes that this result

> seems unconvincing unless we assume foreknowledge of the conditions under which the discovery of B's [the second event's] having happened will be made. [p.1080]

Such foreknowledge I believe obtains in cases in which Bayes uses this result, and I believe therefore that it is correct — but let the reader of the Essay decide for himself.

The other propositions of this Section do not seem excitatory: the definition following Corollary 2 to Proposition 6 has already been mentioned (see §3.2).

3.4 The second section

An examination of the relevant propositions shows that Bayes states the results of this section[7] sometimes in terms of "the probability that the point o should fall [in a certain interval]" and sometimes in terms of "the probability of the event M [is in a certain interval]"[8]. Here, as in §2.4, os denotes the line on which the first ball W comes to rest when it is rolled: the resting of the second ball O between AD and os — see Figure 3.1 — after a single throw is called the happening of M in a single trial. Thus Proposition 8 and its corollary fall in the first category, Proposition 9 falls in both categories, its corollary falls in the second, and Proposition 10 is framed in terms of the probability of the (an?) unknown event.

As Edwards [1978] has noted, these two methods of formulation are, on Bayes's assumptions, identical: however, even if the first ball is *not* uniformly distributed, the distribution of the *probability* will still be uniform. This fact may be demonstrated as follows: suppose the first ball to have the distribution $dF(\cdot)$. Then, by the second part of Bayes's postulate, the associated probability is

$$\theta = \int_0^z dF(x) \ .$$

Thus $d\theta = dF(z)$, and θ has a uniform distribution[9]. Edwards [1974] makes the reasonable deduction that

> probably what happened was that Bayes realised he would need to postulate a uniform prior distribution in order to solve his main problem, and so generated one in his model for that problem. Then he realised that rolling subsequent balls would give the probabilities he wanted for success and failure, but he failed to notice that this would be so even in the event of a non-uniform table. [p.46]

If we interpret all the propositions from the point of view of the probability of the first ball's being in a certain interval, then, with the further assumption of a non-uniform table, the only changes necessitated are the replacements of the limits θ_1 and θ_2 respectively by the integrals $\int_0^{z_1} dF(x)$ and $\int_0^{z_2} dF(x)$, where z_1 and z_2 are the limits of the interval within which the ball lies.

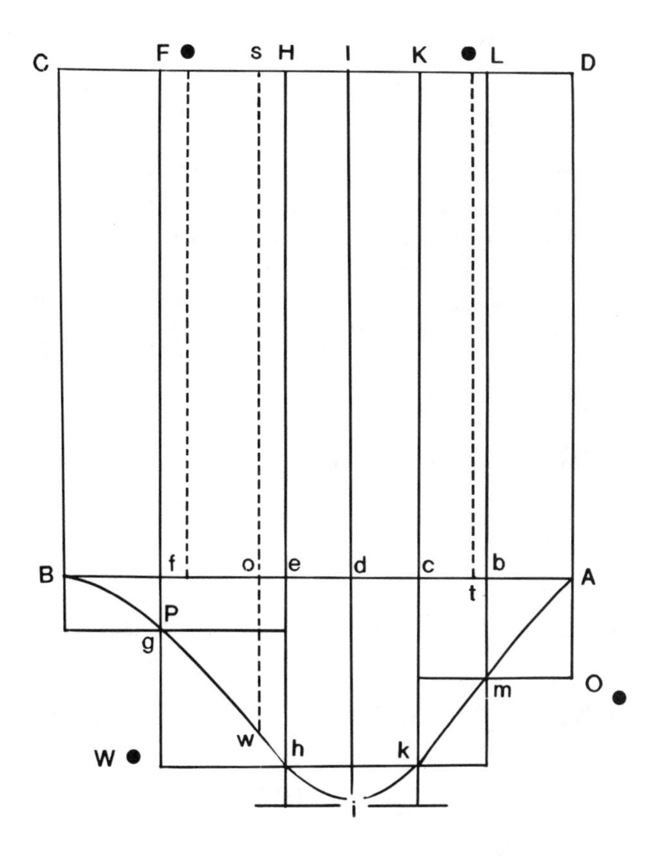

Figure 3.1: The ball W is thrown on the square table $ABCD$ and comes to rest on the line os. A second ball O is then thrown onto the table, its resting on any toss between AD and os being the happening of M.

Edwards [1978, p.117] points out that even if the table is not uniform, the corollary to Proposition 8 is still valid. Indeed, suppose one, and then a further n (distinct) values are drawn from a distribution. Denoting by "success" (S) the event that one of the n values is less than the first one, and by "failure" (F) the event that one of the n values is greater than the first (the respective probabilities now being θ and $1-\theta$ respectively), then, assuming that the values are independently chosen, we have

$$\Pr\left[x\ S\text{'s and } y\ F\text{'s} \mid \theta\right] = \binom{n}{x}\theta^x(1-\theta)^y$$

where $x + y = n$. Hence

$$\Pr[x\ S\text{'s and } y\ F\text{'s}] = \int_0^1 \binom{n}{x} \theta^x (1-\theta)^y \, d\theta = 1/(n+1)\,.$$

There is nothing in Bayes's Essay to say that the square table $ABCD$ is of unit area. This "normalization" has in fact been carried out in the statement of the results in Chapter 2: it might, however, be of some interest to discuss the formulation in more detail. Thus on rewriting the results of this Section in a more modern notation than that adopted by Bayes (and not assuming $ABCD$ to be of unit area) we obtain (see Figure 3.1) the following:

Lemma 1. $\Pr[b < o < f] = (f - b)/AB$.

Lemma 2. $\Pr[M \text{ in a single trial} \mid W]$

$$= \Pr[1 \text{ success} \mid W]$$

$$= Ao/AB.$$

Proposition 8. Let $y = Ex^p r^q$, where $E = \binom{p+q}{p}$. Then

$$\Pr[b < o < f \;\; \& \;\; p, q] = \int_b^f Ex^p r^q \, dx \,/\, \text{area } ABCD.$$

(This proposition will be discussed in more detail later in this section.)

From Bayes's Essay [p.388] we have

$$y = bm/AB, \qquad x = Ab/AB, \qquad r = Bb/AB.$$

Corollary. $\Pr[A < o < B \;\; \& \;\; p, q] = \int_A^B Ex^p r^q \, dx \,/\, \text{area } ABCD.$

On p.393 of the Essay it is pointed out (in a reference to "art.4", which in turn can be found on p.398) that (in essence) in the case of the unit square this corollary yields $1/(n+1)$, independent[10] of x.

Proposition[11] 9.

$$\Pr[b < o < f \mid p, q] \equiv \Pr[Ab/AB < P(M) < Af/AB \mid p, q]$$

$$= \int_b^f Ex^p r^q \, dx \Bigg/ \int_A^B Ex^p r^q \, dx \;,$$

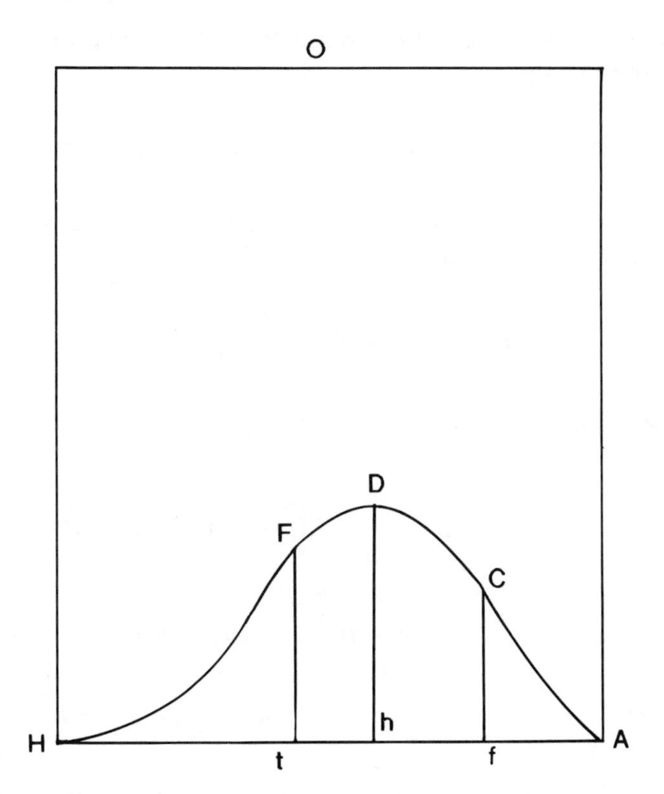

Figure 3.2: The figure used by Bayes in the proof of his Proposition 10.

where $P(M)$ denotes the probability of M.

Corollary.

$$\Pr\left[Ab\,/AB < P(M) < o \mid p, q\right] = \int_b^o Ex^p r^q \, dx \Bigg/ \int_A^B Ex^p r^q \, dx.$$

Proposition 10. Let N be an "unknown event" with probability $P(N)$. Then

$$\Pr\left[Af\,/AH < P(N) < At/\,AH \mid p, q\right] = \int_f^t Ex^p r^q \, dx \Bigg/ \int_A^H Ex^p r^q \, dx \quad .$$

I now propose to examine the "transliteration" of Proposition 8 in more detail. Notice firstly that

$$\Pr[x_1 < x < x_2 \quad \& \quad p,q] = \int_{x_1}^{x_2} f(p,q \mid x) f(x)\, dx.$$

Recalling that $f(x)$ is uniform here, and that $f(p,q \mid x) = Ex^p(1-x)^q$, where $E = \binom{p+q}{p}$, we obtain

$$\Pr[x_1 < x < x_2 \quad \& \quad p,q] = \int_{x_1}^{x_2} E\, x^p(1-x)^q\, dx \ ,$$

the usual result for the unit square. Now let $x = y/B$ in the integrand. Then

$$\Pr[x_1 < x < x_2 \quad \& \quad p,q] = \int_{x_1 B}^{x_2 B} E(y/B)^p(1-y/B)^q\, dy/B$$

$$= \int_b^f E(y/B)^p(1-y/B)^q(1/B)\, dy, \text{ where } b \equiv x_1 B, \quad f \equiv x_2 B$$

$$= \int_b^f (z/B)\, dy \ ,$$

where $z = E(y/B)^p(1-y/B)^q$. Now z being bm/B (see Proposition 8), the integral of z/B from b to f is *not* the area under the curve in Figure 3.1. This area is in fact

$$\int_b^f h\, dy \quad = B\int_b^f (h/B)\, dy$$

$$= B\int_b^f z\, dy$$

(because the height $h = Bz$)[12]. Also h is such that

$$\int_o^B h\, dy = 1 \Rightarrow \int_0^1 z\, dy = 1.$$

Thus

$$\frac{1}{B} \times \text{ area } = \int_b^f z\, dy.$$

Hence finally

$$\Pr[x_1 < x < x_2 \ \& \ p, q] = (1/B^2) \times \text{area under curve from } b \text{ to } f$$

$$= (1/B^2) \int_b^f h\,dy$$

(the latter integral "being" the usual "area under a curve" one)[13]. This is Bayes's result, of which the remaining results are fairly obvious consequences.

We see, then, that there is no loss of generality in considering the square table as being of *unit* area.

3.5 The postulate and the scholium

That Bayes himself presented an argument in defence of (or, better, as justification for) his postulate, although apparently generally ignored, is a fact that has actually frequently been emphasized. One of the most recent to stress this point was Stigler [1982a, p.250], [1986a, p.127 et seqq.], and before him we find the point made by Molina [1930, pp.382-383], [1931, §IV], Savage [1960] and Edwards [1974, p.47].

The positioning of the scholium should be noted: its appearing *after* the corollary to Proposition 9 but *before* Proposition 10 perhaps lends weight to our earlier assertion that the proposition that provides the answer to Bayes's problem is the tenth and not the ninth.

The scholium may be paraphrased as follows: from Proposition 9 (writes Bayes) it is clear that, given the number of times the event M happens and fails in a certain number of trials, "one may give a guess whereabouts it's probability is, and, by the usual methods computing the magnitudes of the areas there mentioned, see the chance that the guess is right" [p.392]. This same rule is to be applied to an event about whose probability we are completely ignorant prior to any trials being made; for "concerning such an event I have no reason to think that, in a certain number of trials, it should rather happen any one possible number of times than another" [p.393]. This being so, one may reason that its probability was at first "unfixed", and then determined in such a way "as to give me no reason to think that, in a certain number of trials, it should rather happen any one possible number of times than another" [p.393]. But this is exactly the case of the event M (see the corollary to Proposition 8). "Hence the model of a uniform prior distribution for p represents complete absence of knowledge about p" [Edwards 1978, p.117]. Finally, Bayes writes[14]

> In what follows therefore I shall take for granted that the rule given concerning the event M in prop. 9. is also the rule to be used in relation to any event concerning the probability of which nothing at all is known antecedently to any trials made or observed concerning it. And such an event I shall call an unknown event. [pp.393-394]

To complete Bayes's argument successfully a converse property must needs be established: viz., none other than the uniform distribution for p has the property of the corollary to Proposition 8. As Murray [1930] has it

> the assumption "all values of p are equally likely" is *equivalent* to the assumption "any number x of successes in n trials is just as likely as any other number y, $x \leq n$, $y \leq n$". [p.129]

In his elegant note Murray verified that part of this quotation which was not proved by Bayes. A shorter proof than his would be provided by noting that the uniform distribution does yield the appropriate sequence of moments, and then using the uniqueness theorem for moment generating functions[15].

Several writers have of course disparaged Bayes's argument[16]: it might be of interest to look at the discussion presented by Hacking [1965], according to whom Bayes argued as follows:

> (i) Before any trials on the billiard table[17], and before the point o is discovered, there is no reason to suppose M will happen any number of times rather than any other possible number — and, he might have added, there is no reason to prefer any value of $P(M)$ [the probability of M] to any other[18]. (ii) Exactly the same is true of the event E, in the case that no parent set-up is known. (iii) Betting rates should be a function of the available data: when all the information in the two situations is formally identical, the betting rates must be identical. (iv) In all that matters, the data in the case of E and M are identical. (v) The initial distribution of betting rates for $P(M)$ is uniform: it assigns equal rates to equal intervals of possible values of $P(M)$. Therefore, (vi) this should also be the initial distribution of betting rates for $P(E)$. [pp.199-200]

To pin-point the fallacy (as he sees it) in Bayes's reasoning, Hacking poses the following dilemma:

> *Interpretation A*: (v) does not follow from (i) directly, but is the consequence of the fact that the table is so made and levelled,

that the long run frequency with which the ball falls in any area is equal to the long run frequency with which it falls in any other area; we infer (v) from this fact plus assumption (3) [viz. when the chance (long run frequency) of getting outcome E on some trial of kind K from some set-up X is known to be p, and when this is all that is known about the occurrence of E on that trial, then the fair rate for betting on E should be $p : 1 - p$].
Interpretation B: (v) does follow from (i) directly. [p.200]

Although "most readers since the time of Laplace have favoured B" [p.200], Hacking believes that Bayes probably meant A — otherwise, why would he have taken such pains in his Essay to compare E to M? If indeed (v) follows from (i) directly (as suggested in Interpretation B), then (vi) follows from (ii) directly, and there would then be no need for any mention of M.

However, under Interpretation A the argument is fallacious. If Hacking's assumption (3) and certain facts about frequency are required for (v), then data about M must be used which are not available for E, and so (iv) must be false (which of course means that the demonstration itself is false).

Interpretation B is similarly discredited, since lack of reason for supposing $P(M)$ to be in one short interval rather than another of the same size should not entail that the betting rate on equal intervals should be in proportion to their size. As an illustration of his point Hacking cites the well-known (though perhaps somewhat shabby) example due to Fisher, in which the assumption that nothing is known about $P(M)$ leads to a similar assumption about $\arcsin P(M)$, and hence to the observation that "betting rates should be proportional to angular size" [Hacking 1965, p.200]. Of course, as Edwards [1978, p.118] notes, "such a change would upset the equal probabilities for all the values of a [the number of successes]". Moreover, it might be disputed whether ignorance of $P(M)$ implies ignorance of $\arcsin P(M)$: indeed, if our interest is in $P(M)$, why should we be at all concerned about whether or no the distribution of $\arcsin P(M)$ is uniform?

I believe that Bayes probably introduced his "table and balls" model for one of two reasons: (a) merely as an example, or (b) because he first gave the result for an unknown event and *then* added his model. The latter interpretation is, I believe, supported by Price's introduction. Salient points from the second paragraph [pp.370-371] of the latter are the following:

(a) Bayes was originally concerned with finding a rule by whose use the probability of an unknown event E could be obtained.

(b) This rule, it appeared to him, must be "to suppose the chance the same that it [i.e. the probability of the unknown event] should lie between any two equidifferent degrees" [p.371].

(c) The quaesitum would then follow by "the common method of proceeding in the doctrine of chances" [p.371].

(d) Bayes in fact gave a proof (suppressed by Price) on these lines.

(e) Second thoughts suggested that not all might regard the *postulate* on which he argued as reasonable.

(f) Bayes therefore " chose to lay down in another form the proposition in which he thought the solution of the problem is contained, and in a *scholium* to subjoin the reasons why he thought so, rather than to take into his mathematical reasoning any thing that might admit dispute" [p.371].

A discussion of the scholium and the postulate would be incomplete without mention of Stigler [1982a], in which paper (*contra* Hacking) it is asserted that "Bayes's actual argument is free from the principal defect it has been charged with" [p.250] (see also Stigler [1986a, pp.126-129]). Stigler's discussion[19] runs as follows: denoting by X the number of successes in $n = p + q$ trials, we may rewrite the corollary to Proposition 8 and the footnote on p.393 as

$$\begin{aligned} \Pr[X = p] &= \int_0^1 \binom{n}{p} x^p (1-x)^{n-p}\, dx \\ &= 1/(n+1) \end{aligned}$$

for all $p \in \{0, 1, ..., n\}$. In terms of this discrete uniform distribution as the marginal distribution of X, Stigler [1982a] constructs Bayes's reasoning as follows:

(i) For the table, $\Pr[X = p] = 1/(n+1)$ for all p.

(ii) In the case of what Bayes describes as "an event concerning the probability [x] of which we absolutely know nothing antecedently to any trials made concerning it" [pp.392-393] [i.e. before X is observed], one should argue that "concerning such an event [success] I have no reason to think that, in a certain number [n] of trials, it should rather happen any one possible number of times than another" [Bayes 1763a, p.393]. (i.e. $\Pr[X = p]$ is constant).

(iii) Since $\Pr[X = p] = 1/(n+1)$ both for the table and for any application in which we are in a state of absolute ignorance, the situations are

> parallel, and x must therefore have a uniform distribution not only on the table, but also in the application. That is, $\Pr[X = p]$ constant *implies* x is uniform.

The second step is characterized by Stigler [1982a, p.253] as "a very distant cousin" of the principle of insufficient reason. Three arguments are advanced in support of this position.

Argument 1. Suppose that before X is observed, we "absolutely know nothing" about x. If $\Pr[X = p]$ were not constant, suppose that there were to exist p and p^* such that $\Pr[X = p^*] > \Pr[X = p]$. A greater expectation would then be attached to p^* than to p, and a future bet ("expectation", in Bayes's terminology) that p^* would occur would be of higher value than a similar one that p would occur. But if we expect one value of X rather than another, then we are *not* in a situation where absolutely nothing is known about x,

> for X/n is an estimate of $[x]$, and we should not describe ourselves as being in a position where we expect this estimate to be one value rather than another. [Stigler 1982a, p.253]

Argument 2. Recalling that Bayes's definition of probability was as an *a priori* expectation, we note that his reluctance to postulate a uniform distribution for x was not a sign of an unwillingness to speak of *a priori* probabilities. Rather, the specification of an *a priori* distribution was removed from "the forever unobservable" x and placed "on the ultimately observable X" [Stigler 1982a, p.253][20]. Thus the second step "makes peculiarly good sense in the context of Bayes's unusual definition of probability (as an expectation)" [Stigler, loc. cit.].

Argument 3. The second step is much more restrictive than the usually invoked principle of insufficient reason: for if knowing absolutely nothing necessitates our taking $\Pr[X = p] = 1/(n+1)$, very few applications will be found in which this requirement is met. Moreover, the argument is strongly linked to the binomial model[21].

The third step in Stigler's reconstruction of Bayes's argument, namely $\Pr[X = p]$ constant implies x is uniformly distributed, while being "intuitively plausible at Bayes's time" [Stigler 1982a, p.253], needs verification. As we have already indicated, however, knowledge of the first n moments, for every n, of a distribution on [0,1] will uniquely determine the distribution. Since Bayes's "certain number of trials" is vague, and since the statement about $\Pr[X = p]$ is *a priori*, "we may be charitable to Bayes and assert that (perhaps inadvertently) he was not actually in error on this point" [Stigler 1982a, p.254].

Stigler [1982a, p.253] and [1986a, p.129] notes further that his interpretation of Bayes's argument shows that, for any strictly monotone function f,

$$\Pr[X = p] = 1/(n+1) \Rightarrow \Pr[f(X) = f(p)] = 1/(n+1).$$

Thus our knowing nothing about X is equivalent to our knowing nothing about $f(X)$, and this observation shows that Bayes's argument is in fact free of the objection raised to it by Fisher and others.

Geisser [1988] proposes three possible versions of Bayes's result. In the first of these a sequence $\{X_i\}_1^N$ of independent and identically distributed random variables taking on values in $\{0, 1\}$ is considered, with

$$\Pr[X_i = 1 \mid \theta] = \theta = 1 - \Pr[X_i = 0 \mid \theta] \ .$$

Setting $R = \sum_{i=1}^N X_i$, we easily find that

$$\Pr[R = r \mid \theta] = \binom{N}{r} \theta^r (1-\theta)^{N-r} \ ,$$

and hence

$$p(\theta \mid r) \propto \theta^r (1-\theta)^{N-r} \ .$$

This, the "Received Version", is contrasted with the "Revised Version" given by Stigler, which we have already discussed.

In the third version, labelled as "Stringent" by Geisser [1988, p.150], it is supposed that the abscissa of the point at which the ball initially rolled comes to rest is a random variable Y. The actual value y of Y is then to be inferred from N further rolls (of a second ball), it being known how often the second ball comes to rest at a position with abscissa less than or equal to y. Assuming that these rolls of the second ball are independent, we have

$$p(y) = 1$$

and

$$\Pr[R = r \mid y] = \binom{N}{y} y^r (1-y)^{N-r} \ .$$

Hence

$$p(y \mid r) \propto y^r (1-y)^{N-r} \ ,$$

an expression independent of any parameters.

3.6 The Appendix

In his appendix "*Containing an Application of the foregoing Rules to some particular Cases*", Price discusses a number of examples illustrating (or purporting to illustrate) the use of the major result of the Essay. I propose to consider this appendix in some detail.

The first illustration runs as follows:

> Let us first suppose, of such an event as that called M in the essay, or an event about the probability of which, antecedently to trials, we know nothing, that it has happened *once*, and that it is enquired what conclusion we may draw from hence with respect to the probability of it's happening on a *second* trial. The answer is that there would be an odds of three to one for somewhat more than an even chance[22] that it would happen on a second trial. [p.405]

Price arrives at his solution by a direct application of Rule I (see §2.4), and then states

> which shews the chance there is that the probability of an event that has happened once lies somewhere between 1 and $\frac{1}{2}$; or (which is the same) the odds that it is somewhat more than an even chance that it will happen on a second trial. [p.405]

Now it is, I think, possible (though perhaps incorrect) to interpret[23] Price's question as requiring an answer given by the rule of succession[24] (a formula obtained by Laplace in 1774), in terms of which the probability of a second occurrence of M is given by

$$\int_0^1 x^2\,dx \Big/ \int_0^1 x\,dx = \frac{2}{3} \quad .$$

This interpretation however does not take account of Price's requirement that there be "more than an even chance that it will happen on a second trial", but this can be incorporated into the solution by taking cognisance of Problem IV, pp.180-183, of Condorcet's[25] *Essai sur l'application de l'analyse à la probabilité des décisions rendues à la pluralité des voix* of 1785. In a slightly different notation to that to be used in our discussion of this problem in Chapter 5, let S_i denote the occurrence on the i-th trial of Price's event M and let $F_{r,s}$ denote the probability that $P(M)$, the probability of M, lies between r and s (with $r < s$). Then by Condorcet's solution, we have

$$\text{(a)} \qquad \Pr\left[F_{\frac{1}{2},1} \mid S_1\right] = \int_{\frac{1}{2}}^{1} x\,dx \Big/ \int_{0}^{1} x\,dx = \tfrac{3}{4} \quad ;$$

$$\text{(b)} \qquad \Pr\left[S_2 \mid S_1 \ \& \ F_{\frac{1}{2},1}\right] = \int_{\frac{1}{2}}^{1} x^2\,dx \bigg/ \int_{\frac{1}{2}}^{1} x\,dx = \tfrac{7}{9} \quad ;$$

$$\text{(c)} \qquad \Pr\left[S_2 \mid S_1\right] = \int_{0}^{1} x^2\,dx \bigg/ \int_{0}^{1} x\,dx = \tfrac{2}{3} \quad ;$$

$$\text{(d)} \qquad \Pr\left[S_2 \ \& \ F_{\frac{1}{2},1} \mid S_1\right] = \int_{\frac{1}{2}}^{1} x^2\,dx \bigg/ \int_{0}^{1} x\,dx = \tfrac{7}{12} \quad .$$

Here part (c) is Laplace's solution, while (a) yields the numerical value determined by Price — and yet there seems to be no mention of a "second trial" in (a)!

However it is possible, by an appropriate interpretation, to obtain Price's result from Bayes's theory. The postulates of §2 of the Essay require that successes and failures be defined referentially to an initial event. Thus the event described by Price as having happened once plays the same rôle as W, the first ball thrown, in the postulates. What is then required by Price is essentially the probability that the next throw results in a "success" (say), inasmuch as it falls in the interval $[\frac{1}{2}, 1]$, the first ball having demarcated the lower limit of this interval. The solution is then immediately given (for *one* success) by

$$\int_{1/2}^{1} x\,dx \bigg/ \int_{0}^{1} x\,dx = \frac{3}{4}$$

as Price showed. This is surely the correct interpretation[26].

Consideration is then given to the odds on the event's happening once again after it has happened twice, thrice,...,p times. Price's answers — odds of $2^{p+1} - 1$ to 1 in the last case — are given similarly by considering, in general,

$$\int_{1/2}^{1} x^p\,dx \bigg/ \int_{0}^{1} x^p\,dx = 1 - 1/2^{p+1} \quad ,$$

and while this is the solution provided by Proposition 10, it is perhaps unfortunate to interpret it[27], as Price does, as the odds "for *more* than an equal chance that it will happen on further trials" [p.405].

Considering next the case of an event which is only known to have happened ten times without failing, Price supposes the

> enquiry to be what reason we shall have to think we are right if we guess that the probability of it's happening in a single trial lies somewhere between 16/17 and 2/3, or that the ratio of the causes of it's happening to those of it's failure is some ratio between that of sixteen to one and two to one. [p.406]

That is, we are trying to find

$$\Pr\left[\tfrac{2}{3} < x < \tfrac{16}{17} \mid 10 \text{ successes and } 0 \text{ failures}\right]$$

or

$$\Pr\left[\tfrac{2}{3} < C(E)\,/C(\overline{E}) < \tfrac{16}{17} \text{ given } 10 \text{ successes and } 0 \text{ failures}\right]$$

(where $C(E)$ denotes the "causes of E"). The former formulation is exactly that of Bayes's Proposition 9: the latter (in terms of causes) has no parallel in the Essay. Price once again uses Bayes's method correctly, obtaining the answer 0.5013 &c.

In discussing his next example, that concerned with the throwing of a die, Price argues in such a manner as to confirm our second interpretation of his first illustration.

> It will appear, therefore, that *after* the first throw and not before, we should be in the circumstances required by the conditions of the present problem, and that the whole effect of this throw would be to bring us into these circumstances. That is: the turning the side first thrown in any subsequent single trial would be an event about the probability or improbability of which we could form no judgement, and of which we should know no more than that it lay somewhere between nothing and certainty. With the second trial then our calculations must begin. [p.407]

Some numerical work follows.

Attention is next given to the famous problem of the probability of the sun's rising[28]. This solar problem, often ignorantly supposed to have originated with Laplace, is in fact to be found, albeit but vaguely expressed, in various forms in Hume's writings[29]. It indeed provides a good illustration of Edgeworth's [1884b] statement that

> the much decried method of Bayes may be employed to deduce from the frequently experienced occurrence of a phenomenon the large probability of its recurrence. [p.228]

In this problem (an entirely similar argument to that given in the die-tossing example mentioned above being advanced) Price explains that the first sinking of the sun a sentient person who has newly arrived in this world would see, leaves him "entirely ignorant whether he should ever see it again" [p.409]. As Pearson [1978] has it

> The first experiment counts nothing because you must know there is a sun or a red ball in a bag before you can argue about the repetition of drawing red balls. [p.368]

Thus, according to Price,

> let him see a second appearance or one *return* of the Sun, and an expectation would be raised in him of a second return, and he might know that there was an odds of 3 to 1 for *some* probability of this. [p.409]

That is,

$$\Pr[(1/2) < x < 1 \mid \text{one return}] = \int_{1/2}^{1} x\,dx \Big/ \int_{0}^{1} x\,dx = 3/4.$$

Next

> let it be supposed that he has seen it return at regular and stated intervals a million of times[30]. The conclusions this would warrant would be such as follow — There would be the odds of the millionth power of 2, to one, that it was likely that it would return again at the end of the usual interval. [pp.409-410]

As Zabell [1988a] has pointed out, there is a slight error here, in that n, the number of occurrences of the event in question, being 1,000,000, the odds should be $2^{1,000,001}$ to 1 on a reappearance. The appropriate exponent of 2 is $(n+1)$ — i.e. the number of risings of the sun — and not n, which is the number of returns of the sun. It was possibly a hasty reading of this section of the Appendix that was responsible for Buffon's incorrectly giving the odds as $2^{m-1} : 1$, where m is the number of risings (see §4.7).

This example is clearly analogous to his first illustration, though it should be noted that while Price is correctly applying Bayes's results, a tendency to apply them to future events seems to be making its presence felt. It is, of course, quite possible that Bayes intended his solution to be applicable to the case of "a single throw" *after* experience: however, this is nowhere explicitly stated *in the Essay*, and, as we shall see in Chapter 6, Bayes's result is in accord with *not* interpreting this "single trial" in the predictive sense (indeed, the actual statements of his ninth and tenth propositions are in the past tense). However, it is not obvious from the first quotation in §3.5 above ("in what follows...") that Bayes intended his result to be used only in a retrodictive sense: in fact, Price writes quite explicitly in his introductory letter that Bayes's intent originally was to find the probability of an event given a number of occurrences and failures.

Price next turns his attention "to cases where an experiment has sometimes succeeded and sometimes failed" [p.411]. To illustrate the general ideas he considers the drawing of blanks and prizes from a lottery, fixing his attention on what is essentially

$$\Pr\left[x_1 < x < x_2 \mid p \text{ blanks and } q \text{ prizes drawn}\right],$$

where x is the (true?) proportion of blanks to prizes in the lottery. Once again this is a straightforward and correct application of Bayes's results.

Price then passes some remarks on the probability of causes[31], and draws towards a conclusion by noting that "The foregoing calculations further shew us the uses and defects of the rules laid down in the essay" [p.417]. These defects seem to be that the second and third rules "do not give us the required chances within such narrow limits as could be wished" [p.417]. However, these limits become narrower as q increases with respect to p, while the exact solution is given by the second rule when $p = q$.

> These two rules therefore afford a direction to our judgement that may be of considerable use till some person shall discover a better approximation to the value of the two series's in the first rule. [pp.417-418]

A footnote (possibly added in proof?) now states that Price had found an improvement of the approximation in the second and third rules, by showing that

$$2\Sigma / (1 + 2Ea^p b^q + 2Ea^p b^q / n)$$

"comes almost as near to the true value wanted as there is reason to desire, only always somewhat less" [p.418]. This too will be reconsidered later.

In his introduction to the Essay Price had commented on de Moivre's rules

> to find the probability there is, that if a very great number of trials be made concerning any event, the proportion of the number of times it will happen, to the number of times it will fail in those trials, should differ less than by small assigned limits from the proportion of the probability of its failing in one single trial. [pp.372-373]

No person, to the best of Price's knowledge, had yet shown how to solve the converse problem, viz.

> the number of times an unknown event has happened and failed being given, to find the chance that the probability of its happening should lie somewhere between any two named degrees of probability. [p.373]

Therefore, de Moivre's work was not sufficient to make consideration of this point unnecessary. Price now concludes the Appendix by noting that

> what most of all recommends the solution in this *Essay* is, that it is compleat in those cases where information is most wanted, and where Mr. De Moivre's solution of the inverse problem can give little or no direction; I mean, in all cases where either p or q are of no considerable magnitude. [p.418]

3.7 Appendix 3.1

In view of the important rôle played by Bayes's Theorem in modern subjective probability, it might be of no little interest briefly to consider the view of de Finetti, a leading exponent of subjective probability, on this result. These views are expressed in §9.2 of his *Probability, Induction and Statistics* of 1972.

After pointing out that Bayes's formulation of his problem is, strictly speaking, unsatisfactory, de Finetti singles out the following assumptions of the Essay for detailed examination:

> (1) The "unknown probability" p has probability dx of being comprised in any interval $(x, x + dx)$ in (0,1).
> (2) The events considered are independent under each hypothesis $p = x$ as to the value of p.
> (3) Therefore, after the observations, the probability that p falls between x and $x+dx$ becomes $Kx^m(1-x)^{n-m}\,dx$. [1972, p.158]

(Here m and $n - m$ denote respectively the numbers of favourable and unfavourable events that have already happened, while K is a normalizing constant.)

Noting that (3) admits of dispute only inasmuch as it concerns "the extent of the domain of applications, which can be narrowed if one wishes to confine the notion of probability to a restrictive meaning" [p.158], de Finetti goes on to point out that the true meaning of the hypothesis in (2) may be clarified to remove reference to the "unknown probability" p.

Turning his attention next to (1), de Finetti points out that a reformulation of the problem might permit the removal of the meaninglessness of the phrase "unknown probability": moreover, this "Bayes' postulate" is "not necessary to the expression of the problem in terms of Bayes' theorem" [p.159] (as we have in fact already seen). The vagueness in the phrase "knowing nothing" leads de Finetti to conclude [p.159] that the postulate is either a tautology (if "knowing nothing" means that a uniform distribution is to be attributed to p), or else a nonsense (if "nothing" is taken literally, for in this case knowing nothing about E_i will mean knowing nothing about

E_iE_j, and hence p^2 will have to have a uniform distribution also).

After some comments on Laplace's more general theorem (in which the initial density need not be uniform), de Finetti recalls some results and applications from the Essay[32]: further details may be found on pp.160-162 of his book cited above.

Chapter 4

Miscellaneous Investigations from 1761 to 1822

As by successive tradition from our forefathers we have received it.

Marcus Aurelius Antoninus.

Of the 54 references cited in Todhunter's "Chronological List of Authors" [1865, pp.619-620] as contributing to probability theory from 1761 to 1822 (and excluding Condorcet and Laplace) only some ten make any contribution to our present topic. The writings to be discussed in this chapter are given, as in others, in order of publication; but once an author is cited, any further pertinent publications of his (although they may well have been written *after* those of another author not yet cited) will be discussed in the same section.

4.1 Moses Mendelssohn (1729-1786)

Moses Mendelssohn[1], born Mendel, and the grandfather of the arguably more famous Jakob Ludwig Felix Mendelssohn-Bartholdy, included an essay on probability in his *Philosophische Schriften.* This appears under the heading "Ueber die Wahrscheinlichkeit" as chapter IV of the second volume of the second edition[2] of 1777.

probability of a causal connexion is $n/(n+1)$. No argument for this value is given, though its use is illustrated by an example concerning repeated onsets of giddiness after drinking coffee.

In a paper on upper and lower probabilities, Dempster [1966, p.369] shows that, when T of the first n sample individuals are observed to fall in a certain category, the upper and lower probabilities ($\overline{P}$ and $\underline{P}$ respectively) that the next sample individual will fall into that category are given by

$$\overline{P} = (T+1)/(n+1)\,, \quad \underline{P} = T/(n+1)\,.$$

(Notice that Laplace's rule of succession would yield a value $(T+1)/(n+2)$ falling between $\underline{P}$ and $\overline{P}$.) On replacing T by n we obtain Mendelssohn's value from $\underline{P}$.

Referring (incorrectly) to "the theorems of Bayes and Laplace" (by which is apparently meant the rule of succession), Todhunter [1865, p.617] notes that the probability that an event, which has already happened n times, will happen a further time, is $(n+1)/(n+2)$: he comments on the close agreement, for large n, of this result with that obtained by Mendelssohn, but the coincidence is more apparent than real, for there is no sign in Mendelssohn's essay of any deep knowledge of probability (or expertise therein). We must conclude with Todhunter (loc. cit.) that "we cannot therefore consider that he [i.e. Mendelssohn] in any way anticipated Bayes".

4.2 Bayes and Price

4.2.1 Bayes's paper on divergent series

Strictly speaking, of course, discussion of this paper has no place here. Nevertheless, in view of Bayes's extremely limited output of writings on mathematical topics, and because of our interest in Bayes in general, I propose to give it some attention, albeit brief[3].

This paper, published in the *Philosophical Transactions* for 1763, pp.269-271, is referred to by Price in a footnote to page 401 of the Essay, in connexion with the evaluation of factorials needed for Rule 2. Price was possibly also responsible for the submission of this paper[4], since, although it bears only the heading "A Letter from the late Reverend Mr. Thomas Bayes, F.R.S. to John Canton, M.A. and F.R.S.", it was read on the 24th November 1763.

Here Bayes considers the expansion of the series for log $z!$ already considered by "some eminent mathematicians" [p.269]. In his introductory note preceding the 1940 reprinting of this paper by Molina, W. Edwards Deming suggests that Bayes had de Moivre and Stirling in mind when using this phrase, and adduces in support of this suggestion the following remarks[5]:

(i) Bayes's use of c for 2π is commonly found in the writings of both de Moivre and Stirling;

(ii) while many series were available to Bayes as illustrations, the one he in fact used is that which de Moivre and Stirling studied extensively;

(iii) Price, Bayes's intimate, refers, on p.401 of the Essay, to "Mr. De Moivre, Mr. Simpson and other eminent mathematicians".

But all this is mere conjecture: let us return to the paper in question. Bayes states that it has been asserted that $\sum_{k=1}^{n} \log k$ is equal to

$$\frac{1}{2}\log c + (z + \frac{1}{2})\log z - S\ ,$$

where c denotes the circumference of a circle whose radius is unity and where

$$S = z - \frac{1}{12z} + \frac{1}{360z^3} - \frac{1}{1260z^5} + \frac{1}{1680z^7} - \frac{1}{1188z^9} + \ \&c.$$

Nowadays, of course, this would be more correctly (and suggestively!) written[6] as

$$\log z! \sim \log\sqrt{2\pi} + (z + \frac{1}{2})\log z - S\ .$$

(The determination of the constant as $\sqrt{2\pi}$ is due to Stirling — see Archibald [1926, p.675], and note the original Latin on p.[2].) Bayes was apparently among the first (if not *the* first) to appreciate the asymptotic character of the series[7], and for this alone he surely deserves some acknowledgement.

4.2.2 The supplement to the Essay

On the 26th of November 1764 Price submitted to John Canton a supplement to the Essay. The first part of this paper was apparently due to Bayes, for shortly before Section 13 we find the words "Thus far have I transcribed Mr. Bayes". Indeed, the fact that Bayes's proof given here is only for Rule 2 perhaps lends weight to my earlier assertion (see §2.5) that the third rule was due to Price.

This supplement is devoted to proofs and some slight elaboration of the Rules of the Essay[8]: we shall content ourselves here with some fairly general remarks (a comparatively detailed discussion may be found in Sheynin [1969]).

After having mentioned the refinements he had proposed to the bounds given by Bayes, Price went on in his letter to Canton to say[9]

> Perhaps, there is no reason about being very anxious about proceeding to further improvements. It would, however, be very agreeable to me to see a yet easier and nearer approximation to the value of the two series's in the first rule: but this I must leave abler persons to seek, chusing now entirely to drop this subject.

Since Price did not die until 1791, it is to be hoped that his desire expressed in this quotation was realized.

Part of the proof given here may be found, in manuscript, in the notebook of Bayes referred to earlier: we shall discuss the relevant passage at a later stage.

It is perhaps worth noting that the Rule as given in the Supplement covers more cases than that of the Essay[10]. In the latter it is stated that, if I guess that the probability of an event's happening in a single trial lies between $p/n + z$ and $p/n - z$, the chance of my being right is greater than

$$2\Sigma\,/(1 + 2Ea^p b^q + 2Ea^p b^q/\,n) \tag{1}$$

and less than

$$2\Sigma\,/(1 - 2Ea^p b^q - 2Ea^p b^q/\,n)\ , \tag{2}$$

while if $p = q$ the chance is exactly 2Σ. (Here a, b, E and Σ are as defined in Rule 2 of the Essay — see §2.4.)

In the Supplement, however, the following six cases are considered:

Case 1. If $q > p$ and I judge that the probability of the event's happening in a single trial lies between p/n and $p/n + z$, the chance that I am correct is greater than Σ and less than

$$\Sigma \times \frac{1 + 2Ea^p b^q + 2Ea^p b^q/n}{1 - 2Ea^p b^q - 2Ea^p b^q/n}\ . \tag{3}$$

Case 2. If $q > p$, and the limits in the previous case are replaced by $p/n - z$ and p/n, my chance of being right is less than Σ and greater than

$$\Sigma \times \frac{1 - 2Ea^p b^q - 2Ea^p b^q/n}{1 + 2Ea^p b^q + 2Ea^p b^q/n}\ . \tag{4}$$

Case 3. If $p > q$, and the limits are p/n and $p/n + z$, the chance of a correct guess is less than Σ and greater than (4).

Case 4. If $p > q$ and the limits are $p/n - z$ and p/n, the chance of my being correct is greater than Σ and less than (3).

Case 5. If $p = q$, my chance is Σ exactly (this must refer to either the interval $(p/n, p/n + z)$ or $(p/n - z, p)$).

Case 6. Whether $p > q$ or $q > p$, and I judge that the probability lies between $p/n - z$ and $p/n + z$, the chance of my being correct is greater than (1) and less than (2). If $p = q$ the chance is 2Σ exactly.
Notice that it is this last case which is given as Rule 2 in the Essay.

It is now that Bayes's contribution ceases and Price proceeds to his own improvements of the bounds, motivating his investigation in the following words:

> It appears, from the Appendix to the Essay, that the rule here demonstrated, though of great use, does not give the required chance within limits sufficiently narrow. It is therefore necessary to look out for a contraction of these limits ...

Articles 13-28 are devoted to this investigation, Price returning in the last of these articles to one of the examples mentioned in the Appendix to the Essay (the fifth case, p.415).

In Article 24 Price concludes that under the conditions of Case 6, the chance of my being right is greater than (1) and less than 2Σ. He next passes on to "determine within still narrower limits whereabouts the required chance must lie" [art.24], his conclusion being summarized in Article 28 as follows [notation slightly altered]:

> If either p or q is greater than 1, the true chance that the probability of an unknown event which has happened p times and failed q in $(p+q)$ or n trials, should lie somewhere between $p/n + z$ and $p/n - z$ is less than 2Σ, and greater than
>
> $$\Sigma + \frac{\Sigma(1 - 2Ea^p b^q - 2Ea^p b^q/n)}{1 + Ea^p b^q + Ea^p b^q/n} .$$
>
> If either p or q is greater than 10, this chance is less than 2Σ, and greater than
>
> $$\Sigma + \frac{\Sigma(1 - 2Ea^p b^q - 2Ea^p b^q/n)}{1 + Ea^p b^q/2 + Ea^p b^q/2n} .$$

To show the improvement effected by his limits, Price returns to an example considered in his Appendix to the Essay: an event, concerning which nothing is known, has happened 100 times and failed 1000 times in 1100 trials. The chance that the probability of this event lies between $\frac{10}{11} + \frac{1}{110}$ and $\frac{10}{11} - \frac{1}{110}$, as computed by Bayes's Rule 2 (art.12) lies between 0.6512 (odds of 186 to 100) and 0.7700 (odds of 334 to 100). (The numbers given originally in the appendix were incorrect, since m^2 was there set equal to n^3/pq instead of $n^3/2pq$.) Using the fact that the improved bounds are 2Σ and

$$\Sigma + \frac{\Sigma(1 - Ea^p b^q - 2Ea^p b^q/n)}{1 + Ea^p b^q/10 + Ea^p b^q/10n} ,$$

Price finds the limits 0.6748 (odds of 207 to 100) and 0.7057 (odds of 239 to 100).

Price's investigations led him to conclude that

> In all cases when z is small, and also whenever the disparity between p and q is not great 2Σ is almost exactly the true chance required. And I have reason to think, that even in all other cases, 2Σ gives the true chance nearer than within the limits now determined. [art.28]

Before leaving the Supplement one might note Price's footnote to the corollary in Article 20. Here he points out that it follows from Article 20 that, in the case in which neither p nor q is very small (or even not less than 10), the probability x of the event satisfies the following:

$$\text{(i)} \quad \Pr\left[\frac{p}{n} - \frac{1}{\sqrt{2}}\gamma < x < \frac{p}{n} + \frac{1}{\sqrt{2}}\gamma\right] \approx \frac{1}{2}$$

$$\text{(ii)} \quad \Pr\left[\frac{p}{n} - \gamma < x < \frac{p}{n} + \gamma\right] \approx \frac{2}{3}$$

$$\text{(iii)} \quad \Pr\left[\frac{p}{n} - \sqrt{2}\gamma < x < \frac{p}{n} + \sqrt{2}\gamma\right] \approx \frac{5}{6}$$

where $\gamma = \sqrt{pq/(n^3 - n^2)}$ is "the point of contrary flexure" [art.26][11]. A numerical example, with $p = 1000$ and $q = 100$, follows.

In the chapter on Laplace we shall show that (in the notation of the present section)[12]

$$\Pr\left[\frac{p}{n} - \frac{\tau\sqrt{2pq}}{n\sqrt{n}} < x < \frac{p}{n} + \frac{\tau\sqrt{2pq}}{n\sqrt{n}}\right] \approx \frac{2}{\sqrt{\pi}}\int_0^\tau e^{-t^2}\,dt . \qquad (5)$$

With $\tau = \frac{1}{2}\sqrt{n/(n-1)}$, the left-hand side of (5) becomes the left-hand side of (i), and hence the latter is approximately equal to

$$\frac{2}{\sqrt{\pi}}\int_0^{\frac{1}{2}\sqrt{n/(n-1)}} e^{-t^2}\,dt = 2\Phi\Big(\sqrt{n/2(n-1)}\Big) - 1 , \qquad (6)$$

where $\Phi(\cdot)$ is the cumulative distribution function of a random variable having the standard Normal distribution. If n is large, then $n/(n-1) \approx 1$, and (6) becomes

$$2\Phi(1/\sqrt{2}) - 1 = 0.5588,$$

which accords reasonably well with (i).

We have already gleaned from our discussion of the Essay that Price probably believed Bayes's results to be applicable in a causal setting. This is given further support by the following passage from the covering letter:

> The solution of the problem enquired after in the papers I have sent you has, I think, been hitherto a *desideratum* in philosophy of some consequence. To this we are now in a great measure helped by the abilities and skill of our late worthy friend; and thus are furnished with a necessary guide in determining the nature and proportions of unknown causes from their effects, and an effectual guard against one great danger to which philosophers are subject; I mean, the danger of founding conclusions on an insufficient induction, and of receiving just conclusions with more assurance than the number of experiments will warrant.

As we have seen, however, there is little (if anything) in Bayes's Essay to warrant such an extension of the results. As it is, at all events, no causal application is called for in this paper, and the matter is accordingly of no importance in the present context.

4.2.3 Bayes's Notebook

In his comments on a paper by Perks [1947], M.E. Ogborn casually mentioned that

> in his own office there was a book which for some time he had not been able to place, but when visiting the Royal Society in connexion with another matter he had realized that the handwriting in the book appeared to be identical with other specimens of Bayes's handwriting. He thought it was, in fact, one of Bayes's notebooks. [Perks 1947, p.318]

No further attention was apparently paid to this remark until Holland [1962] commented on the relic in a biographical note on Bayes, choosing to cite from the contents of the notebook

> a method of "finding the time and place" of the conjunction of two planets, some notes on weights and measures, on a method of differentiation and a note on logarithms. [p.457]

Holland also mentioned the note on an electrifying machine and drew attention to the shorthand (a modification by Elisha Coles of one of Thomas Shelton's systems) used by Bayes[13]. (In addition to the shorthand, Bayes used English, French and Latin.)

Only one passage in the notebook pertains to probability: it is concerned with a proof of one of the Rules in Bayes's Essay, and since it is probably not readily available I present, from here to the end of this section, a free translation of the original Latin text (for further details see Dale [1986]). Some of the formulae are given in a more modern notation, and certain obvious *lapsus calami* have been corrected. Bayes's first paragraph is unlabelled.

Firstly, let $S/V = x$. Then $\dot{x} = (S/V)(\dot{S}/S - \dot{V}/V)$. Thus if $\dot{S}/S > \dot{V}/V$ and S and V are both increasing (and of the same sign), $\dot{x} > 0$, and so x is increasing. Similarly, if $y = V/S$ and S and V are both decreasing (and of opposite sign) then V/S is increasing.

Art. 2. Let

$$A = (1 - nz/p)^p(1 + nz/q)^q$$

$$B = (1 + nz/p)^p(1 - nz/q)^q$$

where $n = p + q$. Then

$$\int_{-q/n}^{p/n} A\,dz = n^n\left[(n+1)\binom{n}{p}p^p q^q\right]^{-1} .$$

The integral of B from $z = -p/n$ to q/n reduces to the same expression.

Art. 3. With A and B as given above, and D and Δ defined by

$$D = (1 - n^2z^2/q^2)^{nq/2p}, \quad \Delta = (1 - n^2z^2/p^2)^{np/2q} ,$$

we find that

$$\begin{aligned} \dot{B}/B : \dot{D}/D \ &:: \ (1 - n^2z^2/q^2) : (1 - nz/q)(1 + nz/p) \\ &:: \ (1 + nz/q) : (1 + nz/p). \end{aligned}$$

Since $q > p$ and $\dot{D}/D$ is negative, it follows that $\dot{B}/B > \dot{D}/D$. (Examination of the subsequent Articles shows that attention is restricted to $z \geq 0$.) Hence, since B and D are both decreasing functions of z, we find from the first Article that B/D is increasing; and since $B = D = 1$ when $z = 0$, we may conclude that $B > D$ — and similarly that $A < \Delta$. Moreover, since

$$\Delta^2 = (1 - n^2z^2/p^2)^{np/q} \text{ and } AB = (1 - n^2z^2/p^2)^p(1 - n^2z^2/q^2)^q,$$

it follows that

$$\Delta^2 : AB :: (1 - n^2z^2/p^2)^{p^2/q} : (1 - n^2z^2/q^2)^q .$$

Thus

$$\Delta^{2q} : (AB)^q :: (1 - n^2z^2/p^2)^{p^2} : (1 - n^2z^2/q^2)^{q^2}$$

and hence $\Delta^2 < AB$ and $2\Delta < A + B$.

Art. 4. Let A be an unknown event with prior probability x. By Proposition 10 of the Essay it follows that, for $z \geq 0$,

$$\begin{aligned} P_1 &\equiv \Pr[p/n - z \leq x \leq p/n \mid A \text{ has happened exactly } p \text{ times} \\ &\qquad\qquad \text{in } n = p + q \text{ trials}] \\ &= \int_{p/n-z}^{p/n} \binom{n}{p} x^p(1-x)^q\,dx \Big/ \int_0^1 \binom{n}{p} x^p(1-x)^q\,dx \\ &= (n+1)\binom{n}{p} \int_0^z (p/n - u)^p(q/n + u)^q\,du \\ &= (n+1)\binom{n}{p}(p^pq^q/n^n) \int_0^z (1 - nu/p)^p(1 + nu/q)^q\,du \ . \end{aligned} \tag{7}$$

Having noticed in Article 3 that

$$(1 - nu/p)^p(1 + nu/q)^q < (1 - n^2u^2/p^2)^{np/2q},$$

we see that

$$P_1 \leq (n+1)\binom{n}{p}(p^pq^q/n^n) \int_0^z (1 - n^2u^2/p^2)^{np/2q}\,du \ .$$

2°. Under the same hypotheses it follows that

$$\begin{aligned} P_2 &\equiv \Pr[p/n \leq x \leq p/n + z \mid A \text{ has happened exactly } p \text{ times in} \\ &\qquad\qquad n = p + q \text{ trials}] \\ &= (n+1)\binom{n}{p}(p^pq^q/n^n) \int_0^z (1 - nu/q)^q(1 + nu/p)^p\,du \qquad (8) \\ &\geq (n+1)\binom{n}{p}(p^pq^q/n^n) \int_0^z (1 - n^2u^2/p^2)^{nq/2p}\,du \ . \end{aligned}$$

3°. Finally,

$$P_3 \equiv \Pr[p/n - z \leq x \leq p/n + z \mid A \text{ has happened exactly } p \text{ times in } n = p + q \text{ trials}]$$

$$= (n+1)\binom{n}{p}(p^p q^q/n^n)\int_0^z \left[\frac{(1-nu/p)^p}{(1+nu/q)^q} + \frac{(1-nu/q)^q}{(1+nu/p)^p}\right] du$$

$$\geq 2(n+1)\binom{n}{p}(p^p q^q/n^n)\int_0^z (1-n^2u^2/p^2)^{np/2q}\, du \quad (\text{since } A + B > 2\Delta).$$

In subsequent Articles certain approximations to these integrals are obtained.

Art. 5. From the expression

$$n! = \sqrt{2\pi}\, n^{n+\frac{1}{2}} e^{-n} e^{\theta/12n} \quad , \quad 0 < \theta < 1$$

one finds that

$$\binom{n}{p} p^p q^q / n^n = \sqrt{\frac{n}{2\pi pq}}\, e^{\theta(pq - np - nq)/12npq} \quad .$$

Since $(pq - n^2) < 0$, we may conclude that

$$N\sqrt{\frac{n}{2\pi pq}} < \binom{n}{p} p^p q^q / n^n < \sqrt{\frac{n}{2\pi pq}}$$

where $\ln N = (pq - n^2)/12npq$.
(Bayes does not give details of his derivation of this last expression: I have tried to reconstruct his argument.)

Art. 6.

$$\int_{-p/n}^{p/n} (1 - n^2u^2/p^2)^{np/2q}\, du = \frac{2^{(np+q)/q}\, p\, [\Gamma((np/2q)+1)]^2}{n\,\Gamma((np/q)+2)} \quad .$$

On our using the approximation

$$\Gamma(z+1) \sim \sqrt{2\pi}\, z^{z+\frac{1}{2}} e^{-z} \quad ,$$

we obtain

$$\int_{-p/n}^{p/n} (1 - n^2u^2/p^2)^{np/2q}\, du \sim \frac{ep}{np+q}\sqrt{\frac{2\pi pq}{n}}\,(1+q/np)^{-(np/q+\frac{1}{2})}$$

$$> \frac{p}{2(np+q)}\sqrt{\frac{2\pi pq}{n}} \ .$$

Similarly

$$\int_{-q/n}^{q/n} (1 - n^2u^2/q^2)^{nq/2p}\, du > \frac{q}{2(nq+p)}\sqrt{\frac{2\pi pq}{n}} \ .$$

Art. 7. From Articles 6 and 4.3 it follows that

$$\Pr\left[0 \le x \le 2p/n \mid A \text{ has happened exactly } p \text{ times in } n = p+q \text{ trials}\right]$$

$$> (n+1)\binom{n}{p}(p^pq^q/n^n)\int_0^{p/n} (1 - n^2u^2/p^2)^{np/2q}\, du$$

$$> N(n+1)p/(np+q) \ .$$

If p and q are both large, and $q > p$, then the right-hand side of this last expression is approximately 1. Furthermore, under such conditions on p and q,

$$\Pr\left[p/n - z \le x \le p/n + z \mid A \text{ has happened exactly } p \text{ times in } n = p+q \text{ trials}\right]$$

$$\sim 2(n+1)\sqrt{n/(2\pi pq)}\int_{-p/n}^{p/n} (1 - n^2u^2/p^2)^{np/2q}\, du$$

without appreciable error.

In the last two Articles of his work Bayes turns his attention to the evaluation of the integrals in (7) and (8).

Art. 8. If $x : r :: p : q$ and $(x+r)^n$ is expanded, then

$$\binom{n}{q}x^pr^q \ : \ \binom{n}{q-1}x^{p+1}r^{q-1} \ :: \ 1 \ : \ (q/(p+1))(p/q)$$

$$\binom{n}{q}x^pr^q \ : \ \binom{n}{q-2}x^{p+2}r^{q-2} \ :: \ 1 \ : \ \frac{q(q-1)}{(p+1)(p+2)}(p/q)^2 \ \ \&c.$$

Similarly

$$\binom{n}{q}x^pr^q \ : \ \binom{n}{q+1}x^{p-1}r^{q+1} \ :: \ 1 \ : (p/(q+1))(q/p)$$

$$\binom{n}{q}x^pr^q \ : \ \binom{n}{q+2}x^{p-2}r^{q+2} \ :: \ 1 \ : \ \frac{p(p-1)}{(q+1)(q+2)}(q/p)^2 \ \ \&c.$$

Thus

$$\sum_{k\le q}\binom{n}{k}x^{n-k}r^k \Big/ \sum_{k>q}\binom{n}{k}x^{n-k}r^k$$

$$= \left[1+\frac{q}{p+1}\left(\frac{p}{q}\right)+\frac{q(q-1)}{(p+1)(p+2)}\left(\frac{p}{q}\right)^2 + \frac{q(q-1)(q-2)}{(p+1)(p+2)(p+3)}\left(\frac{p}{q}\right)^3+\cdots\right]\Big/$$

$$\left[\frac{p}{q+1}\left(\frac{q}{p}\right)+\frac{p(p-1)}{(q+1)(q+2)}\left(\frac{q}{p}\right)^2 + \frac{p(p-1)(p-2)}{(q+1)(q+2)(q+3)}\left(\frac{q}{p}\right)^3+\cdots\right].$$

Now if $q>p$ the k-th term of the numerator on the right-hand side of this last expression is greater than the corresponding term of the denominator. For small p the series in the denominator terminates first, wherefore the series in the numerator is greater than that in the denominator. If in fact 1 is subtracted from the series in the numerator, the series in the denominator will be larger than that in the new numerator.

Art. 9. Attention is next turned to the evaluation of the integral

$$\begin{aligned} I &\equiv \int_{-q/n}^{p/n}(n+1)\binom{n}{p}(p^p q^q/n^n)(1-nz/p)^p(1+nz/q)^q\,dz \\ &= \int_{-q/n}^{p/n}\left[(n+1)\binom{n}{p}(p/n-z)^p(q/n+z)^q\right]dz \\ &= \int_{-q/n}^{0}(n+1)\binom{n}{p}x^p r^q\,dz + \int_0^{p/n}(n+1)\binom{n}{p}x^p r^q\,dz \\ &\equiv I_1+I_2\ , \end{aligned}$$

say, where $x=p/n-z$, $r=q/n+z$.

To evaluate I_1, consider the series

$$V = r^{n+1}+\binom{n+1}{1}r^n x+\cdots+Fr^{q+1}x^p\ .$$

Since $\dot{r} = \dot{z} = -\dot{x}$,

$$\begin{aligned}\dot{V} &= (n+1)r^n\dot{r} + \binom{n+1}{1}\left[nr^{n-1}\dot{r}\,x + r^n\,\dot{x}\right] + \binom{n+1}{2}\left[(n-1)r^{n-2}\dot{r}\,x^2\right.\\ &\quad \left.+r^{n-1}2x\dot{x}\right] + \cdots + F\left[(q+1)r^q\dot{r}\,x^p + r^{q+1}px^{p-1}\dot{x}\right]\\ &= (n+1)r^n\dot{r} + \binom{n+1}{1}\left[nr^{n-1}\dot{r}\,x - r^n\dot{r}\right] + \binom{n+1}{2}\left[(n-1)r^{n-2}\dot{r}\,x^2\right.\\ &\quad \left.-r^{n-1}2x\dot{r}\right] + \cdots + F\left[(q+1)r^q\dot{r}\,x^p - r^{q+1}px^{p-1}\dot{r}\right]\\ &= (q+1)Fr^qx^p\dot{r}\ .\end{aligned}$$

Noting that F is the coefficient of $r^{q+1}x^p$ in the expansion of $(x+r)^{n+1}$, we find that $(q+1)F = (n+1)\binom{n}{p}$, and hence V can be written in the form

$$V = r^{n+1} + \binom{n+1}{1}r^nx + \cdots + \left[(n+1)/(q+1)\right]\binom{n}{p}r^{q+1}x^p\ ,$$

a series which reduces, when $z = 0$, to

$$B \equiv \frac{(n+1)}{n}\frac{q}{q+1}\binom{n}{p}\frac{p^pq^q}{n^n}\left[1 + \frac{pq}{(q+2)p} + \frac{p(p-1)q^2}{(q+2)(q+3)p^2} + \cdots\right]\ .$$

Thus

$$\begin{aligned}I_1 &= \int_{-q/n}^{0}(n+1)\binom{n}{p}x^pr^q\,dz\\ &= \int_0^{q/n}(n+1)\binom{n}{p}(1-r)^pr^q\,dr \qquad (= B)\\ &= (n+1)\binom{n}{p}I_{q/n}(q+1,p+1)\ ,\end{aligned}$$

$I_a(b,c)$ denoting the incomplete beta-function[14]. A similar procedure shows that

$$\begin{aligned}I_2 &= \int_0^{p/n}(n+1)\binom{n}{p}x^pr^q\,dz\\ &= (n+1)\binom{n}{p}I_{p/n}(p+1,q+1).\end{aligned}$$

4.2.4 Price's Four Dissertations

In 1767 Richard Price published[15] a volume entitled *Four Dissertations*, these being the following:

1. On Providence.
2. On Prayer.
3. On the Reasons for expecting that virtuous Men shall meet after Death in a State of Happiness.
4. On the Importance of Christianity, the Nature of Historical Evidence, and Miracles.

Only the fourth of these essays contains anything pertinent to our topic: in the second section, entitled "The Nature and Grounds of the Regard due to Experience and to the Evidence of Testimony, stated and compared", we find some discussion of probability. Although no direct use of Bayes's Theorem is made, Price does quote examples illustrating the results he had given in the Appendix to Bayes's Essay[16].

But before considering these examples, it might be of interest to note Price's illustration of the influence of knowledge on future observation. After a long quotation from Hume's *Essay on Miracles*[17] Price turns to a consideration of the assurance, given by experience, of the laws of nature. "This assurance", he says,

> is nothing but the conviction we have, that future events will be agreeable to what we have hitherto found to be the course of nature, or the *expectation* arising in us, upon having observed that an event has happened in former experiments, that it will happen again in *future* experiments. [pp.389-390]

This is then illustrated by the following example:

> if I was to draw a slip of paper out of a wheel, where I knew there were more white than black papers, I should intuitively see, that there was a probability of drawing a white paper, and therefore should *expect* this; and he who should make a mystery of such an expectation, or apprehend any difficulty in accounting for it, would not deserve to be seriously argued with. — In like manner; if, out of a wheel, the particular contents of which I am ignorant of, I should draw a white paper a hundred times together, I should see that it was probable, that it had in it more white papers than black, and therefore should expect to draw a white paper the next trial. There is no more difficulty in this

> case than in the former; and it is equally absurd in both cases to ascribe the *expectation*, not to *knowledge*, but to *instinct*. [pp.390-391]

Similar examples, concerned with the tossing of a die and with the happening of an event in every trial a million times, are also cited to show that an observed frequency should be used as a reasonable predictor for future occurrences.

In a long footnote (stretching over four pages) Price proceeds to the examples mentioned above. Although some of these are somewhat similar to those given in his Appendix to the Essay, I choose to give them all here in detail as they are seldom cited.

> In an essay published in vol. 53d of the *Philosophical Transactions*, what is said here and in the last note, is proved by mathematical demonstration, and a method shewn of determining the exact probability of all conclusions founded on induction. — This is plainly a curious and important problem, and it has so near a relation to the subject of this dissertation, that it will be proper just to mention the results of the solution of it in a few particular cases.
> Suppose, 1st, all we know of an event to be, that it has happened ten times without failing, and that it is inquired, what reason we shall have for thinking ourselves right, if we judge, that the probability of its happening in a single trial, lies somewhere between sixteen to one and two to one. — The answer is, that the chance for being right, would be .5013, or very nearly an equal chance. — Take next, the particular case mentioned above, and suppose, that a solid or dye of whose number of sides and constitution we know nothing, except from experiments made in throwing it, has turned constantly the same face in a million of trials. — In these circumstances, it would be improbable, that it had *less* than 1,400,000 more of these sides or faces than of all others; and it would be also *improbable*, that it had above 1,600,000 more. The chance for the latter is .4647, and for the former .4895. There would, therefore, be no reason for thinking, that it would never turn any other side. On the contrary, it would be likely that this would happen in 1,600,000 trials. — In like manner, with respect to any event in nature, suppose the flowing of the tide, if it has flowed at the end of a certain interval a million of times, there would be the probability expressed by .5105, that the odds for its flowing again at the usual period was *greater* than 1,400,000 to 1, and the probability expressed

> by .5352, that the odds was *less* than 1,600,000 to one.
> Such are the conclusions which *uniform* experience warrants. — What follows is a *specimen* of the expectations, which it is reasonable to entertain in the case of *interrupted* or *variable* experience. — If we know no more of an event than that it has happened ten times in eleven trials, and failed once, and we should conclude from hence, that the probability of its happening in a single trial lies between the odds of nine to one and eleven to one, there would be twelve to one *against* being right. — If it has happened a hundred times, and failed ten times, there would also be the odds of near three to one *against* being right in such a conclusion. — If it has happened a thousand times and failed a hundred, there would be an odds *for* being right of a little more than two to one. And, supposing the same *ratio* preserved of the number of happenings to the number of failures, and the same guess made, this odds will go on increasing for ever, as the number of trials is increased. — He who would see this explained and proved at large may consult the essay in the Philosophical Transactions, to which I have referred; and also the supplement to it in the 54th volume. — The specimen now given is enough to shew how very inaccurately we are apt to speak and judge on this subject, previously to calculation. ... It also demonstrates, that the order of events in nature is derived from permanent causes established by an intelligent Being in the constitution of nature, and not from any of the powers of chance. And it further proves, that so far is it from being true, that the understanding is not the faculty which teaches us to rely on experience, that it is capable of determining, *in all cases*, what conclusions ought to be drawn from it, and what *precise degree* of confidence should be placed in it. [pp.395-398]

In a further footnote [pp.440-452], Price provides two definitions and two propositions concerned with probability. These are as follows:

> Definition 1st. An event is *probable*, when the odds *for* its happening are greater than those *against* its happening; *improbable*, when the odds *against* are greater than those *for*; and neither *probable* nor *improbable* when these odds are equal. — This is the proper sense of these words; but the writers on the *doctrine of chances* use the word *probable* in a more general sense.
> Definition 2nd. Two events are *independent*, when the happening of one of them has no influence on the other.

> Proposition 1st. The improbabilities of *independent* events are the same whether they are considered *jointly* or *separately*. That is; the improbability of an event remains the same, whether any other event which has no influence upon it happens at the same time with it, or not. This is self-evident.
> Proposition 2nd. The *improbability* that two independent events, each of them not improbable, should both happen, cannot be greater than the odds of *three* to *one*; this being the odds that two equal chances shall not both happen; and an equal chance being the lowest event of which it can be said that it is not improbable.

We shall not discuss other illustrations given in this Dissertation; enough has been said to show the flavour of the probability used here.

4.3 John Michell (1724-1793)

Three years after the posthumous publication of Bayes's Essay, Michell[18] published a paper entitled *An Inquiry into the probable Parallax, and Magnitude of the fixed Stars, from the Quantity of Light which they afford us, and the particular Circumstances of their Situation.* Although Michell's argument is markedly similar to that used by Arbuthnott[19] in 1710 in his essay in which an argument for divine providence is put forward on the basis of an observed constant regularity in the birth rates of the two sexes, and to that of Daniel Bernoulli in his prize-winning essay of 1734 on the attribution to chance of the inclinations to the ecliptic of the planetary orbits, inasmuch as it can perhaps be interpreted as a significance test, many of those who examined Michell's memoir in the nineteenth century found in it an application of inverse probability. Thus it is expedient to pay some attention to the memoir here, the particularly relevant section being found on pages 243-250.

The assertion Michell proposes to prove is the following:

> that, from the apparent situation of the stars in the heavens, there is the highest probability, that, either by the original act of the Creator, or in consequence of some general law (such perhaps as gravity) they are collected together in great numbers in some parts of space, whilest in others there are either few or none. [p.243]

The method to be used in order to prove this assertion

> is of that kind, which infers either design, or some general law, from a general analogy, and the greatness of the odds against things having been in the present situation, if it was not owing to some such cause. [p.243]

The first thing to be examined is "what it is probable would have been the least apparent distance of any two or more stars, any where in the whole heavens", it being always supposed that "they had been scattered by mere chance, as it might happen" [p.243]. Consider firstly two stars A and B: the probability that B will be within a distance of one degree of A is the ratio of the area of a circle of one degree angular radius to the area of the sphere (of radius R) of fixed stars, i.e. working in radians[20],

$$\pi(2\pi R/360)^2/4\pi R^2 \ ,$$

which reduces to 0.000076154 or 1/13,131. Thus the probability that B is not found within one degree of A is 13,130/13,131. Furthermore,

> because there is the same chance for any one star to be within the distance of one degree from any given star, as for every other [p.244]

the probability that none of n stars will lie within one degree of A is $(13{,}130/13{,}131)^n$, while the complement of this quantity to unity is the probability that at least one of the n stars is within the given distance of A.

Wishing now to abandon the significance given to the star A, Michell states that

> because the same event is equally likely to happen to any one star as to any other, and therefore any one of the whole number of stars n might as well have been taken for the given star as any other [p.244]

it follows that the probability that no two of the n stars are within one degree of each other is $[(13,130)^n/(13,131)^n]^n$: we shall comment on the correctness of this statement later.

It follows similarly that to find the probability that, of n stars, no two stars should be one within the distance x and the other within the distance z of a given star, one must firstly consider the fractions

$$\alpha = \left[\frac{(6875.5')^2 - x^2}{(6875.5')^2}\right]^n \quad \text{and} \quad \beta = \left[\frac{(6875.5')^2 - z^2}{(6875.5')^2}\right]^n$$

(the denominators being the square of 2 radians, in minutes) which give the probabilities that no star is within the distances x and z of the given star. Since

> the probability that two events shall both happen, is the product of the respective probabilities of those two events multiplied together [p.245]

it follows that the probability that one star is within a distance x of the given star, and that another is within a distance z of that same star is $(1-\alpha)(1-\beta)$. And finally, the probability that of n stars, no two exist which are within respective distances x and z of the same star, is $[1-(1-\alpha)(1-\beta)]^n$.

Two examples follow. In the first of these Michell calculates the probability

> that no two stars, in the whole heavens, should have been within so small a distance from each other, as the two stars β Capricorni, to which I shall suppose about 230 stars only to be equal in brightness. [p.246]

Under the supposition that the distance between these stars is something less than $3\frac{1}{3}'$, the required probability is found to be

$$\left[1-\pi(2\pi R3\tfrac{1}{3}\,/360\times 60)^2/\,4\pi R^2\right]^{230\times 230} \quad ,$$

or 80/81.

In the second example Michell considers the six brightest stars of the Pleiades, the stars Taygeta, Electra, Merope, Alcyone and Atlas being respectively at distances 11, $19\frac{1}{2}$, $24\frac{1}{2}$, 27 and 49 minutes from Maia. Supposing the number of stars "which are equal in splendor to the faintest of these" [p.246] to be 1,500, Michell finds the odds to be almost[21] 500,000 to 1

> that no six stars, ...scattered at random, in the whole heavens, should be within so small a distance from each other as the Pleiades are. [p.246]

Michell states further that the same argument will be found to be "still infinitely more conclusive" [p.249] if extended to smaller stars and those in clusters.

> We may from hence, therefore, with the highest probability conclude (the odds against the contrary opinion being many million millions to one) that the stars are really collected together

> in clusters in some places, where they form a kind of systems, whilst in others there are either few or none of them, to whatever cause this may be owing, whether to their mutual gravitation, or to some other law or appointment of the Creator. And the natural conclusion from hence is, that it is highly probable in particular, and next to a certainty in general, that such double stars, &c. as appear to consist of two or more stars placed very near together, do really consist of stars placed near together, and under the influence of some general law, whenever the probability is very great, that there would not have been any such stars so near together, if all those, that are not less bright than themselves, had been scattered at random through the whole heavens. [pp.249-250]

Thus far the relevant work.

Had Michell contented himself with stopping before the last quotation, his work would in all probability have been seen as an early significance test, and we should have been spared much of the ensuing controversy. But the passage quoted above suggests strongly that Michell thought the strength of his argument to be measurable, and his work came to be seen as an application of inverse probability.

In 1827 Struve proposed a completely different argument, which ran as follows. The number of possible binary combinations of n stars being $\binom{n}{2}$, the chance that any pair falls within a small circle of area s is $\binom{n}{2}s/S$, where S is a given area of the celestial sphere. As a special case Struve considered the surface from -15° declination to the north pole (so $S = 4\pi \sin^2 52\frac{1}{2}^o$), with $n = 10229$ and $x = 4''$, where x is the radius of the small circle. He evaluated the above expression as 0.007814. Struve also considered the cases in which $x = 8, 16$ and 32 seconds, and discussed similar results for the triple star problem.

For β Capricorni Michell takes $n = 230$ and $x = 3\frac{1}{3}'$. In this case[22] $s/S = 1/4254517$, which Michell in fact takes as $1/4254603$. Application of Struve's formula to Michell's figures yields

$$1 - \frac{230 \times 229}{2} \times \frac{1}{4254603} = 160.6/161.6$$

(rather than Michell's $80/81$), a figure which Lupton [1888, p.273] interprets as "the probability that no two such stars fall within the given area."

An endorsement of Michell's argument appeared in 1849 in J.F.W. Herschel's *Outlines of Astronomy.* Here the example of the Pleiades is rehearsed, though Herschel finds Michell's estimate of 1500 stars to be "considerably too small" [1873, art.833]. Citing also Struve's *Catalogus novus stellarum duplicium et multiplicium* of 1827, Herschel finds[23] that "The conclusion of a physical connexion of some kind or other is therefore unavoidable."

Comment on Herschel's work followed swiftly. In the same year in a short letter to the editors of the *Philosophical Magazine and Journal of Science*, J.D. Forbes[24] wrote

> Now I confess my inability to attach any idea to what would be the distribution of stars or of anything else, if "fortuitously scattered," much more must I regard with doubt and hesitation an attempt to assign a numerical value to the antecedent probability of any given arrangement or grouping whatever. An equable spacing of the stars over the sky would seem to me to be far more inconsistent with a total absence of Law or Principle, than the existence of spaces of comparative condensation, including binary or more numerous groups, as well as of regions of great paucity of stars. [pp.132-133]

In his 1850 review of Quetelet's *Lettres à S.A.R. le Duc régnant de Saxe-Cobourg et Gotha sur la Théorie des Probabilités appliquée aux Sciences Morales et Politiques* Herschel, mentioning neither Michell nor Forbes, in an attempt to clear up

> a singular misconception of the true incidence of the argument from probability which has prevailed in a quarter where we should least have expected to meet it [p.36],

indicated the inductive nature of the argument for a physical connexion between stars and its independence of any calculations. It seems, however, that Herschel's argument was misaimed, and, reasonable though it was, it did not invalidate Forbes's reasoning. As Gower [1982] has pointed out, the difference between the two revolved around the meaning of terms like "random scattering".

Forbes could not let this go unremarked, and on the 6th of August 1850 he read a paper on the matter before the Physical Section of the British Association, an expanded version being published in the *Philosophical Magazine and Journal of Science* in the same year. The aim of this paper is expressly stated in the sixth article as follows:

> the argument which I have to state is not meant to controvert the truth of the general result at which Mitchell [sic] and Struve arrive, namely, that the proximity of many stars to one spot, or the occurrence of many close binary stars distributed over the

> heavens, raises a *probability*, or rather we would call it an *inductive argument*, feeble perhaps, but still real, that such proximity may be actual, not merely apparent; but I deny *that such probable argument is capable of being expressed numerically at all.* [p.403]

Two main objections are raised to Michell's work, these being summarized as follows:

> *First*, a confusion between the *expectation* of a given event in the mind of a person speculating about its occurrence, and an *inherent improbability* of an event happening in one particular way when there are many ways equally possible. *Secondly*, a too limited and arbitrary conception of the utterly vague premiss of stars being "scattered by mere chance, as it might happen;" — a statement void of any condition whatever. [pp.421-422]

In a Note to his paper Forbes takes exception to Michell's expression $[(13130)^n/(13131)^n]^n$ for the probability that no two of n stars are within one degree of each other. With the assistance of "a mathematical friend, whose skill in these matters gives the utmost attainable assurance of his accuracy" [p.425], Forbes proposed to consider n (the number of stars) dice, each having p sides. Then the chance of doublets when the dice are thrown simultaneously is equal to that of two stars "being found at a less distance than the radius of a small circle of the sphere which includes an area $1/p$-th of the entire surface of the sphere" [p.425]. The total number of arrangements, without repetition, being $p(p-1)\ldots(p-n+1)$, and the total number of outcomes being p^n, the probability of an outcome without repetition is $p(p-1)\ldots(p-n+1)/p^n$, and the chance that two or more dice show the same face is

$$1 - p(p-1)\ldots(p-n+1)/p^n \quad .$$

Using Michell's figures for β Capricorni, with $p = 4254603$ and $n = 230$, and approximating this last expression by

$$1 - \frac{1}{e^n}\left(\frac{p}{p-n}\right)^{p-n+\frac{1}{2}} \quad , \tag{9}$$

Forbes obtains, for the required probability, a value of 0.00617 (a modern calculation yields 0.00625709), or approximately[25] 1/160. This agrees closely with the value 0.00618977 = 1/161.6 obtainable from Struve's formula.

That Forbes's work excited much discussion is shown by the letters reprinted in Shairp et al. [1873]. In a letter to Forbes dated 5th September 1850, Kelland pointed out that the approximation of Forbes's

$$1 - p(p-1)\dots(p-n+1)/p^n$$

by

$$1 - (p - n/2)^n/p^n$$

was unsatisfactory. Further letters, both in support of (from Terrot and Ellis) and against (from Airy) Forbes's argument, are well-worth reading, though the controversy is perhaps most fairly expressed in a chapter written by Tait in Shairp et al. [1873], where we find the words

> Forbes ... hit upon a real blot in Mitchell's argument, and rightly denounced its revival in Sir John Herschel's justly celebrated text-book. But they [the extracts quoted by Tait] also show that in dealing with the subject, he fell, at first at least, into mistakes quite as grave as those he was endeavouring to expose. [p.485]

In 1851 Boole entered the controversy. As he saw it, the statement of Michell's problem in relation to β Capricorni was as follows:

> 1. Upon the hypothesis that a given number of stars have been distributed over the heavens according to a law or manner whose consequences we should be altogether unable to foretell, what is the probability that such a star as β Capricorni would nowhere be found?
> 2. Such a star as β Capricorni having been found, what is the probability that the law or manner of distribution was not one whose consequences we should be altogether unable to foretell? [1952, p.249]

Denoting the two probabilities by p and P respectively, Boole finds p to be a determined number, and finds the fallacy to lie in the identification of p and P. (The same observation had earlier been made to Forbes by Bishop Terrot — see Shairp et al. [1873, p.476].) Rewriting p and P as $\Pr[\overline{B} \mid A]$ and $\Pr[\overline{A} \mid B]$, one has, by the discrete form of Bayes's Theorem,

$$P = \Pr[B \mid \overline{A}] \Pr[\overline{A}] \big/ \big(\Pr[B \mid \overline{A}] \Pr[\overline{A}] + \Pr[B \mid A] \Pr[A]\big) \quad .$$

As a special case Boole considers $p = 159/160$ (which he considers to be the correct value, rather than 80/81), $\Pr[A] = \frac{1}{2} = \Pr[B \mid \overline{A}]$. It then follows that $P = 80/81$, as Michell in fact found — but for p rather than P!

Boole reconsiders the problem, though without making any further specific comments, in Chapter XX of *An Investigation of The Laws of Thought* of 1854.

In the second part of his paper of 1851 Donkin presents a Bayesian approach to Michell's problem. He supposes that there are n visible stars of a certain class, for no two of which, were they within a certain angular distance of each other, could any conclusion be drawn from their apparent brightness as to whether they were merely optically double or actually formed a true binary system. Suppose further that there are in fact m pairs of stars within these angular limits, the other $n-2m$ being single, and let p denote the *a priori* probability that a proposed system is binary (a *system* is defined to be either a *single star* or a *binary system*). Then, all systems but single and binary being excluded, $1-p$ is the *a priori* probability of a single system. Donkin explains p as follows:

> Suppose a person to be perfectly acquainted with the mode in which the stars are produced; he would be able, setting aside difficulties of calculation, to assign the probability that a system *about to be produced* would turn out to be binary, and this would be the value of p. [pp.462-463]

It is assumed further that p is uniformly distributed over the unit interval.

Now let P_i^n denote the *a priori* probability that there are i binary systems among n stars, and let Q_r^s denote the *a priori* probability that there are r optically double pairs among s single stars "whose configurations were *accidental*" [p.463]. The aim is to determine the posterior probability of i. Donkin's reasoning is somewhat loose, no clear distinction between joint and conditional probabilities being observed. In an attempt to put things on a firmer footing, let us denote by A_m^n the event that there are m pairs among the n stars, by B_i^m the event that there are i binary stars among m stars, and by C_i^m the event that there are i optically double stars among m. Furthermore let us replace Donkin's p by P. Then

$$\begin{aligned}\Pr[A_m^n \mid P=p] &= \sum_{i=0}^{m} \Pr\left[B_i^m C_{m-i}^{n-2i} \mid P=p\right]\\ &= \sum_{i=0}^{m} \Pr[B_i^m \mid P=p] \Pr\left[C_{m-i}^{n-2i} \mid B_i^m \,\&\, P=p\right]\\ &= \sum_{i=0}^{m} P_i^m\, Q_{m-i}^{n-2i}\\ &= \Phi(p) \quad , \text{ say} .\end{aligned}$$

Therefore

$$\Pr[A_m^n] = \int_0^1 \Phi(p) f(p)\, dp = \omega \ ,$$

where $f(\cdot)$ denotes the (uniform) density of P, and hence

$$\Pr[p < P < p + dp \mid A_m^n] = [\Phi(p)/\omega]\, dp \quad .$$

Notice further that

$$\Pr[B_i^m \,\&\, p < P < p + dp \mid A_m^n] = \Pr[p < P < p + dp \mid A_m^n] \times \Pr[B_i^m \mid A_m^n \,\&\, p < P < p + dp]$$

$$= \frac{\Phi(p)\, dp}{\omega} \Pr[B_i^m A_m^n \mid p < P < p + dp] / \Pr[A_m^n \mid p < P < p + dp]$$

$$= \frac{\Phi(p)\, dp}{\omega} \Pr\left[B_i^m C_{m-i}^{n-2i} \mid p < P < p + dp\right] / \Pr[A_m^n \mid p < P < p + dp]$$

$$= \frac{\Phi(p)\, dp}{\omega} P_i^m Q_{m-i}^{n-2i} \frac{1}{\Phi(p)} \quad .$$

Thus

$$\Pr[B_i^m \mid A_m^n] = \frac{1}{\omega} \int_0^1 P_i^m Q_{m-i}^{n-2i}\, dp \quad .$$

Denoting this last integral by $\varphi(i)$, Donkin points out that one may equivalently write

$$\Pr[B_i^m \mid A_m^n] = \varphi(i) \Big/ \sum_{i=0}^{m} \varphi(i) \quad .$$

Turning now to the evaluation of P_i^n and Q_r^s, Donkin notes firstly that, were k systems about to be produced, the probability that i would turn out to be binary would be $p^i q^{k-i}$, where $q = 1 - p$. If n stars have been produced, and if no knowledge of the division into systems is available, the probability of i binary stars will be proportional to $\binom{n-i}{i} p^i q^{n-i}$, and hence

$$P_i^n = \binom{n-i}{i} p^i q^{n-i} \Big/ \sum_{i=0}^{\nu} \binom{n-i}{i} p^i q^{n-i}$$

where $\nu = n/2$ or $(n-1)/2$ according as n is even or odd.

Secondly he points out that

> The *à priori* probability that two given stars, whose positions were accidental, would be within a given angular distance θ of one another, is $\sin^2(\theta/2)$ [p.465]

though it appears, from Michell's paper, that this factor should be divided by four. Donkin then, like Forbes, considers s dice, each having t faces, where t is the nearest integer to $1/\sin^2(\theta/2)$. The probability of getting doublets with a given pair of dice is then $\sin^2(\theta/2)$, and it is then suggested that Q_r^s be approximated by the probability of getting, in one trial with the s dice, r different doublets and $s-2r$ different numbers. From an earlier article of his paper (not discussed here), this probability is found to be

$$\frac{t(t-1)\dots(t-(s-r)+1)}{t^s} \times \frac{s!}{(1.2)^r\, r!\, (s-2r)!} \quad .$$

Notice that this expression reduces to that given by Forbes [1850] when $r=0$, i.e. when there are no doublets.

Donkin now concludes by saying

> I should consider it a great waste of time and labour to attempt anything like a numerical result in the actual case. All that I have aimed at is to show that there is no real difficulty of principle in applying the theory of probabilities to this and similar questions, however impracticable it may be to obtain a complete numerical solution. [p.466]

In 1859 and 1860 Newcomb published a series of notes on probability in the *Mathematical Monthly*. In the fourth of these he discusses the Poisson distribution and applies it

> to the determination of the probability that, if the stars were scattered at random over the heavens, any small space selected at random would contain s stars. [p.137]

Taking N as the whole number of stars, h as the number of units of space and l as "the extent of space selected at random" [p.137], Newcomb finds the desired probability P to be given by

$$P = \frac{N^s l^s}{h^s s!} e^{-Nl/h} \quad .$$

A specific numerical example, with which we shall not concern ourselves, then follows.

Newcomb's general conclusion is, however, that despite the vagueness and uncertainty present in the problem, Michell's "general method is ... better applicable to this particular problem than that given above" [p.138].

For further comments on Michell's work, and the remarks of Forbes, Herschel and Donkin, the reader may be referred to Jevons [1877], where Michell's investigations are described as "admirable speculations" [p.212] and where it is noted that "The conclusions of Michell have been entirely verified by the discovery that many double stars are connected by gravitation" [pp.247-248]: Jevons also concludes that any error there may be in Michell's work lies in his methods of calculation and "not in the general validity of his reasoning and conclusions" [p.248]. Proctor [1872, pp.314-316] discusses a similar problem, as does Bertrand [1907, art.135], while Porter [1986, p.79] proffers some general comments on Donkin, Forbes and Herschel. Venn [1888, chap.XX, §§21-23] is also pertinent.

In 1888 a detailed investigation was undertaken by Lupton of the arguments of Michell, Struve and Forbes, it being concluded that the latter's methods were the least open to objection. Kleiber [1887], [1888] dissented sharply from this view, finding on the contrary that Forbes's experiments in fact *supported* Michell's argument: Lupton was not altogether convinced, as his further letter of 1888 showed. Keynes [1921], in a careful discussion, found that "Michell's argument owes more, perhaps, to Daniel Bernoulli than to Bayes" [chap.XVI, footnote to §11] and concluded further [chap.XXV] that Michell's argument was in part invalid and elsewhere less conclusive than he had supposed. An excellent modern discussion is provided by Gower [1982].

Before we leave Michell's essay, it might be of interest more closely to examine some of the alternative formulae proposed. There can be no doubt that Michell's formula is wrong: as Gower [1982, p.148] has pointed out, the probability found does not reduce, as it should, to zero for $n \geq 13,131$. The error clearly arises from the tacit assumption that the events whose probabilities are multiplied together are independent, whereas in fact the event that star A is more than one degree from any other star is not independent of the event that star B is more than one degree from any other star.

Turning next to Struve's work, we recall that he found

$$\begin{aligned}\pi_1 &\equiv \Pr[\text{any binary pair falls in a small circle of area } s] \\ &= \binom{n}{2} p ,\end{aligned}$$

where $p = s/S$. It thus follows that

$$\begin{aligned}\pi_2 &\equiv \Pr[\text{no binary pair falls in a small circle of area } s] \\ &= 1 - \binom{n}{2} p.\end{aligned}$$

Forbes's argument, on the other hand, yields

$$\begin{aligned}\pi_3 &\equiv \Pr\,[\text{all dice show different faces}]\\ &= \Pr\,[\text{no two stars are in the same small circle}]\\ &= v(v-1)\ldots(v-n+1)/v^n \ ,\end{aligned}$$

and thus

$$\begin{aligned}\pi_4 &\equiv \Pr\,[\text{at least two dice show the same face}]\\ &= \Pr\,[\text{at least two stars are in the same small circle}]\\ &= 1 - v!/[(v-n)!\,v^n].\end{aligned}$$

Using Michell's figures for β Capricorni, and (9) where necessary, one obtains

$$\begin{aligned}\pi_1 &= 6.189766706 \times 10^{-3}\\ \pi_2 &= 9.938102333 \times 10^{-1}\\ \pi_3 &= 9.937429075 \times 10^{-1}\\ \pi_4 &= 6.257092500 \times 10^{-3}\ .\end{aligned}$$

A comparison of π_2 and π_3 (or π_1 and π_4) shows that, even though the numerical values are markedly similar, these probabilities are in fact answers to different questions. That the numerical answers coincide is a consequence of the fact that, for large n and very much larger p $(= 1/v)$,

$$1 - \binom{n}{2}\frac{1}{v} \approx 1 - \frac{(n-2)^2}{2v}$$

$$\frac{v!}{(v-n)!\,v^n} \approx 1 - \frac{n(2n-1)}{2v}\ .$$

In conclusion, let us see whether Michell is in fact guilty of some of the charges levelled against him. Recall that the method he advocated consisted of two parts, viz.

(i) the inferring of design, or some general law, from a general analogy, and

(ii) the greatness of the odds against things having been in the present situation, were it not for some such cause.

If one denotes by D the event that a certain group of stars (e.g. those in β Capricorni or the Pleiades) has a certain physical distribution, and by R the event that the stars are randomly scattered, then one sees that Michell has in each of his examples calculated $\Pr\,[D \mid R]$. Further, in the case of β Capricorni he states

> If we now compute ... what the probability is, that no two stars ... should have been within so small a distance from each other, as the two stars β Capricorni, ... we shall find it to be about 80 to 1 [p.246]

while in that of the Pleiades he writes

> we shall find the odds to be near 500 000 to 1, that no six stars, ... scattered at random,... would be within so small a distance from each other as the Pleiades are. [p.246]

Thus "odds" and "probability" are used in an apparently synonymous manner.

What Michell is in fact concluding, then, is that $\Pr[\overline{R} \mid D]$ is large, or equivalently that $\Pr[R \mid D]$ is small. Since

$$\Pr[R \mid D] = \Pr[D \mid R] \Pr[R] / \Pr[D] \tag{10}$$

and since $\Pr[D \mid R]$ has been found to be small (1/80 for β Capricorni, and 1/496000 for the Pleiades), it is "clear" that $\Pr[R \mid D]$ will indeed be small — provided, of course, that the other terms in (10) are of appropriate size. Thus Michell has clearly made use of part (ii) of his method.

As regards part (i), notice that, after considering in detail the cases of β Capricorni and the Pleiades, Michell writes

> If, besides these examples that are obvious to the naked eye, we extend the same argument to the smaller stars, as well those that are collected together in clusters, such for example, as the Præcepe Cancri, the nebula in the hilt of Perseus's sword, &c. as to those stars, which appear double, treble, &c. when seen through telescopes, we shall find it still infinitely more conclusive, both in the particular instances, and in the general analogy, arising from the frequency of them. [pp.247-249]

This "analogy" argument may perhaps also be seen as being implied by the long quotation given above from pages 249-250 of Michell's memoir.

4.4 Nicolas de Beguelin (1714-1789)

The only memoir by Beguelin[26] which has any bearing on our subject (and that bearing, let it be admitted, is but slight) is entitled *Sur l'usage du principe de la raison suffisante dans le calcul des probabilités*, a memoir published in the volume of 1767 of the *Histoire de l'Académie royale des Sciences et Belles-Lettres, Berlin* (published in 1769), pp.382-412.

In a reference to an earlier memoir[27] Beguelin stresses the importance that prior information has in probability calculations:

> j'ai montré dans un Mémoire précédent que la doctrine des probabilités étoit uniquement fondée sur le principe de la raison suffisante; il ne seroit donc pas surprenant que les Mathématicians ne suffent pas d'accord entr'eux dans la solution des problemes qui ont la probabilité pour objet; leurs calculs sont de vérité nécessaire, mais la nature du sujet auquel ils les appliquent ne l'est pas. Les vérités contingentes ne peuvent être démontrées qu'en partant d'une supposition; & quelque plausible qu'une supposition est, elle n'en exclut pas nécessairement d'autres, qui peuvent servir de base à d'autres calculs, & donner par conséquent des resultats différents. [p.382]

He goes on next to distinguish between the possibility and the probability of an event:

> toute combinaison qui n'implique pas contradiction est possible, & comme on ne sauroit impliquer à demi, toutes les combinaisons possibles sont également possibles; ce n'est qu'improprement qu'on diroit d'un événement possible, qu'il est plus ou moins possible qu'un autre; il n'y a point de milieu, ni de degrés à concevoir, entre ce qui peut exister, & ce qui répugne à l'existence. Mais la simple possibilité ne suffit pas pour donner l'existence à un événement; il faut de plus qu'il y ait une raison suffisante qui détermine l'événement à être plutôt celui qu'il est, qu'un des autres également possibles: & c'est ici que commence la probabilité. [p.383]

Then follows a clear definition of "sufficient reason", viz.

> la raison suffisante de la probabilité d'un événement, c'est la preponderance des raisons de s'attendre à cet événement sur celles de s'attendre à l'événement contraire. [p.383]

Todhunter is perhaps a little harsh in writing "the memoir does not appear of any value whatever" [1865, art.616]: certainly the emphasis on the bearing of prior knowledge on probability calculations is important, though little else seems relevant here.

4.5 Joseph Louis de la Grange (1736-1813)

Of this famous mathematician's many writings, the only one at all pertinent to our subject is his first memoir on probability, viz. *Mémoire sur l'utilité de la méthode de prendre le milieu entre les résultats de plusiers observations, dans lequel on examine les avantages de cette méthode par le calcul des probabilités, et où l'on résout différents problèmes relatifs à cette matière.* This was published[28] in volume 5 of the *Miscellanea Taurinensia* (1770-1773), pp.167-232. Todhunter [1865] remarks on the merit of this memoir in the following words:

> The memoir at the time of its appearance must have been extremely valuable and interesting, as being devoted to a most important subject; and even now it may be read with advantage. [art.556]

Of the ten problems[29] considered in this memoir, the sixth is pertinent to our work. Because it is both an early example in "inverse probability" and a precursor of Pearson's important investigations of the (P, χ^2) problem[30], we have chosen to discuss the question in some detail. The problem is posed by Lagrange as follows[31]:

> Je suppose qu'on ait vérifié un instrument quelconque, et qu'ayant réitéré plusiers fois la même vérification on ait trouvé différentes erreurs, dont chacune se trouve répétée un certain nombre de fois; on demande quelle est l'erreur qu'il faudra prendre pour la correction de l'instrument. [p.200]

Supposing errors $p, q, r, \ldots$ to be made $\alpha, \beta, \gamma, \ldots$ times respectively in n observations, Lagrange assumes the unknown frequencies to be $a, b, c, \ldots$, and considers the polynomial $(ax^p + bx^q + cx^r + \cdots)^n$, with general term $N(ax^p)^\alpha(bx^q)^\beta(cx^r)^\gamma \ldots$ Now the coefficient $Na^\alpha b^\beta c^\gamma \ldots$ of $x^{p\alpha+q\beta+r\gamma+\cdots}$ divided by $(a+b+c+\cdots)^n$ gives the probability that the errors $p, q, r, \ldots$ will be found together in such a way that p occurs α times, q β times, r γ times, &c. From an earlier problem (viz. the fifth) it is known that $N = n!/(\alpha!\,\beta!\,\gamma!\ldots)$. The most probable value is then (correctly) taken to be the highest term in the multinomial, which yields

$$\alpha = \frac{na}{a+b+c+\cdots} \;, \quad \beta = \frac{nb}{a+b+c+\cdots} \;, \quad \gamma = \frac{nc}{a+b+c+\cdots} \;, \ldots$$

from which the unknowns $a, b, c, \ldots$ may be determined. Again by Problem V, it follows that the correction to be made is $(\alpha p + \beta q + \gamma r + \cdots)/n$, "c'est-à-dire égale à l'erreur moyenne entre toutes les erreurs particulières que les n vérifications ont données" [p.201].

Now, as Pearson has noted[32], the $\alpha, \beta, \gamma, \ldots$ which give the maximum term in the multinomial are taken by Lagrange as being the *observed* $\alpha, \beta, \gamma, \ldots$: this may well be reasonable, but no discussion of the point is essayed.

Following a corollary (which does not concern us at the moment) may be found two *Remarques*, in which Lagrange turns to a problem of inverse probability[33]. These remarks Todhunter dismisses as follows:

> Lagrange proposes further to estimate the probability that the values of $a, b, c, \ldots$ thus determined from observation do not differ from the true values by more than assigned quantities. This is an investigation of a different character from the others in the memoir; it belongs to what is usually called the theory of inverse probability, and is a difficult problem.
> Lagrange finds the analytical difficulties too great to be overcome; and he is obliged to be content with a rude approximation. [art.562]

Condemning Todhunter for his myopia, Pearson [1978, p.599] notes that Lagrange came "within an ace" of solving the (P, χ^2) problem, a tough nut cracked by Pearson himself[34] in 1900. However, one might plead in mitigation that Todhunter was writing a history, and not a statistical text. Thus while he was perhaps a little brusque in his dismissal of what has proved to be statistically feracious, it is a bit harsh to judge him for lacking the foresight to appreciate its value.

As the Normal distribution[35] was reached by de Moivre as a limit to the skew binomial[36] in 1733 so, using the multinomial, Lagrange arrived at the multivariate Normal distribution. Let us examine the derivation. The problem posed is the following:

> ...on voulait savoir de plus quelle est la probabilité que ces mêmes valeurs [viz. $a, b, c, \ldots$] ne s'écarteront pas de la vérité d'une quantité quelconque $\pm(rs/n)$ [p.202]

where $s = a + b + c + \cdots$. (Notice that the true values are now assumed unknown.)

Noting that $a, b, c, \ldots$ are proportional to $\alpha, \beta, \gamma, \ldots$ only when one is working with the most probable value of the multinomial, Lagrange considers now

$$a = \frac{s(\alpha + x)}{n} \ , \qquad b = \frac{s(\beta + y)}{n} \ , \qquad c = \frac{s(\gamma + z)}{n} \ , \ \ldots$$

taking $x, y, z, \ldots$ equal to $\pm 1, \pm 2, \ldots, \pm r$ successively, subject to the constraint that $x + y + z + \cdots = 0$ (since, by hypothesis, $\alpha + \beta + \gamma + \cdots = n$ and $a + b + c + \cdots = s$). If P is the probability that $a = s\alpha/n$, $b = s\beta/n$, $c = s\gamma/n, \ldots$, then substitution of these values in an earlier result (Problem V) yields

$$P = \frac{n!}{n^n} \frac{\alpha^\alpha}{\alpha!} \frac{\beta^\beta}{\beta!} \cdots$$

Similarly, if Q is the probability that one has

$$a = \frac{s(\alpha + x)}{n}, \quad b = \frac{s(\beta + y)}{n}, \quad c = \frac{s(\gamma + z)}{n}, \ldots$$

then

$$\begin{aligned} Q &= P(1 + x/\alpha)^\alpha (1 + y/\beta)^\beta (1 + z/\gamma)^\gamma \ldots \\ &\equiv PV, \quad \text{say.} \end{aligned}$$

The desired probability will then be $P \int V$.

Noting the difficulty of evaluating this integral in general, Lagrange remarks that it can be evaluated by multiplying the mean value of V by the number of all the values of V entering into the integral, "et la difficulté ne consistera qu'à trouver ce nombre" [p.203]. Denoting by m the number of the quantities $\alpha, \beta, \gamma, \ldots$, he points out that the number required will be the coefficient T of u^0 in the expansion of

$$\left(u^{-r} + u^{-r+1} + \cdots + u^{-1} + u^0 + u^1 + \cdots + u^{r-1} + u^r\right)^m,$$

whence, in fact [p.203],

$$\begin{aligned} T = {} & \frac{(mr+1)(mr+2)(mr+3)\ldots(mr+m-1)}{1.2.3\ldots(m-1)} \\ & - m\frac{[(m-2)r]\,[(m-2)r+1]\,[(m-2)r+2]\ldots[(m-2)r+m-2]}{1.2.3\ldots(m-1)} \\ & + \frac{m(m-1)}{2}\frac{[(m-4)r-1]\,[(m-4)r]\,[(m-4)r+1]\ldots[(m-4)r+m-3]}{1.2.3\ldots(m-1)} - \cdots \end{aligned}$$

If W denotes the mean value of V, then $\int V$ is to be approximated by TW, and the desired probability is then approximately PTW.

If, however, one were to take the smallest value of V, rather than the mean value W, one would necessarily underestimate the true value of $\int V$, and hence the desired probability. Thus one may advantageously wager PTW to $1 - PTW$ that in taking

$$\frac{a}{s} = \frac{\alpha}{n}, \quad \frac{b}{s} = \frac{\beta}{n}, \quad \frac{c}{s} = \frac{\gamma}{n}, \ldots,$$

one does not make a mistake of an amount greater in absolute value than r/n.

In his *Remarque II*, Lagrange essentially "passes to the limit": that is, he supposes n (and consequently $\alpha, \beta, \gamma, \ldots$) to be very large. Proceeding from what is essentially the Stirling-de Moivre theorem, he deduces that

$$\frac{1.2.3\ldots u}{u^u} = \frac{\sqrt{\pi u}}{e^u} \quad .$$

His "π" being what one would nowadays call "2π", we shall change to the modern notation. It follows that

$$\begin{aligned} P &= \frac{n!}{n^n} \frac{\alpha^\alpha}{\alpha!} \frac{\beta^\beta}{\beta!} \cdots \\ &= \sqrt{\frac{2\pi n}{(2\pi\alpha)(2\pi\beta)(2\pi\gamma)\ldots}} \quad . \end{aligned}$$

Turning next to the expression V Lagrange shows that

$$\begin{aligned} \log V &= \alpha \log(1 + \frac{x}{\alpha}) + \beta \log(1 + \frac{y}{\beta}) + \gamma \log(1 + \frac{z}{\gamma}) + \cdots \\ &= \alpha\frac{x}{\alpha} + \beta\frac{y}{\beta} + \gamma\frac{z}{\gamma} + \cdots - \tfrac{1}{2}\left(\frac{x^2}{\alpha} + \frac{y^2}{\beta} + \frac{z^2}{\gamma} + \cdots\right) \\ &\qquad + \tfrac{1}{3}\left(\frac{x^3}{\alpha^2} + \frac{y^3}{\beta^2} + \frac{z^3}{\gamma^2} + \cdots\right) + \cdots \\ &= -\tfrac{1}{2}\left(\frac{x^2}{\alpha} + \frac{y^2}{\beta} + \frac{z^2}{\gamma} + \cdots\right) , \end{aligned} \tag{11}$$

since $x + y + z + \cdots = 0$, and the cubic term in (11) above (given by Pearson [1978, p.600] but *not* by Lagrange) is negligible in comparison with the quadratic.

On defining $x = \xi\sqrt{n}$, $y = \eta\sqrt{n}$, $z = \zeta\sqrt{n}, \ldots$ and $\alpha/n = A$, $\beta/n = B$, $\gamma/n = C, \ldots$, one deduces that $\xi + \eta + \zeta + \cdots = 0$, $A + B + C + \cdots = 1$, and

$$PV = \left[(2\pi n)^{m-1} ABC\ldots\right]^{-\frac{1}{2}} \exp\left[-\tfrac{1}{2}(\xi^2/A + \eta^2/B + \zeta^2/C + \cdots)\right] \quad .$$

Now, when the increment or the difference of the quantities $x, y, z, \ldots$ is 1, the difference of the variables $\xi, \eta, \zeta, \ldots$ will be $1/\sqrt{n}$ (and hence infinitely small). Denoting this difference by $d\theta$, one will have

$$PV = \left[(2\pi)^{m-1} ABC\ldots\right]^{-\frac{1}{2}} \exp\left[-\tfrac{1}{2}(\xi^2/A + \eta^2/B + \zeta^2/C + \cdots)\right] d\theta^{m-1} \quad .$$

This result, incidentally, Pearson [1978, p.600] finds "extraordinarily brilliant", in particular for the following reasons[37]

(i) a measure of the terms we are neglecting;

(ii) it deduces the probability that the true values differ from the observed values and not the inverse relation;

(iii) it involves precisely the P and the χ^2 that I obtained by a most troublesome algebraic process in 1900.

Lagrange next turns his attention to the $(m-1)$-fold integration of $\exp[-\frac{1}{2}(\xi^2/A+\eta^2/B+\zeta^2/C+\cdots)]\,d\theta^{m-1}$, and takes note that there are only $m-1$ independent variables, which results in his substituting for ξ, $-\eta-\zeta-\ldots$ The solution of the general problem being only obtainable by tables[38], Lagrange restricts his attention to the case in which only two errors are present. Pearson [1978, p.602] has pointed out that certain numerical errors present in this discussion suggest that Lagrange copied de Moivre's results in places. Nevertheless, the right answer for the approximate evaluation of

$$P = (2\pi ABn)^{-\frac{1}{2}} \exp(-\tfrac{1}{2}\,\xi^2/AB)$$

is obtained — viz. 0.682688. The section concludes with further discussion, not relevant to the present study, of the multivariate case.

4.6 William Emerson (1701-1782)

In 1776 a treatise entitled *Miscellanies, or a Miscellaneous Treatise; containing several Mathematical Subjects*, and published by J. Nourse of London, appeared under the name of Emerson[39].

The first article [pp.1-48] of this treatise is devoted to the laws of chance. The treatment is fairly standard: indeed, one must agree with Todhunter [1865] that

> There is nothing remarkable about the work except the fact that in many cases instead of exact solutions of the problems Emerson gives only rude general reasoning which he considers may serve for approximate reasoning. [art.641]

In Emerson's own words

> It may be observed, that in many of these problems, to avoid more intricate methods of calculation, I have contented myself with a more lax method of calculating, by which I only approach near the truth. [1776, p.47]

That Emerson expected criticism of his essay (perhaps even welcomed it) is shown by one of his introductory paragraphs, in which he writes

> Therefore my readers may please to take notice, that if any envious, abusive, dirty Scribbler, shall hereafter take it into his head to creep into a hole like an Assassin, and lie lurking there on purpose to scandalize and rail at me; and dare not shew his face like a Man; I shall give myself no manner of trouble about such an Animal, but look upon him as even below contempt. [p.v]

Harsh words, but perhaps not out of character for one who could decline an F.R.S.[40]!

The only part of the Essay that might possibly be of interest is Article 1, *The Laws of Chance* (pp.1-48). Here Emerson sets out the following definitions and axioms[41]:

> Definition I. *Chance* is an event, or something that happens without the design or direction of any agent; and is directed or brought about by nothing but the laws of nature.
> Def. II. The *probability* or *improbability* of an event happening, is the judgement we form of it, by comparing the number of Chances there are for its happening, with the number of Chances for its failing.
> Def. III. *Expectation* in play, is the value of a man's Chance; that is, of the thing played for, considered with the probability of gaining it; and therefore is the product of its value multiplied by the probability of obtaining the prize.
> Def. IV. *Risk* is the value of the stake considered with the probability of losing it; & therefore is the product of its value multiplied by the probability of losing it.
> Def. V. Events are *independent* when they have no manner of connection with one another; or when the happening of one neither forwards nor obstructs the happening of any other of them.
> Def. VI. An event is *dependent* when the probability of its happening is altered by the happening of some other.
> Axiom I. In computing the number of Chances, it is supposed that all Chances are equal, or made with equal facility.
> Axiom II. The whole expectation for any prize, is the sum of all the expectations upon the particulars.
> Axiom III. The value of any Chance or expectation is what would purchase the like Chance or expectation, in a fair game. [pp.2-3]

4.7 George Louis Leclerc, Comte de Buffon (1707-1788)

From the pen (or quill) of this distinguished naturalist there flowed a memoir entitled *Essai d'Arithmétique Morale*, which work, published in 1777, constitutes part of the *Supplément* to the *Histoire Naturelle*, Tome IV. Exactly when this memoir[42] was written is uncertain, though Gouraud says

> Cet ouvrage, dont la composition remonte à 1760 environ, ne parut qu'en 1777 dans le tome IV du *Supplement à l'Histoire naturelle.* [1848, p.54]

Most of this long essay has little (if indeed any) bearing on our subject. However, after distinguishing three kinds of truths (viz. geometrical truths known by reasoning, physical truths known by experience, and truths believed on testimony), Buffon illustrates those of the second kind by considering the question of the sun's rising. Like Price, Buffon stresses that, to the man who has only once seen the rising and the setting of the sun, the second rising will be

> une première experiénce, qui doit produire en lui l'espérance de revoir le soleil, & il commence à croire qu'il pourrait revenir, cependant il en doute beaucoup. [1778, p.76]

With the repeated returns of the sun the observer's doubt diminishes, until

> il croira être certain qu'il le verra toujours paroître, disparoître & se mouvoir de la même façon. [1778, p.77]

Buffon then concludes that the probabilities of subsequent risings increase like the sequence $1,2,4,\ldots,2^{n-1}$, the meaning of this becoming clear only later in the *Essai*, where we read

> $\ldots 2^{13} = 8192, \ldots$ & par conséquent lorsque cet effet est arrivé treize fois, il y a 8192 à parier contre 1, qu'il arrivera une quatorzième fois ... [pp.85-86]

that is, a probability of 2^{n-1} is to be interpreted as odds of 2^{n-1} to 1 in favour of the event in question[43].

As a numerical example, it is supposed that the age of the earth is 6,000 years, with leap years being neglected. Buffon then asserts that, if one knows that the sun has risen 2,190,000 times, the probability of its rising once more is $2^{2,189,999}$ (or, as we have seen, $2^{2,189,999}$ to 1). This is plainly inconsistent with Laplace's expression $(n+1)/(n+2)$, though, as Sheynin

[1969] and Zabell [1988a] have noted, it is more in line with that given by Price — if we gloss over a confusion between "number of risings" and "number of returns" (see §3.6).

4.8 Jean Trembley (1749-1811)

Only one work by this author contains matter directly pertinent to our topic, viz. the memoir *De probabilitate causarum ab effectibus oriunda: disquisitio mathematica*, published in volume 13 of the *Commentationes Societatis Regiae Scientiarum Gottingensis*, 1795-1798, pp.64-119 of *Commentationes mathematica*[44] (published 1799).

The scope of this work is clearly delineated in the opening paragraph[45]:

> Hanc materiam pertractarunt eximii Geometrae, ac potissimum Cel. la Place in Commentariis Academiae Parisinensis. Cum autem in hujusce generis Problematibus solvendis sublimior et ardua analysis fuerit adhibita, easdem quaestiones methodo elementari ac idoneo usu doctrinae serierum aggredi operae pretium duxi. Qua ratione haec altera pars calculi Probabilium ad theoriam combinationum reduceretur, sicut et primam reduxi in dissertatione ad Regiam Societatem transmissa. Primarias quaestiones hic breviter attingere conabor, methodo dilucidandae imprimis intentus [§1]

— though as Todhunter has noted, the claims of "lucidness" and "rigour" are perhaps a little exaggerated[46].

The first problem Trembley[47] considers is the following: let there be an urn containing an infinite number of white and black balls[48] in unknown proportion. Let p white and q black balls be withdrawn from the urn: we seek the probability of drawing m white and n black balls in future drawings (all drawings being made with replacement). The solution to this problem is, as we shall see in our chapter on Laplace, given by

$$\binom{m+n}{m} \int_0^1 x^{m+p}(1-x)^{n+q}\,dx \Big/ \int_0^1 x^p(1-x)^q\,dx,$$

though Trembley does not give his solution in this form.

After discussing the problem thus far, Todhunter goes on to say

> the investigations are only approximate, the error being however inappreciable when the number of balls is infinite. If each ball is *replaced* after being drawn we can obtain an *exact* solution of

> the problem by ordinary Algebra ... and of course if the number of balls is supposed infinite it will be indifferent whether we replace each ball or not, so that we obtain indirectly an exact elementary demonstration of the important result which Trembley establishes approximately. [art.766]

It seems to me that Todhunter has missed, in the original, the sentence "Schedulae eductae supponuntur rursus conicii in vas" — or is the emphasis merely on an expert use of algebra to solve the problem?

Certain other problems, involving balls and urns, are considered by Trembley: in each case, however, he relates them to work by Laplace, and we shall therefore postpone consideration of Trembley's transcriptions to the appropriate place in Chapter 6. The treatment of the Problem of Points, considered by Laplace in his *Mémoire sur la probabilité des causes par les évènemens*, is extended slightly by Trembley: to this we shall likewise return.

His preceding discussion, Trembley states, leads to the conclusion that the probability of causes, generated by effects, requires a method which consists of two parts[49]:

> In prima parte assignantur formulae quae repraesentant hanc Probabilitatem; in altera parte indicantur approximationes quae possibilem reddant usum harum formularum ubi ingentes adsunt numeri. [§14]

The example (again one from Laplace) adduced to illustrate this assertion is that concerning the observed difference between the ratio of the number of boys born to the number of girls born (in a certain time period) in London, and the similar ratio in Paris. As we shall see in the discussion on Laplace, one is led to consideration of the ratio

$$\frac{\displaystyle\int_{x=0}^{1}\int_{x'=0}^{x} x^p(1-x)^q x'^{p'}(1-x')^{q'}\,dx'\,dx}{\displaystyle\int_{x=0}^{1}\int_{x'=0}^{1} x^p(1-x)^q x'^{p'}(1-x')^{q'}\,dx'\,dx},$$

which Trembley evaluates by expansion of the integrands and term-by- term integration: an alternative way of reaching his final result is given by Todhunter [1865, art.773].

4.9 Pierre Prevost (1751-1839) & Simon Antoine Jean Lhuilier (1750-1840)

There are three memoirs by these authors which have some bearing on our subject. The first of these, and, of the three, the only technical one, is entitled *Sur les probabilités.* It occupies pp.117-142 of the *Classe de Mathématique* of the *Mémoires de l'Académie royale des Sciences et Belles-Lettres, Berlin* 1796 (published 1799), and was read before the Academy on the 12th November 1795.

In this essay Prevost and Lhuilier propose to consider the following problem:

> Soit une urne contenant des billets de deux espèces (que j'appellerai blancs et noirs), dans un rapport inconnu. Soit tiré successivement un certain nombre de ces billets, sans remettre dans l'urne, à chaque extraction, le billet tiré. Connoissant le nombre des billets de chaque espèce qui ont été tirés, on demande la probabilité que tirant de la même manière de nouveaux billets, en nombre donné, il y a en aura des nombres donnés de ces deux espèces. [p.117]

As Todhunter [1865, art.849] has noted, this memoir is the first in which the urn-sampling problem when the balls sampled are *not* replaced[50], is considered.

The solution of this problem requires the following principle:

> *Principe étiologique.* Si un événement peut être produit par un nombre n de causes différentes, les probabilités de l'existence de ces causes prises de l'événement, sont entr'elles comme les probabilités de l'événement prises de ces causes. Et (par consequent) la probabilité de l'existence de chacune d'elles est égale à la probabilité de l'événement prise de cette cause, divisée par la somme de toutes les probabilités de l'événement prises de chacune de ces causes. [p.125]

This principle, "fécond en consequences" [p.125], is copied *verbatim* from Laplace's memoir of 1774, though for once an appropriate reference as to the source is made in the memoir itself. We shall postpone discussion of this principle to the chapter on Laplace.

The perhaps slightly general statement of the problem as initially posed is now refined as follows:

> Probléme. Soit une urne contenant un nombre n de billets; on a tiré $p+q$ billets, dont p sont blancs & q non-blancs (que

> j'appellerai noirs). On demande les probabilités que les billets blancs & les billets noirs de l'urne étoient des nombres données, dans la supposition qu'á chaque tirage on n'a pas remis dans l'urne le billet tiré [p.126]

and this in turn is further sharpened to

> Probléme. Tout étant posé comme dans le §4 [i.e. the preceding version]. On demande les probabilités d'amener dans un nombre donné r de nouveaux tirages faits de la même manière, des nombres donnés $r - m$, & m de billets blancs & noirs. [p.129]

Immediately following this last problem is the principle of solution; the probabilities of the event sought, corresponding to assumptions as to its causes, are made up in proportion to the probabilities of these causes and to the probabilities of the event depending on these causes, the probability of the event being the sum of these probabilities (clearly the principle follows from the *Principe étiologique* mentioned above).

All solutions are given in product form: full details may be found in Todhunter [1865, art.843]. All we shall do here, to give the flavour of the original presentation, is to present the *récapitulation* of §7, viz.

> On a tiré d'urne p billets blancs, & q billets noirs, en ne remettant dans l'urne à aucun des tirages le billet extrait. On tiré de nouveau r billets de la même manière. On obtient les expressions suivantes des probabilités que les nombres des billets blancs & noirs seront comme il suit.

Nombres des billets blancs	... billets noirs	Probabilités
r	0	$1 \times \frac{p+1.p+2. \ldots p+r}{p+q+2.p+q+3. \ldots p+q+r+1}$
$r-1$	1	$\frac{r}{1} \times \frac{p+1.p+2. \ldots p+r-1.q+1}{p+q+2.p+q+3. \ldots p+q+r+1}$
$r-2$	2	$\frac{r.r-1}{1.2} \times \frac{p+1.p+2. \ldots p+r-2.q+1.q+2}{p+q+2.p+q+3. \ldots p+q+r+1}$
&c.		

It is clear from this that the desired probability of drawing r white and s black balls can be expressed, more compactly, as

$$\frac{r!}{s!\,(r-s)!}\;\frac{(p+r-s)!}{p!}\;\frac{(q+s)!}{q!}\;\frac{(p+q+1)!}{(p+q+r+1)!}$$

or

$$\binom{p+r-s}{p}\binom{q+s}{q}\Big/\binom{p+q+r+1}{r} \ ,$$

an expression which the authors note, in their ninth section, is independent of the number of balls initially in the urn.

So far there is little, if indeed anything, that seems pertinent to our work. However, the authors go on to point out that the conclusion noted at the end of the preceding paragraph will *not* hold if sampling is effected *with* replacement. They state that a future memoir would consider this latter problem when the number of balls is infinite, but such observations apparently did not see the light of day. However Todhunter has considered the possible contents of such a memoir, and his thoughts run as follows (we present them here as an interesting example of a non-futile speculation): suppose that, from an urn with an infinite number of balls, p white and q black are chosen (without replacement). The probability that the next $r+s$ draws will result in r white and s black is then, by the Laplace theorem[51],

$$\binom{r+s}{r}\int_0^1 x^{p+r}(1-x)^{q+s}\,dx \Big/ \int_0^1 x^p(1-x)^q\,dx \ ,$$

evaluation of which results in the answer given above for the finite case. The coincidence appears to Todhunter to be "remarkable" [art.847]: but when we consider that the result for the finite case is independent of the number m of balls initially in the urn, should we not expect the same answer to hold "in the limit as $m \to \infty$", so to speak?

The remaining two memoirs, which were published in the same volume of the *Mémoires de l'Académie royale des Sciences et des Belles-Lettres*, being less mathematical in nature, are published in the *Classe de Philosophie Spéculative*, the second memoir occupying pp.3-25, and the third pp.25-41.

The second memoir, entitled *Mémoire sur l'art d'estimer la probabilité des causes par les effets*, is divided into two sections, of which only the first (*Des principes de cette partie de l'art de conjecturer*) need be considered here (the second part, *Précis de la marche des applications*, consists in some simple applications of the principle propounded in the first part to some die problems)[52].

Two early definitions, given at the start of the first section of this, the second, memoir, are, I think, of interest. They are the following:

> La Stochastique, ou l'art de conjecturer avec rigeur, ayant en pour premier objet d'estimer les hasards du jeu, est fondée sur des principes relatifs à cette origin. [p.3]
>
> La Stochastique entière repose sur cette hypothèse que je vais

> maintenant énoncer sous une forme plus générale. Hypothèse Stochastique. Lorsqu'en vertu d'une certaine détermination des causes, plusiers événemens nous paroissent également possibles; nous feignons que tous ces événemens ont lieu successivement tour-à-tour & sans répétition. [p.6]

Here we find strongly stated the opinion that "la stochastique" (dare we translate this by the archaic noun "stochastic"?) has, as its *fons et origo* (and also its prime purpose), games of chance. The "hypothèse stochastique" is also of interest, stating as it does that a judgement of equipossibility is, in a sense, basic, and that it is on the grounds of such a judgement that we suppose events occur in turn and without repetition.

This is one of the few French papers in which reference to earlier authors is specifically made. We read further in the memoir, in fact,

> MM. JAC. BERNOULLI, MOYVRE, BAYES & PRICE ont successivement appliqué le calcul à la recherche des causes. Mais le principe sur lequel repose la justesse de leurs résultats, n'étant pas énoncé, laisse un vide qui nuit à la clarté: & ce défaut, trés-sensible à tout lecteur attentif, a rendu timides ces auteurs mêmes; en sorte que leurs résultats n'ont ni l'étendu ni l'utilité qu'ils auroient pu leur donner. Et si une sage défiance les a garantis de l'erreur, l'incertitude de leur marche a laissé des hasards à courir à ceux qui tenteroient des les suivres. §9. M. de la Place le premier a posé disertement le principes sur lequel repose toute cette partie de la théorie des probabilités. Voice comme il l'a énoncé:
> Principe. Si un événement peut être produit par un nombre n de causes différentes, les probabilités de l'existence de ces causes prises de l'événement, sont entre elles comme les probabilités de l'événement prises de ces causes. [p.8]

Here I believe Prevost and Lhuilier are unjust to Bayes: it is, I trust, quite clear from what has already been said that his presentation and solution of the problem were perfectly satisfactory. On Price they are perhaps more correct, while their opinions on Bernoulli and de Moivre do not concern us. They are, however, quite correct in attributing to Laplace the first announcement of the principle.

The authors then restate this fundamental principle as their *principe étiologique* (in a slightly different form to that given in the first memoir). After that, we read[53]

> Tel est le principe reconnu par M. de la Place, lequel a rendu claire & sûre l'estimation de la probabilité des causes par les

> effets, & que, par cette raison, j'ai cru devoir appeler *Principe étiologique.* [p.8]

Prevost and Lhuilier now prove Laplace's principle (their statement of the *principe étiologique* here is framed in terms of dice-throwing), and deduce the discrete "Bayes's theorem" from it.

The third memoir is entitled *Remarques sur l'utilité & l'étendue du principe par lequel on estime la probabilité des causes*, and it also deals with Laplace's fundamental principle. Again there is a reference to Bayes — as Bayer! The first section is on the utility of the principle, the second on its extent, and the third on the comparison of some results of the (probability) calculus to the judgements of common sense. Of interest to us is the start of Section 19:

> Enfin la théorie de l'estimation des probabilités *a posteriori* fournit une conséquence nouvelle & remarquable: c'est que l'hypothèse de l'ignorance des causes, & l'hypothèse de la connaissance de leur nature, ne donnent la mêmes résultats que dans le cas où on estime une probabilité simple,

this being illustrated by a die-tossing example. The fourth, and final, section is devoted to some mathematical developments.

4.10 Carl Friedrich Gauss (1777-1855)

Gauss's works, although legendary, contain relatively little pertinent to our topic[54]. Indeed the relevant writings are limited to two: an 1815 review of Laplace's *Sur les comètes* and a passage from the 1809 opus *Theoria Motus Corporum Coelestium in Sectionibus Conicis Solem Ambientium.* The former will be considered in §6.12; we turn our attention immediately to the latter.

In Article 176 of the third Section of the second Book Gauss[55] cites the following result[56]:

> Si posita hypothesi aliqua H probabilitas alicuius eventus determinati E est $= h$, posita autem hypothesi alia H' illam excludente et per se aeque probabili eiusdem eventus probabilitas est $= h'$: tum dico, quando eventus E revera apparuerit, probabilitatem, quod H fuerit vera hypothesis, fore ad probabilitatem, quod H' fuerit hypothesis vera, ut h ad h'.

That is, $\Pr[H \mid E]/\Pr[H' \mid E] = \Pr[E \mid H]/\Pr[E \mid H']$ under the assumption that $\Pr[H] = \Pr[H']$. Arguing from numbers of equally-likely cases

Gauss demonstrates this theorem, and goes on to apply it in the following case: suppose there are $\mu(>\nu)$ functions[57] $V, V', V'' \ldots$ of the ν unknown quantities $p, q, r, s, \ldots$. Suppose further that the values of the functions found by direct observation are $V = M$, $V' = M'$, $V'' = M''$ etc. Expressing by $\varphi(M - V)$ the probability that observation yields the value M for V, and substituting in V a determinate system of values for $p, q, r, s, \ldots$, we find, under the assumption of independent observations, that the probability ("or expectation") that all these values will result together from observation is

$$\Omega = \varphi(M - V)\,\varphi(M' - V')\,\varphi(M'' - V'')\ \ldots$$

Using the theorem cited above one finds that[58]

$$\Pr\left[p < P < p + dp,\ q < Q < q + dQ, \ldots \mid V = M, V' = M', \ldots\right]$$

$$= \lambda\Omega\, dp\, dq\ \ \ldots$$

where $1/\lambda = \int_{-\infty}^{\infty} \cdots \int_{-\infty}^{\infty} \Omega\, dp\, dq\ \cdots$. This result of course obtains under the assumption that "omnia systemata valorum harum incognitarum ante illas observationes aeque probabilia fuisse" [art.176].

Gauss now concludes that the most probable system of values of the quantities p, q, r, s, etc. is that which maximizes Ω, whence he deduces that the probability to be assigned to an error Δ should be given by

$$\varphi(\Delta) = \frac{h}{\sqrt{\pi}}\, e^{-h^2\Delta^2} \quad ,$$

h being "considered as the measure of precision of the observations" (Davis [1857, p.259]).

4.11 William Morgan (1750-1833)

William Morgan, a nephew of Richard Price[59], was by profession an actuary, and contributed himself nothing to our subject — although his 1783 paper "Probability of Survivorship" was excellent enough to win him the gold medal of the Royal Society, and a fellowship followed soon thereafter.

However Morgan also wrote a small monograph bearing the title "Memoirs of the Life of the Rev. Richard Price, D.D. F.R.S." in which reference was made to Price's involvement with Bayes's Essay. William had not intended to write this memoir: in his foreword he in fact states that his brother George[60] had

> undertaken to write a very circumstantial history of his uncle's life, and had made a considerable progress in it, when, towards the close of the year 1798, a fatal disorder put a final period to this and all his other pursuits.
> The confused state in which his papers were found, and the indistinct short hand in which they were written, rendered it impossible either to arrange or to understand them properly; and therefore, after many fruitless attempts, I was reluctantly obliged to give up the investigation, and to take upon myself the task of writing a new, but more concise account ... [1815, pp.vi-vii].

The rôle Richard Price played in communicating Bayes's Essay to the Royal Society is succinctly summarised as follows by Morgan (the quotation is long, but I think worthy of inclusion).

> On the death of his friend Mr. Bayes of Tunbridge Wells in the year 1761, he was requested by the relatives of that truly ingenious man, to examine the papers which he had written on different subjects, and which his own modesty would never suffer him to make public. Among these Mr. Price found an imperfect solution of one of the most difficult problems in the doctrine of chances, for "determining from the number of times in which an unknown event has happened and failed, the chance that the probability of its happening in a single trial lies somewhere between any two degrees of probability that can be named." The important purposes to which this problem might be applied, induced him to undertake the task of completing Mr. Bayes's solution; but at this period of his life, conceiving his duty to require that he should be very sparing of the time which he allotted to any other studies than those immediately connected with his profession as a dissenting minister, he proceeded very slowly with the investigation, and did not finish it till after two years; when it was presented by Mr. Canton to the Royal Society, and published in their Transactions in 1763.
> — Having sent a copy of his paper to Dr. Franklin, who was then in America, he had the satisfaction of witnessing its insertion the following year in the American Philosophical Transactions[61]. — But not withstanding the pains he had taken with the solution of this problem, Mr. Price still found reason to be dissatisfied with it, and in consequence added a supplement to his former paper; which being in like manner presented by Mr. Canton to the Royal Society, was published in the Philosophical Transac-

tions in the year 1764. In a note to his Dissertation on Miracles, he has availed himself of this problem to confute an argument of Mr. Hume against the evidence of testimony when compared with the regard due to experience; and it is certain that it might be applied to other subjects no less interesting and important. By these two communications to the Royal Society, Mr. Price had proved himself not unworthy the honour of being admitted a member of that learned body, and he was accordingly elected in a few months after the publication of his second paper. [1815, pp.24-27]

4.12 Sylvestre François Lacroix (1765-1843)

In his *Traité Élémentaire du Calcul des Probabilités* [1816][62] Lacroix has this to say on the probability of causes:

> C'est ainsi qu'on a posé pour principe que les probabilités des causes (ou des hypothèses) sont proportionelles aux probabilités que ces causes donnent pour les événemens observés. [p.133]

In a footnote to this passage he writes

> Cet énoncé se trouve dans le tome VI des *Savans étrangers*[63], p.263. Bayes, dans les *Transactions philosophiques* de 1763, et Price, dans celle de 1764 (p.296), s'étaient déjà occupés de ce sujet; mais M. Laplace l'a réduit le premier à la forme analytique sous laquelle on le traite maintenant, qui en facilite et en généralise beaucoup les applications. [p.133]

Once again it is doubtful whether the animadversion to Bayes as having been concerned with causes is correct. Other pertinent passages are the following:

> Enfin il faut remarquer encore que ces fractions, ou les probabilités des diverses hypothèses, se forment en divisant la probabilité de l'événement composé, calculée dans chaque hypothèse, par la somme de ses probabilités dans toutes les hypothèses [p.143]

and

> On trouverait de même, pour tout autre example, que la probabilité d'un nouvel événement simple s'obtient en calcul, d'après

> les événemens passés, la probabilité des diverses hypothèses possibles, et faisant la somme des produits de ces probabilités par celles de l'événement, prises dans chaque hypothèse. [pp.135-136]

These statements lead, in a manner that is by now perhaps all too familiar, to expressions of the form

$$\alpha x^{m+1}(1-x)^n \Big/ \alpha \int x^m(1-x)^n\,dx,$$

and

$$\frac{p(p-1)\dots(p-q+1)}{1.2.\ \dots\ q} \times \frac{\int_0^1 x^{m+p-q}(1-x)^{n+q}\,dx}{\int_0^1 x^m(1-x)^n\,dx}$$

where [0,1] is divided into small parts, denoted by α.

4.13 Conclusions and Summary

In the half-century following the publication of Bayes's Essay, there seems to have been little published which might be regarded not only as pertinent but also as original — excluding, of course, the works of Condorcet and Laplace, to which we shall turn in the following chapters.

Hard on the heels of the Essay came a paper communicated by Price to the Royal Society, in which Bayes's proofs of the rules of the Essay were detailed and developed. Much of the refinement was due to Price himself.

It is possible to find in Mendelssohn's writings a precursor of Laplace's rule of succession, though hindsight and charity are probably required for such a discovery. The expression that Buffon advances for the solution of similar problems bears no resemblance to Laplace's, though it is (more or less) in accord with Price's.

It seems clear to me that the Bayes integrals to be found in some of the papers discussed here are in fact due to Laplace, and that a number of the results we have noted are but application or development of Laplace's work.

More noteworthy is the discussion we find here, by Lagrange, of a problem in inverse probability — perhaps the first in print. This discussion appeared in 1770, a scant six years after publication of Bayes's results. Perhaps one should consider Lagrange, rather than Bayes, as the father (albeit unwittingly) of inverse probability.

4.14 Appendix 4.1

In 1774 the collected works of Guillaume Jacob 'sGravesande appeared. Here, in Part II of the *Introduction a la Philosophie, contenant la Metaphysique, et la Logique*, may be found, in chapter XVII, "De la probabilité", what in effect is an example of an inverse to Bernoulli's theorem (though it amounts to little more than the advocating of the approximation of a probability by an observed frequency, and the mentioning that the error involved in such an approximation decreases as the number of trials increases). Since this work was apparently first printed[64] in 1736, however, it falls outside the ambit of the present study.

Chapter 5

Condorcet

The Productions of an exalted Genius are very liable to Misconstruction and Cavil, as the Subject is often clouded with some natural Intricacy.

Francis Blake.

5.1 Introduction

Marie Jean Nicolas Caritat, Marquis de Condorcet (1734-1794) was a man of polymathic, if not polyhistoric, proportions. Pearson [1978] has described him as follows:

> there have been better mathematicians, better economists, better historians, better philosophers and better politicians than Condorcet, but scarcely any man has been at the same time as good a mathematician, as good an economist, as good an historian, as good a philosopher and as good a politician as he was. [p.425]

Of the some half-dozen writings by Condorcet considered in this chapter, two, a memoir and an essay, outstrip the others in importance[1]. Although the memoir was published in a number of parts (almost as separate papers) over a number of years, and although the essay was published during this period, we shall consider the former as a unit and discuss it *in toto* (where relevant).

5.2 Unpublished manuscripts

The existence of two early probabilistic works by Condorcet, presently housed in the *Bibliothèque de l'Institut de France*, has been noted by Baker [1975, p.436]. The first of these, MS883, ff.216-221, was probably written in 1772: it contains nothing pertinent to the present study. The second, MS875, ff.84-99 (copy 100-109) dates from 1774, and bears the title "Histoire abrigée de le calcul". It is clear from the manuscript that the work was revised at some stage, and it is in one of these revisions that the only reference to Bayes (a reference not repeated in the fair copy) is to be found, to wit,

> Les principes de les calculs se trouverant dans les Transactions Philosophiques annee 1764 No. LIII dans différens morceaux de Mrs Bayes et Price.

The reference which this sentence replaced was to a memoir by Laplace "imprime dans le Tome VI": this is clearly a reference to Laplace's paper of 1774, and suggests that Condorcet became aware of Bayes's work after the publication of this paper of Laplace's.

Crepel [1987] has recently pointed out that the first of the manuscripts mentioned above, viz. MS883, ff.216-221, is really only the first part of a longer work, the second part of which, Z30, ff.1-6, is housed in the *Bureau des Longitudes*, while the third, MS875, ff.132-133, is to be found in the *Bibliothèque de l'Institut de France.* An outline of the contents of these fragments is given by Crepel (op. cit.): it does not appear that anything germane to the present work is to be found there.

5.3 The Memoir

This memoir[2], in six parts, was published in the *Histoire de l'Académie royale des Sciences* for the years 1781, 1782, 1783 & 1784, although the dates of publication are usually later than these dates.

The first part of the *Mémoire sur le calcul des probabilités* is entitled *Réflexions sur la règle générale qui prescrit de prendre pour valeur d'un évènement incertain, la probabilité de cet évènement, multipliée par la valeur de l'évènement en lui-même*; and it occupies pp.707-720 of the volume for 1781 (although it was read on the 4th August, 1784). This part contains nothing pertinent: the second, however, filling pp.720-728 of the same volume and entitled *Application de l'analyse à cette question: Déterminer la probabilité qu'un arrangement régulier est l'effet d'une intention de le produire*, contains some observations that are at least slightly relevant.

The first noteworthy detail concerns n possible combinations, of which only one is regular.

> Je suppose qu'il y ait n combinaisons possibles, & qu'une seule d'elles soit régulière. Si une cause a eu l'intention de produire cette combinaison, elle a eu lieu nécessairement, & sa probabilité sera 1; si, au contraire, elle a été l'effet du hasard, sa probabilité sera $1/n$. [Condorcet 1781, p.720]

Applying what Pearson [1978, p.454] describes as "inverse probability" — though an argument framed in terms of odds might perhaps be more readily understood — Condorcet says that cause and chance are then in the ratio of $1 : 1/n$, and hence the chance of a cause and the chance of a chance are $1/(1 + 1/n)$ and $(1/n)/(1 + 1/n)$ (i.e. $n/(n + 1)$) respectively. As we have already seen (§4.1 above), these are the values given by Mendelssohn, the first edition [1761] of whose *Philosophische Schriften* antedates the publication of Bayes's Essay by three years.

The second pertinent detail concerns sequences of regularities; specifically, the two series

$$\begin{array}{cccccccccc} 1, & 2, & 3, & 4, & 5, & 6, & 7, & 8, & 9, & 10 \\ 1, & 3, & 2, & 1, & 7, & 13, & 23, & 44, & 87, & 167 \end{array}$$

or respectively,

$$a_n = 2a_{n-1} - a_{n-2} \ , \ \ n \in \{2, 3, \dots, 10\} \ , \ \ \& \text{ given } \ a_0 = 1, \ a_1 = 2 \ ;$$

$$a_n = a_{n-1} + a_{n-2} + a_{n-3} + a_{n-4} \ , \ n \in \{4, 5, \dots, 10\} \ ,$$

$$\& \text{ given } \ a_0 = 1, \ a_1 = 3, \ a_2 = 2, \ a_3 = 1 \ .$$

These symbolic formulations are in accord with what Condorcet himself wrote; however, the first series could of course have been obtained in many different ways (e.g. $a_0 = 1$ and $a_n = a_{n-1} + 1$, $n \in \{1, 2, 3, \dots, 10\}$), and any other method of obtaining it would change Condorcet's solution. But we shall not worry about this point: rather let us examine how Condorcet continues his example.

Keeping e terms of the first sequence and e' of the second, one is assured that the probability that the law of formation of the sequence will be continued q times is

$$(e + 1)\Big/(e + q + 1) \qquad \text{and} \qquad (e' + 1)\Big/(e' + q + 1) \qquad (1)$$

respectively for the two sequences. This is essentially Pearson's exposition [1978, p.455]: the original reads as follows:

> Soit donc pour une de ces suites e le nombre des termes assujettis à une loi, et e' le nombre correspondant pour une autre suite, et qu'on cherche la probabilité que pour un nombre q de termes suivans, la même loi continuera d'être observée. La première probabilité sera exprimée par $(e+1)/(e+q+1)$, la seconde par $(e'+1)/(e'+q+1)$, et le rapport de la seconde à la première par $(e'+1)(e+q+1)/(e+1)(e'+q+1)$.

Although a numerical example is given, no argument is presented for the derivation of (1), the values in which are certainly those which would arise from an application of the rule of succession. Todhunter considers this example in some detail in his Article 724; but in view of the arbitrariness of the assumptions on which it is based, there seems little point in pursuing the matter further.

The third part of the memoir appeared in the volume for 1782 (published 1785), pp.674-691, and is entitled *Sur l'évaluation des droits éventuels.* Writing of this part, Todhunter [1865] says that it is

> neither important nor interesting, and it is disfigured by the contradiction and obscurity which we have noticed in Condorcet's Essay. [art.728]

However, Todhunter devotes some three pages [arts 726-732] to a discussion of this trivial and tedious tractate, while Pearson found it (or at least parts of it) worthy of fairly detailed comment in his historical lectures [1978, pp.455-457]. For us, the importance of this memoir lies in its use of multiple Bayes's integrals, introduced (as we shall see in Chapter 6) by Laplace[3] in 1778.

Condorcet begins by examining the case in which the cause (or event) by which the right is produced necessarily happens in a certain length of time ("as, for example, when the right accrues on every succession to the property" [Todhunter 1865, art.728]), this case being followed by one in which the event does not necessarily happen ("as, for example, when the right accrues on a sale of the property, or on a particular kind of succession" (Todhunter loc. cit.)). Three methods are given for the first case: we detail two of them here.

The first method proceeds as follows[4]: let $a_1, a_2, \ldots, a_n$ be the number of years elapsing between two transfers ("mutations observees")[5] and $b_1, b_2, \ldots, b_n$ the number of transfers corresponding to those intervals. (This is somewhat vague: what is perhaps meant is that, starting from a contingency realized in year $a_1 = 1$, b_1 further contingencies have become realized, in the second year ($a_2 = 2$), b_2 became realized, &c.) Further, let 1 be the value of the right for any property whatsoever at the moment of its transfer, and $1/m$ the annual interest of right 1. The problem is to determine the total value of the right, as much as for the actual transfer as for all future transfers, this value being reported at the present time. One knows that the right 1 which will only be due at the end of z years will then be given by $(m/(m+1))^z$, or abbreviated, by c^z.

If we then consider p successive transfers, of which p_1 occur at the end of a_1 years, p_2 at the end of a_2 years,..., p_n at the end of a_n years, it is clear that, in whatever order these transfers succeed each other, the last will happen at the end of $p_1a_1 + p_2a_2 + \cdots + p_na_n$ years; so that the sum due for this transfer will always be

$$c^{a_1p_1+a_2p_2+\cdots+a_np_n} \quad .$$

If, in the next place ("ensuite"), one denotes by x_1 the probability of the transfer after a_1 years, x_2 the probability after a_2 years,... and finally $1 - x_1 - x_2 - \cdots - x_{n-1}$ the probability after a_n years, the probability of this p-th transfer which we are considering will be expressed by

$$\frac{p!}{p_1!\, p_2! \ldots p_n!} x_1^{p_1} x_2^{p_2} \ldots (1 - x_1 - x_2 - \cdots - x_{n-1})^{p_n}$$

so that the value of all the p-th transfers, each multiplied by its respective probability, will be

$$[c^{a_1}x_1 + c^{a_2}x_2 + \cdots + c^{a_n}(1 - x_1 - x_2 - \cdots - x_{n-1})]^p$$

which represents the mean value of the right of this transfer.

Noting that here x is neither given nor constant, Condorcet goes on to say that one knows only that the event whose probability is expressed by x_1 has happened b_1 times, that whose probability is expressed by x_2, b_2 times, &c. The mean value of the right for the p-th transfer will then be expressed by

$$\frac{\int \left\{x_1^{b_1} x_2^{b_2} \ldots y^{b_n} [c^{a_1}x_1 + c^{a_2}x_2 + \cdots + c^{a_n}y]^p \, dx_1 \, dx_2 \ldots dx_{n-1}\right\}^{n-1}}{\int \left\{x_1^{b_1} x_2^{b_2} \ldots y^{b_n} \, dx_1 \ldots dx_{n-1}\right\}^{n-1}}$$

where $y = (1 - x_1 - \cdots - x_{n-1})$, the integration being repeated $n-1$ times and the integrals[6] being taken from $x_{n-1} = 0$ to $x_{n-1} = 1 - x_1 - \cdots - x_{n-2}$, from $x_{n-2} = 0$ to $x_{n-2} = 1 - x_1 - \cdots - x_{n-3}, \ldots$, from $x_1 = 0$ to $x_1 = 1$.

Following this development of a multiple Bayes's integral, Condorcet remarks ("somewhat naively", according to Pearson [1978, p.457])

> Nous ne dirons rien de plus de ces formules, si n'est qu'elles s'intègrent par les méthodes connues, & que d'ailleurs on eut auroit des valeurs très-approchées, soit par la méthode donnée par M. Euler, soit par celles que M. de la Place a exposées dans ce même volume.[7] [p.682]

By now it will probably be quite clear that Todhunter has not erred in frequently drawing his readers' attention to Condorcet's obscure and often obnubilated oratory. We choose at this stage, therefore, to present a second method of approaching this problem, following Todhunter [1865, arts 729-730] and Pearson [1978, p.457].

Suppose, then, that the right is equally likely to occur in any year (e.g. change by sale, rather than death of present holder). If c is the present value of the fee to be paid in the event of the right being realized, the value of the whole right is

$$x\left(c + c^2 + \cdots\right) = xc\,/(1-c) \quad .$$

If during the past $m+n$ years the event happened m times and failed to happen n times, one might well estimate x by $m/(m+n)$, in which case the whole value of the right becomes $[c/(1-c)]\,[m/(m+n)]$. Since, however, Condorcet views x as unknown, the whole value of the right must rather be taken as

$$\int_0^1 x^m(1-x)^n xc(1-c)^{-1}\,dx \Bigg/ \int_0^1 x^m(1-x)^n\,dx$$

$$= \frac{c}{1-c}\cdot\frac{B(m+2,n+1)}{B(m+1,n+1)}$$

$$= \frac{c}{1-c}\cdot\frac{m+1}{m+n+2} \ ,$$

a result which differs from the preceding estimate by the replacement of m and n by $m+1$ and $n+1$ respectively (a substitution of little moment if m is large).

In his Article 730, Todhunter criticizes this second method on two accounts, viz.

(i) Condorcet asserts that this method is applicable to his first case, that is, one in which the event must happen in a given number of years. In an example such as he mentions, namely one where the right would accrue on the death of the present holder of the property, the method is clearly inapplicable, since the probability of the event concerned may well vary from year to year. This method would, however, be applicable in the second case — i.e. when the right is supposed to accrue from a sale (as we have in fact supposed in our discussion of this method), the probability of which latter event might well be supposed to be constant from year to year.

(ii) The use of Bayes's theorem here adds very little to our knowledge when $m+n$ is large; and when it is small, "our knowledge of the past would be insufficient to justify any confidence in our anticipations of the future." [Todhunter 1865, art.730]

Finishing off his discussion of the second method detailed above, Pearson [1978, p.457] writes "Todhunter not unjustly calls Condorcet's method 'an extravagant extension and abuse of Bayes' Theorem' ". In writing this the worthy biometrician has erred: the quotation from Todhunter is in fact a reference to a later part of the memoir in which the total value arising from two different rights is investigated.

The fourth part of the memoir, published in 1786 (i.e. after the *Essai*) in the volume for 1783, is entitled *Réflexions sur la méthode de déterminer la probabilité des évènemens futurs, d'après l'observation des évènemens passés*, and occupies pp.539-553. The purpose of the work is summarized succinctly in the opening words as follows:

> Cette partie de l'Analyse qui enseigne à déterminer la probabilité des évènemens futurs, d'après l'ordre qu'ont suive les évènemens passés du même genre que l'on a observés, est susceptible d'un grand nombre d'applications utiles & curieuses; j'ai cru en conséquence qu'il pourroit n'être pas inutile d'examiner les principes sur lesquels cette Analyse est fondée; tel est l'objet des Réflexions suivantes. [p.539]

Despite the fact that Condorcet was a personal friend of Price's[8], there is mention neither of the latter nor of Bayes[9]. Writing in the 1920's Pearson [1978] says

> It is simply the French custom[10], which never cites authorities, so that it is impossible to say of a French work or memoir how much or how little is original. Of course it is a very bad custom, which has lasted from 1700 to the present day in France. [p.457]

We shall however find later mention of Bayes in Condorcet's work.

Condorcet begins by supposing that there are only two events A and N, of a nature which we should today describe as "mutually exclusive and only possible", and that these two events have occurred m and n times respectively. The probability, then, of having, in $p+q$ trials[11], p events (or occurrences of the event) A and q events N, will be

$$\frac{(p+q)\ldots(q+1)}{1.2\ldots q}\int_0^1 x^{m+p}(1-x)^{n+q}\,dx\Big/\int_0^1 x^m(1-x)^n\,dx \qquad (2)$$

"telle est la règle générale" [Condorcet 1783, p.539]. (This rule also occurs in the *Essai* — cf. Todhunter [1865, art.704] and Dinges [1983, p.74]: we shall take up this point later on.)

It is, I believe, important to consider Pearson's [1978] comments on this formula: he says

> This is the generalised Bayes' Theorem; it is the generalisation which is due to Condorcet. Bayes took $p=1$ and $q=0$. But Bayes is more correct than Condorcet, for he shows why he puts the 'dx' in on his hypothesis of first ball determining the chance of success or failure. Condorcet does not explain where the dx comes from. I think it can only be explained by the Euler-Maclaurin bridge and in this case, we must suppose the differential coefficients finite at the terminals. The point is, I think, an important one, because Condorcet starts from ball drawing in urns, and thus his x is really the ratio of two numbers and not continuous unless the total number of balls in the urn be infinite. x would go by stages, and it may be just possible that for small m, n, p and q the terminal conditions do become of some importance. [p.458]

Now one must bear in mind that Pearson's *History* is composed of a series of *lectures* and was not designed by him for publication[12]. It is quite possible, therefore, that any criticism one may level against this work might well have been removed had his intentions been otherwise. Nevertheless, it is, I feel, necessary to comment briefly on this passage.

(i) The first, and perhaps the most important, remark is that Bayes's result is *not* that given here with $p=1$ and $q=0$. We have already hinted (and shall say more on the matter in the chapter on Laplace) that there is no reference to (the occurrence of) any future event in Bayes's Essay *per se* (although such an extension is of course made by Price).

(ii) The "generalization", if such we may call it, is not due to Condorcet: it was in fact given by Laplace in 1774 in his[13] *Mémoire sur la probabilité des causes par les évènements.*

(iii) As minor comments, we might mention two points: firstly, there is in fact no "dx" in Bayes's work (he did not use integral notation). Secondly, I have not managed to find, either in Todhunter's discussion or in the pertinent part of the original, any reference to the drawing of balls from an urn. Such reference is however made in Laplace's memoir cited in (ii) above, and we shall return to this in the appropriate chapter.

After presenting (and I use this word purposely, for no further argument is given) this formula, Condorcet points out that it really expresses the probability only in the case of the following two hypotheses:

> 1. Si la probabilité des évènemens A & N reste la même dans toute la suite des évènemens; cela est évident par la formule même qui exprime la loi.
> 2. Dans le cas où cette même probabilité est variable, mais où l'on supposeroit en même temps que la valeur de la probabilité, quoique pouvant être différente pour chaque évènement, est cependant prise au hasard pour chacun, d'après une certaine probabilité générale x pour A, & $1 - x$ pour N. [p.540]

After some discussion of these hypotheses, Condorcet gives his definition of "probability" as

> n'est que le rapport du nombre des combinaisons qui amènent un évènement à celui des combinaisons qui ne l'amènent pas; combinaisons que notre ignorance nous fait regarder comme également possibles [p.540]

and then relates this definition to the two hypotheses. He stresses that for any other hypothesis the formula cited should not be regarded as giving accurate results: in such a case a procedure advocated in the *Essai* [p.179] may be adopted.

Having noted that the same formula holds for any ordering of the m A's and n N's, Condorcet points out that when the assumption that x is constant contradicts that which reason indicates, one ought perhaps to use some method in which the probability depends on the order of the events: two cases involving variable x are briefly considered in the fourth article.

In Article 5 Condorcet considers the case in which the probability may differ from one event to another, although it is independent of the order in which the events occur. Let t denote the total number of events, past or future, let $t_1 = m + n$ be the number of past events, and let $t_2 = p + q$ be the number of future events. Denoting by $x_1, x_2, \ldots, x_t$ the different probabilities in favour of A, he gives, for the probability of p events A and q events N in t_2 future events (instead of the earlier formula) the expression

$$\frac{t_2(t_2-1)\dots(p+1)}{q!}\frac{\int\cdots\int s_t^{m+p}/t\ [1-s_t/t]^{n+q}\,dx_1\dots dx_t}{\int\cdots\int s_t^m/t\ [1-s_t/t]^n\,dx_1\dots dx_t} \tag{3}$$

where $s_k=\sum_{i=1}^k x_i$ and each integral is taken from 0 to 1. He then proceeds to evaluate the integrals in the usual manner, noting the correspondence with the customary result if $t=1$. An expression is also given for the probability that, in an unlimited sequence of events, more A's than B's will occur.

In the sixth article Condorcet supposes the probability to be variable, but possibly dependent on the order of events. He is once again rather confusing, and I shall therefore quote the original:

> soit x' la probabilité du premier A, & $1-x'$ celle du premier N; $(x'+x'')/2$ & $(2-x'-x'')/2$ pourront exprimer les probabilités du second A ou du second N, $(x'+x''+x''')/3$ & $(3-x'-x''-x''')/3$ celles du troisième A ou du troisième N; & celles des r^{es} A ou N pourront l'être par $(x'+x''+x'''+\dots+x'''^r)/r$ & $(r-x'-x''-x'''-\dots-x'''^r)/r$, où l'on voit que x' est la probabilité de A au premier coup, x'' celle de A au second si elle est différente de celle du premier, x''' celle de A au troisième si elle est différente de celle des deux autres, & ainsi de suite. [p.545]

Noting the difficulties that can arise when future occurrences are to be taken into account, Condorcet restricts his subsequent attention to the case in which future events occur in the same order as that which has already been observed (event E, say). If we let n be the number of constantly occurring events, and p the number of future events, the probability that that event (E) will occur (or that that law will be observed during the time of p revolutions) will be expressed by

$$\frac{\int\cdots\int s_1(s_2/2)\dots(s_{n+p}/(n+p))\,dx_1\dots dx_{n+p}}{\int\cdots\int s_1(s_2/2)\dots(s_n/n)\,dx_1\dots dx_n}\ . \tag{4}$$

Todhunter, writing of formula (2) above, says

> Condorcet quotes this result; he thinks however that better formulae may be given, and he proposes two. But these seem quite arbitrary, and we do not perceive any reason for preferring them to the usual formula. [1865, art.734]

However, as Pearson [1978, p.459] has noted, Condorcet is in fact considering three distinct problems, formulated in the seventh section of this part of the memoir as follows:

> 1^0. celle où la probabilité est constante, c'est-à-dire, où l'on suppose chaque évènement également probable, ou du moins la probabilité moyenne pour chacun, déterminée d'une manière semblable; 2^0. celle où l'on suppose cette probabilité variable, mais indépendante du temps où les évènemens sont arrivés, & de l'ordre dans lequel ils ont été observés; 3^0. celle où on les suppose dépendans, ou plutôt pouvant dépendre de cet ordre. [pp.548-549]

The solutions to these problems are those respectively given by formulae (2) - (4) above. In his comments on this section, Pearson writes

> In (i) Condorcet agrees with and generalises Bayes. This is an advance, but no more than Bayes has he any hesitation about the equal distribution of ignorance. In (ii) he takes a mean value of all the unknown chances and integrates with regard to each of them. If he had integrated solely with regard to the mean chance he would have really fallen back on Bayes. I think to be accurate he ought to have recorded the success or failure at each trial and integrated the resulting products, and this would give the answer in the same manner as Bayes. If this be done it seems to me that we should get precisely the same result for (ii) and (iii) unless in (iii) we make some hypothesis as to the correlation between successive x's. [1978, p.459]

Let us now examine this quotation:

(a) We have already commented on the claim that (2) is a generalization of Bayes's result. Further, Pearson is perhaps a little too ready to say that both Condorcet and Bayes had no hesitation in using the "equal distribution of ignorance" assumption. We have previously discussed Bayes's argument for this prior postulate, an argument which one must agree is singularly lacking in Condorcet's work.

(b) As regards the sentence starting "if he had integrated solely ...", this is clearly true.

(c) In the sentence starting "I think to be accurate ...", is Pearson suggesting merely the integration, in the usual manner, of some product $\prod_1^n x_i^{\alpha_i}(1-x_i)^{\beta_i}$?

(d) One must agree with Pearson as regards the hypothesis of correlation; and the hypothesis that Condorcet has in fact chosen is, like those on which other formulae presented in the memoir and of similar type to those already mentioned, are based, rather arbitrary.

As a final example from this part of the memoir we instance that presented in the ninth section. Here Condorcet supposes that two sequences S and S' of events A and N have been observed, with A and N occurring m and n times respectively in S, and m' and n' times respectively in S'. In addition it is supposed that the ratio $m : n$ differs sufficiently from the ratio $m' : n'$ that one may assume that the probability of A is not the same in the two sequences. It is required to find the probability of getting p A's and q N's in $(p+q)$ future events. Letting x and $1-x=z$ (x' and $1-x'=z'$) be the probabilities of A and N respectively in the first (second) sequence, Condorcet defines X and X' by $X = x^m(1-x)^n$ and $X' = x'^{m'}(1-x')^{n'}$. He then considers in $(x+z+x'+z')^{p+q}$ the sequence of all terms in which the sum of the exponents of x and x' is p, and that of z and z' is q. On our letting $A\, x^a x'^b z^{a'} z'^{b'}$ be one of these terms, the resultant probability is found to be

$$\frac{A\,.\int X\, x^a z^{a'}\, dx.\int X' x'^b z'^{b'}\, dx'}{\int X\, dx\ .\ \int X'\, dx'}\ ,$$

the required probability being the sum of all the terms thus formed, provided that it is equally probable that a future event belongs to either S or S'.

If, contrariwise, one supposes that this same probability depends on the order observed in the two sequences, then the term given must be multiplied by

$$\binom{p+q}{a+a'}\int (X\, dx)^{a+a'}\int (X'\, dx')^{b+b'}\ ,$$

the required probability being found by summing all such terms and dividing by

$$\int (X\, dx + X'\, dx')^{p+q}\ .$$

Finally one may suppose this probability ordered in accordance with the number of terms of each sequence, in which case the same term must be multiplied by

$$\binom{p+q}{a+a'}\int x_2^{m+m'+a+b}(1-x_2)^{n+n'+a'+b'}\, dx_2\ ,$$

taking the sum of all such terms and dividing by

$$\int x_2^{m+m'} (1 - x_2)^{n+n'} \, dx_2 \quad .$$

In the fifth part of his memoir, *Sur la probabilité des faits extraordinaires*, published on pp.553-559 of the same volume of the *Histoire de l'Académie* as the fourth, Condorcet devotes no little attention to the question of testimony[14]. In doing so, he presents in the second section the following argument:

> Supposons que u désigne la probabilité d'un évènement A, & e celle d'un évènement N, que u' & e' désignent les probabilités de deux autres évènemens A' & N'; $uu'/(uu' + ee')$ exprimera la probabilité de la combinaison des évènemens A, A'; & $ee'/(uu' + ee')$ la probabilité de celle des évènemens N, N'. [p.554]

An example involving the drawing of coins from an urn follows, and this in turn is followed by a testimonial example, in which the use of the discrete Bayes's Theorem is perhaps more clearly expressed. The relevant passage runs as follows:

> Supposons maintenant que u & e représentent les probabilités de la vérité d'un évènement extraordinaire & dela fausseté du même évènement, & qu'en même-temps u' & e' expriment la probabilité qu'un témoignage sera ou non conforme à la verité, & qu'un témoin ait assuré de la vérité de cet évènement. ... ainsi la probabilité que l'évènement extraordinaire déclaré vrai l'est réellement, sera $uu'/(uu' + ee')$, & celle qu'il est faux $ee'/(uu' + ee')$. [pp.554-555]

If we let E denote the truth of the extraordinary event, and E^* the conforming of the testimony to the truth of E, then $u = \Pr[E] = 1 - e, u' = \Pr[E^* \mid E]$ and $e' = \Pr[E^* \mid \overline{E}]$. Thus

$$\begin{aligned} \frac{uu'}{uu'+ee'} &= \frac{\Pr[E]\ \Pr[E^*|E]}{\Pr[E]\ \Pr[E^*|E]+\Pr[\overline{E}]\ \Pr[E^*|\overline{E}]} \\ &= \Pr[E \mid E^*] \ , \end{aligned}$$

that is, the probability that an event declared to be true is really so[15].

While much of the rest of this part of the memoir is devoted to amplification of the above formula, the main use of it is made in the sixth part, *Application des principes de l'article précédent à quelques questions de critique*, published in the *Histoire de l'Académie* for 1784, pp.454-468. It seems unnecessary to rehearse these applications here[16].

We have had occasion, in the course of this section, frequently to comment on the "obscurity and inutility" [Todhunter 1865, art.753] in Condorcet's writing. Others' comments on this score are reported in Todhunter, Article 753: the last sentence of this article is well-worth repeating:

> Condorcet seems really to have fancied that valuable results could be obtained from any data, however imperfect, by using formulae with an adequate supply of signs of integration.

Gouraud's opinion of the memoir is more glowing[17]. Speaking of the first four parts, in preparation for the writing of which Condorcet had spent three years in familiarizing himself with the calculus, in studying the general rules and methods and the principal kinds of application, Gouraud [1848] says that these researches

> produisirent de 1781 à 1783 les quatre premières parties d'un vaste et beau mémoire où l'ingénieux géomètre déposa les résultats de longues réflexions sur tout le passé de la théorie des hasards, résultats précieux, dont la découverte faisait également honneur au philosophe et à l'analyste. [p.91]

A similar comment is made (op. cit.) on the last two parts of the memoir, viz.

> A la fin de 1783 et dans le courant de 1784, il montra dans une cinquième et dernière partie du mémoire qui l'occupait déjà depuis trois ans, que ces premiers travaux n'étaient que les préliminaires d'une publication plus originale et plus hardie. [p.92]

5.4 *Probabilité*, from the *Encyclopédie Méthodique*

The mathematical part of the *Encyclopédie Méthodique, ou par ordre de matières* was published in three volumes in 1784, 1785 and 1789, the second of these having two articles entitled "Probabilité". The first of these articles, pp.640-649, is a reprint of the article under the same title from the earlier *Encyclopédie ou Dictionnaire Raisonné*; it is apparently by Diderot[18], and contains nothing useful to our purpose. The second article, pp.649-663, is unsigned, but the last sentence makes it clear that the author was

Condorcet. Devoted to general principles of the calculus of probabilities, the article is divided into three parts, only the third of which concerns us here.

Condorcet's aim in this third section is stated at the outset as follows:

> Jusqu'ici nous avons regardé le nombre des combinaisons qui donnent chaque évènement comme déterminé & connu. Nous allons maintenant supposer ce nombre inconnu & variable, en sorte qu'il n'y ait plus une *probabilité* déterminée des évènemens, mais seulement une *probabilité* moyenne d'après laquelle on puisse déterminer celle de leur production. [p.657]

In the second article of this section he supposes that from an urn containing black balls and white, n white and m black balls have been drawn. What will then be the probability of drawing p white and q black balls? Under the further assumption that the urn contains an infinite number of balls, "afin que le rapport des boules blanches, au nombre total, puisse avoir toutes les valeurs depuis 1 jusqu'à 0" [p.657], Condorcet finds the required probability to be

$$\binom{p+q}{p} \int_0^1 x^{n+p}(1-x)^{m+q}\,dx \Big/ \int_0^1 x^n(1-x)^m\,dx$$

$$= \frac{m+n+1}{m+q+n+p+1}\binom{p+q}{p}\binom{m+n}{n} \Big/ \binom{m+n+p+q}{n+p} .$$

Supposing next that $n > m$, Condorcet asks what the probability will be that in the sequence of events the number of white balls will exceed that of black by a given amount. Three conclusions about this probability present themselves, viz.

> 1^0. que cette *probabilité* ne peut jamais approcher indéfiniment de 1; 2^0. que, suivant les hypothèses de pluralité, elle peut, après avoir été croissante, devenir décroissante; 3^0. qu'après un certain terme, elle continuera indéfiniment d'approcher de la fonction
>
> $$\overline{\int x^n \,.\, \overline{1-x}^m\,dx}^{\frac{1}{2}} \Big/ \overline{\int x^n \overline{1-x}^m\,dx} \; > \; 1/2 \;,$$
>
> la formule $\overline{\int x^n \,.\, \overline{1-x}^m\,dx}^{\frac{1}{2}}$ indiquant que l'intégrale est prise seulement depuis $x = 1$, jusqu'à $x = 1/2$. [p.657]

The following is an attempt at an explanation of the above passage.

Let W and B denote the numbers of white and black balls in the sequence, with $W + B = N$. Then $W > B \Rightarrow W/N > 1/2$. Moreover, if $W = B + \delta$, with $\delta > 0$, then $W/N = (1/2) + \delta/2N$, and hence

$$\Pr[W > B + \delta] = \Pr[W/N > (1/2) + \delta/2N] \ .$$

Clearly this probability increases with increasing N and decreases with increasing δ, provided that the ratio W/N is unchanged. Furthermore, if δ is fixed, this probability will decrease as $N \to \infty$– i.e. the probability does not tend to 1. Finally, note that

$$\begin{aligned} J \equiv \Pr[W/N > (1/2)] &= \int_{1/2}^{1} x^n(1-x)^m\,dx \Big/ \int_0^1 x^n(1-x)^m\,dx \\ &= \int_0^{1/2} \frac{1}{B(m+1,n+1)} x^m(1-x)^n\,dx \ . \end{aligned} \tag{5}$$

If, as stated at the outset, $n > m$, then $m/(m+n) < 1/2$. Recognizing that $m/(m+n)$ is the mode of the beta density in (5), we find that

$$n > m \Rightarrow \ \text{mode} \ < 1/2 \ .$$

It thus follows from (5) that $J > 1/2$.

Condorcet next considers the case in which $n < m$ (though this is mistakenly printed in the original as $m < n$), and concludes that in this case $J < 1/2$. Similarly it follows that, in an infinite number of future draws,

$$\begin{aligned} \Pr[W/N > p/(p+q)] &= \int_{\alpha}^{1} x^n(1-x)^m\,dx \Big/ \int_0^1 x^n(1-x)^m\,dx \\ &= \int_0^{1-\alpha} \frac{1}{B(m+1,n+1)} x^m(1-x)^n\,dx \ , \end{aligned}$$

(where $\alpha = p/(p+q)$), a probability which exceeds, or is less than, 1/2 according as the mode $m/(m+n)$ is less than, or greater than, $q/(p+q)$. Finally (at least in this subsection), it is shown that, for $p' > p$,

$$\Pr[\alpha < W/N < \beta] = I_\beta(m+1,n+1) - I_\alpha(m+1,n+1)$$

where $\alpha = p/(p+q)$ and $\beta = p'/(p+q)$.

Condorcet next addresses himself to considering “s'il n'est question que d'une pluralité absolue ou proportionelle, observée entre les évènemens” [p.658] what the probability of indefinite continuation of this plurality may be. The answer in the case of absolute plurality is given as

$$\int_{1/2}^{1} x^{a+b}(1-x)^{a}\,dx \Big/ \int_{0}^{1} x^{a+b}(1-x)^{a}\,dx \quad ,$$

while for proportional plurality we have

$$\int_{\gamma}^{1} x^{ca}(1-x)^{a}\,dx \Big/ \int_{0}^{1} x^{ca}(1-x)^{a}\,dx$$

with $\gamma = c/(1+c)$. No argument for these solutions is presented: Condorcet is apparently assuming, in these two cases, that (in our previous notation), $W = B + b$ and $W = cB$, with $B = a$ in each case. He also derives an expression, in the case of proportional plurality, for the probability that W/N lies between two given functions of c.

In the next subsection Condorcet applies the preceding theory to the question of births, showing that "tout restant dans le même état" the probability that in an indefinite period there will be more boys born than girls is

$$\int_{1/2}^{1} x^{a+b}(1-x)^{a}\,dx \Big/ \int_{0}^{1} x^{a+b}(1-x)^{a}\,dx$$

where $a + b$ is the number of boys and a is the number of girls. Further applications follow to problems of life annuities and contingent rights.

Recalling in the tenth subsection that the probability has hitherto been regarded as constant in a sequence of events of the same type, Condorcet notes that this assumption may in some cases appear gratuitous. He supposes now that the events are independent of one another, keeping the same probability. In the notation introduced earlier, the probability of obtaining the event A, after A and N have been observed n and m times respectively, is $(n+1)/(n+m+2)$. But if the events are independent, this same probability will be

$$\int_{0}^{1} x\,dx = \frac{1}{2} \ .$$

Further, the probability of n A's and m N's is

$$\binom{n+m}{m} \int_{0}^{1} x^{n}(1-x)^{m}\,dx$$

under the first hypothesis and

$$\binom{n+m}{m} \left(\int_{0}^{1} x\,dx\right)^{n} \left(\int_{0}^{1} (1-x)\,dx\right)^{m}$$

under the second. These two probabilities are then in the ratio

$$\frac{m!\,n!}{(m+n+1)!} : \frac{1}{2^{n+m}} ,$$

and consequently "la probabilité moyenne A" will be

$$\left[\frac{(m+1)!\,n!}{(m+n+2)!} + \frac{1}{2^{n+m+1}}\right] \Big/ \left[\frac{m!\,n!}{(m+n+1)!} + \frac{1}{2^{n+m}}\right] .$$

An application to $(n+m)$ tosses of a coin is then given (see Problem III of the *Essai* for further detail).

Condorcet now focusses his attention on the first hypothesis used above, finding that it is legitimate in only two cases:

> 1^0. lorsque la *probabilité* de chaque évènement est toujours la même, comme lorsqu'on tire des boules noires ou blanches toujours d'une même urne; 2^0. lorsque les tirant d'urnes différentes, on suppose que ces urnes ont été remplies en prenant des boulles dans une masse commune, où elles étoient dans un certain rapport. [p.660]

In the first case he asserts that it is the probability itself that is constant, while in the second it is the mean probability[19]. An application to the drawing of cards from packs follows.

A further modification is made in the twelfth subsection, where the following assertion is made:

> On doit donc en général, & si l'on n'a pas *à priori* quelque raison d'adopter une autre hypothèse, regarder la *probabilité* non-seulement comme dépendante des évènemens, mais aussi comme dépendante de l'ordre qu'ils suivent entr'eux. [p.661]

The probability of successive occurrences of events of types A and B are then given respectively by the two sequences

$$x, \quad (x+x')/2\,, \quad (x+x'+x'')/3, \ldots$$

and

$$(1-x), \quad [(1-x)+(1-x')]/2, \quad [(1-x)+(1-x')+(1-x'')]/3, \ldots$$

The probability of a specified sequence of future events is then a fraction whose numerator is the repeated integral of the products of the probabilities of the events already observed and those expected, and whose denominator is the repeated integral of the products of the probabilities of the observed events: all integrals are taken over the unit interval. Further ramifications of typical Condorcetian character follow. Many of the results of this article are given in more detail in the *Essai*, and we shall consider them in due course. The article concludes with the following historical observations:

> La théorie exposée dans ce troisième article est encore peu connue. MM. Price & Bayes en ont donné les principes fondamentaux dans les *Transactions philosophiques* des années 1764 & 1765. M. Delaplace l'a traitée le premier analytiquement, & en a fait plusiers savantes applications dans les *Mémoires de l'académie des sciences.* On trouvera aussi quelques réflexions sur le même sujet dans l'ouvrage que j'ai publie sur le *probabilité* des décisions, & dans quelques mémoires insérés dans les volumes de l'académie, années 1781, 1782 & 1783. [p.663]

It is this last sentence, as we mentioned at the outset, that identifies Condorcet as the author of this article.

5.5 The Essay

The work entitled *Essai sur l'application de l'analyse à la probabilité des décisions rendues à la pluralité des voix* was published in Paris in 1785. Like so much of Condorcet's work, this essay is fraught with difficulty. Todhunter is particularly severe on Condorcet in respect of this work[20]: in his 1865 history he writes

> the difficulty does not lie in the mathematical investigations, but in the expressions which are employed to introduce these investigations and to state their results: it is in many cases almost impossible to discover what Condorcet means to say. The obscurity and self contradiction are without any parallel, so far as our experience of mathematical works extends; some examples will be given in the course of our analysis, but no amount of examples can convey an adequate impression of the extent of the evils. We believe that the work has been very little studied, for we have not observed any recognition of the repulsive peculiarities by which it is so undesirably distinguished. [art.660]

Gouraud's praise, on the other hand, is as fulsome as usual; he writes

> cette remarquable composition, le traité de la plus longue haleine et du plus ambitieux dessein qui jusque-là, dans les cent cinquante ans d'existence de la théorie des hasards, eût attiré

l'attention publique, par la nature des materières que l'auter entreprend d'y soummettre au calcul, l'adresse des hypothèses auxquelles il se livre dans cet objet, la nouveauté des méthodes analytiques dont il faut usage, les vues immenses qu'il découvre à la géométrie, et, par-dessus tout cela, la sécurité sans égale avec laquelle il travaille à la conquête de la terre vierge encore où il aborde le premier, restera dans l'histoire de l'intelligence de l'homme comme un des plus naïfs et des plus éclatants témoignages de l'insatiable avidité de ses désirs et de ses espérances. [1848, pp.94-95]

Even when criticizing Condorcet Gouraud is incapable of suppressing his favourable views. Further on in the same work we find the following:

Un style embarrassé, dénué de justesse et de coloris, une philosophie souvent obscure ou bizarre, une analyse que les meilleurs juges ont trouvée confuse, tels sont, sans préjuger d'ailleurs la légitimité de l'innovation de Condorcet, les défauts de l'ouvrage où il en a consigné les principes: des idées ingénieuses et neuves, des méthodes originales, quelques traits d'une véritable éloquence, en font le mérite et les beautés. [p.99]

The essay consists basically of two parts: a *Discours Préliminaire* of cxci pages, and the *Essai* proper of 304 pages. We shall discuss these *seriatim.*

Opinions on the usefulness of the preliminary discussion vary. Todhunter [1865, art.661] writes

We shall not delay on the Preliminary Discourse, because it is little more than a statement of the results obtained in the Essay. The Preliminary Discourse is in fact superfluous to any person who is sufficiently acquainted with Mathematics to study the Essay, and it would be scarcely intelligible to any other person.

Pearson, on the other hand, in writing of Condorcet's mathematical treatment, says "much light on these matters can be obtained from the preliminary discourse" [1978, p.469]. We shall content ourselves here with discussing only those parts of the introduction which are particularly pertinent.

The *Essai* being divided into five parts (plus a short introduction), the preliminary discourse is similarly partitioned. The aim of this discourse is clearly stated:

ainsi j'ai cru devoir y joindre un Discours, où, après avoir exposé les principes fondamentaux du Calcul des probabilités, je me

> propose de développer les principales questions que j'ai essayé de résoudre & les résultats auxquels le calcul m'a conduit. Les Lecteurs qui ne sont pas Géomètres, n'auront besoin, pour juger de l'ouvrage, que d'admettre comme vrai ce qui est donné pour prouvé par le calcul. [p.ij]

Basic to his theory is the following general principle[21]:

> si sur un nombre donné de combinaisons également possibles, il y en a un certain nombre qui donnent un évènement, & un autre nombre qui donnent l'évènement contraire, la probabilité de chacun des deux évènemens sera égale au nombre des combinaisons qui l'amènent, divisé par le nombre total. [p.v]

A similar sentiment is expressed on p.lxxxvj.

Condorcet next gives various results, which we can express as follows:

(i) for any event A, $\Pr[A] + \Pr[\overline{A}] = 1$;

(ii) if S denotes the certain event, $\Pr[S] = 1$;

(iii) $\Pr[A \cup \overline{A}] = 1$;

(iv) probability is expressed by a fraction, certitude by 1.

He also considers the case in which the combinations are not equally possible: if one combination is twice as possible as another, the former should be viewed as two similar equipossible combinations.

Condorcet goes on to say that one should not regard the above principle as limiting the definition of the probability of an event to an appropriate ratio of numbers of combinations. Rather, he believes it should include belief in the following sense[22]:

(i) if one knows the number of combinations which occasion an event, and the number which do not occasion it, and if the former exceeds the latter, then there is reason to believe that the event will happen rather than that it will not happen;

(ii) this reason for belief increases as the ratio of the number of favourable combinations to the total number increases; and finally

(iii) that it increases proportionally in the same ratio.

He cites as a source of the proof of the last two statements Bernoulli's *Ars Conjectandi*[23]: both of them are, he states, consequences of the first, the latter being proved in the following way: however small the excess of the probability of one event may be over that of another, in a sequence of similar events one will find that the event of the greater of these two probabilities will occur more often than the other (a result proved in the *Essai*). Thus, by hypothesis, one will have reason to believe it will happen more often than the other, and consequently reason to believe that it will happen rather than fail to occur.

In view of the attention we shall give later to Condorcet's treatment of the rule of succession, it seems wise at this stage to give his definition of a future event, viz. "un évènement futur n'est pour nous qu'un évènement inconnu" [p.x]. A clear distinction is also drawn between certainty and probability:

> nous donnons le nom de certitude mathématique à la probabilité, lorsqu'elle se fonde sur la constance des loix observées dans les opérations de notre entendement. Nous appelons certitude physique la probabilité qui suppose de plus la même constance dans un ordre de phénomènes indépendans de nous, & nous conservons le nom de probabilité pour les jugemens exposés de plus à d'autres sources d'incertitude. [p.xiv]

After discussing various matters concerned with voting, Condorcet turns in his *Analyse de la troisième Partie* [pp.lxxxij-cxxviij] to matters which directly concern us. The object of this part he describes as follows[24]:

> nous nous proposons dans cette troisième Partie de donner les moyens, 1^0. de déterminer par l'observation la probabilité de la vérité ou de la fausseté de la voix d'un homme ou de la décision d'un Tribunal; 2^0. de déterminer également, pour les différentes espèces de questions qu'on peut avoir à résoudre, la probabilité que l'on peut regarder comme donnant une assurance suffisante, c'est-à-dire, la plus petite probabilité dont la justice ou la prudence puisse permettre de se contenter. [p.lxxxij]

The first of these questions he proposes to answer in two different ways:

(a) by determining the probability of a future judgement, from the knowledge of the truth or falsity of judgements already delivered, and

(b) by determining the probability of a future judgement, from those of judgements delivered, using only the hypothesis that the probability that one opts rather for truth than for error, is at least $1/2$.

He also states that it is to be assumed in such calculations that the law of the events is constant.

He passes on next to the rule of succession, phrasing it as follows:

> que pour avoir la probabilité d'un évènement futur, d'après la loi que suivent les évènemens passés, il faut prendre, 1^0. la probabilité de cet évènement dans l'hypothèse que la production en est assujettie à des loix constantes; 2^0. la probabilité du même évènement dans le cas où la production n'est assujettie à aucune loi; multiplier chacune de ces probabilités par celle de la supposition en vertu de laquelle on l'a déterminée, & diviser la somme des produits par celle des probabilités des deux hypothèses. [p.lxxxiv]

This is illustrated by a numerical example: we shall postpone any discussion of this point until the pertinent part of the *Essai* proper.

Condorcet passes on next to what we recognize as a discrete form of Bayes's Theorem[25], one which we can write as

$$\Pr[H_i \mid E] = \Pr[E \mid H_i] \,/ \sum_j \Pr[E \mid H_j] \quad .$$

This is in turn followed by a verbal statement of what is essentially the theorem of total probabilities, i.e. $\Pr[E] = \sum \Pr[EH_i]$, which in turn is followed by the curious remark that

> ce n'est donc pas la probabilité réelle que l'on peut obtenir par ce moyen, mais une probabilité moyenne. [p.lxxxvj]

In his *Analyse de la quatrième Partie* Condorcet discusses the application of the methods of his third part to certain voting situations. He emphasizes that, when one has past data to consider, it is only the *pertinent* information which must be taken into account[26]:

> lorsqu'il s'agiroit de déterminer la probabilité d'une nouvelle décision, on emploîroit, non la totalité des décisions passées, mais seulement le système de celles où le rapport de la pluralité au nombre des Votans est à peu-près le même que dans la nouvelle décision. [p.cxxx]

The two methods discussed in the third Part, while both being usable in the questions of the fourth Part, may be appropriate in different cases: indeed,

> si au lieu considérer la distribution des voix dans les décisions, on considéroit les décisions en elles-mêmes, alors il faudroit préférer la première méthode, la seconde ne pouvant s'appliquer à cette dernière question qu'avec difficulté, & ne pouvant même conduire alors qu'à des résultats hypothétiques. [p.cxxxviij]

Condorcet next provides an example to distinguish between the real probability of the truth of a proposition and the probability that this same proposition has a certain degree of absolute or mean probability. The example concerns withdrawals from an urn (or urns) containing white and black balls, under the following conditions:

(i) there are two urns, the numbers of white and black balls present being known to the drawer, who also knows from which urn the ball is taken;

(ii) one or more witnesses testify as to which urn the ball comes from (such testimony having a certain probability of being true);

(iii) the witnesses have concluded on the basis of past drawings, which of the urns contains more white balls;

(iv) the drawer is completely ignorant of the composition of the urns (in this case only a mean probability is available).

So much for the *Discours Préliminaire*: we pass on now to the *Essai* proper[27].

The *Essai* opens with a two-page introduction summarizing the contents of its five parts. Earlier parts of the essay not being pertinent, let us turn our attention immediately to the paragraph in the introduction that is connected with the third part:

> dans le troisième, on cherchera une méthode pour s'assurer à *posteriori* du degré de probabilité d'un suffrage ou de la décision d'une assemblée, & pour déterminer les dégres de probabilité que doivent avoir les différentes espèces de décisions. [p.2]

The problems to be discussed in this third part, Condorcet states, require firstly

> qu'on ait établi en général les principes d'apres lesquels on peut déterminer la probabilité d'un évènement futur ou inconnu, non par la connoissance du nombre des combinaisons possibles qui donnent cet évènement, ou l'évènement opposé, mais seulement par la connoissance de l'ordre des évènemens connus ou passés de la même espèce. [p.176]

To this end Condorcet discusses thirteen problems, in which both the rule of succession and Bayes's Theorem are illustrated[28]: we shall consider these problems *seriatim*.

Problem 1

> Soient deux évènemens seuls possibles A & N, dont on ignore la probabilité, & qu'on sache seulement que A est arrivé m fois, & N, n fois. On suppose l'un des deux évènemens arrivés, & on demande la probabilité que c'est l'évènement A, ou que c'est l'évènement N, dans l'hypothèse que la probabilité de chacun des deux évènemens est constamment la même. [p.176]

Let H_1 denote this hypothesis, and let x denote the probability of A. Then the probability of m A's and n N's (event E, say)[29], is $\binom{m+n}{n}x^m(1-x)^n$. Hence the probability of E "pour toutes valeurs de x depuis zéro jusqu'à 1" [p.177] will be given by

$$\Pr[E \mid H_1] = \int_0^1 \binom{m+n}{n} x^m (1-x)^n \, dx \ .$$

Proceeding similarly we can show that[30]

$$\begin{aligned}\Pr[A \mid EH_1] &= \int_0^1 x^{m+1}(1-x)^n \, dx \Big/ \int_0^1 x^m (1-x)^n \, dx \\ &= (m+1)/(m+n+2) \ ,\end{aligned}$$

a similar result holding for $\Pr[N \mid EH_1]$.

Problem 2

> On suppose dans ce Problème, que la probabilité de A & de N n'est pas la même dans tous les évènemens [hypothesis H_2, say], mais qu'elle peut avoir pour chacun une valeur quelconque depuis zéro jusqu'à l'unité. [p.177]

In this case, asserts Condorcet (and in the same notation as in the previous problem)[31],

$$\begin{aligned}\Pr[E \mid H_2] &= \binom{m+n}{n}\left[\int_0^1 x\, dx\right]^m \left[\int_0^1 (1-x)\, dx\right]^n \\ &= \binom{m+n}{n} 2^{-(m+n)} \ .\end{aligned}$$

Thus

$$\begin{aligned}\Pr[AE \mid H_2] &= \binom{m+n}{n}\left[\int_0^1 x\, dx\right]^{m+1} \left[\int_0^1 (1-x)\, dx\right]^n \\ &= \binom{m+n}{n} 2^{-(m+n+1)} \ ,\end{aligned}$$

and hence

$$\Pr[A \mid EH_2] = 1/2$$

(and similarly for $\Pr[N \mid EH_2]$). Noting that this is the same as the result we would obtain on taking $\Pr[A] = 1/2 = \Pr[N]$, we see that Condorcet seems to have confused the sentiment[32] "suppose that the probabilities are not constant" with "do not suppose that the probabilities are constant".

Problem 3

> On suppose dans ce problème que l'on ignore si à chaque fois la probabilité d'avoir A ou N reste la même, ou si elle varie à chaque fois, de manière qu'elle puisse avoir une valeur quelconque depuis zéro jusqu'à l'unité, & l'on demande, sachant que l'on a eu m évènemens A, & n évènemens N, quelle est la probabilité d'amener A ou N. [p.178]

Two cases are considered here:

(i) if the probability is constant (hypothesis H_1),

$$\Pr[E \mid H_1] = \binom{m+n}{n} m!\, n! \,/(m+n+1)!$$

(ii) if the probability is not constant (hypothesis H_2),

$$\Pr[E \mid H_2] = \binom{m+n}{n} 2^{-(n+m)} \quad .$$

Thus, under the implicit assumption of equal initial probabilities for H_1 and H_2, and using a discrete form of Bayes's Theorem, we see that

$$\begin{aligned}
\Pr[H_1 \mid E] &= \frac{m!\, n!}{(m+n+1)!} \Big/ \left[\frac{m!\, n!}{(m+n+1)!} + \frac{1}{2^{n+m}}\right] \\
\Pr[H_2 \mid E] &= 2^{-(n+m)} \Big/ \left[\frac{m!\, n!}{(m+n+1)!} + \frac{1}{2^{n+m}}\right] \quad .
\end{aligned}$$

Recalling that

$$\Pr[A \mid EH_1] = (m+1)\,/(m+n+2)$$

$$\Pr[N \mid EH_1] = (n+1)\,/(m+n+2)$$

$$\Pr[A \mid EH_2] = 1/2 = \Pr[N \mid EH_2] \quad ,$$

we see finally that

$$\begin{aligned}
\Pr[A \mid E] &= \Pr[A \mid EH_1]\Pr[H_1 \mid E] + \Pr[A \mid EH_2]\Pr[H_2 \mid E] \\
&= \frac{m+1}{m+n+2}\,\frac{m!\,n!}{(m+n+1)!}\Big/\left[\frac{m!\,n!}{(m+n+1)!}+\frac{1}{2^{n+m}}\right] \\
&\qquad +(1/2)2^{-(n+m)}\Big/\left[\frac{m!\,n!}{(m+n+1)!}+\frac{1}{2^{n+m}}\right] \\
&= \left[\frac{(m+1)!\,n!}{(m+n+2)!}+\frac{1}{2^{n+m+1}}\right]\Big/\left[\frac{m!\,n!}{(m+n+1)!}+\frac{1}{2^{n+m}}\right]
\end{aligned}$$

a similar expression holding for $\Pr[N \mid E]$.

As a remark Condorcet considers the ratio of the terms $m!\,n!/(m+n+1)!$ and $2^{-(n+m)}$ when $m = an$ and $n \to \infty$. If $a = 1$, it follows from the Stirling-de Moivre approximation that, as $n \to \infty$,

$$\frac{m!\,n!}{(m+n+1)!}\Big/ 2^{-(n+m)} \to 0 \;.$$

Furthermore, if $a \neq 1$, the ratio tends to infinity as $n \to \infty$. Condorcet then goes on to expand verbally on this result (for criticism see Todhunter [1865, art.700]).

Problem 4

> On suppose ici un évènement A arrivé m fois, & un évènement N arrivé n fois; que l'on sache que la probabilité inconnue d'un des évènemens soit depuis 1 jusqu'à $\frac{1}{2}$, & celle de l'autre depuis $\frac{1}{2}$ jusqu'à zéro, & l'on demande, dans les trois hypothèses des trois problèmes précédens, 1^0. la probabilité que c'est A ou N dont la probabilité est depuis 1 jusqu'à $\frac{1}{2}$; 2^0. la probabilité d'avoir A ou N dans le cas d'un nouvel évènement; 3^0. la probabilité d'avoir un évènement dont la probabilité soit depuis 1 jusqu'à $\frac{1}{2}$. [p.180]

Condorcet supposes firstly that the (unknown) probability is constant (hypothesis H_1). Denoting by p_A and p_N the probabilities of A and N we have[33]

$$\Pr[E \;\&\; 0 \le p_A \le 1/2 \mid H_1] = \binom{m+n}{n}\int_0^{1/2} x^m(1-x)^n\,dx$$

$$\Pr[E \;\&\; 1/2 \le p_A \le 1 \mid H_1] = \binom{m+n}{n}\int_{1/2}^{1} x^m(1-x)^n\,dx \;,$$

where E denotes the event that A and N have occurred m and n times respectively. Again by a tacit application of Bayes's Theorem Condorcet deduces that

$$\begin{aligned}
&\Pr[1/2 \leq p_A \leq 1 \mid EH_1] \\
&\quad = \frac{\Pr[1/2 \leq p_A \leq 1 \,\&\, EH_1]}{\Pr[0 \leq p_A \leq 1/2 \,\&\, EH_1] + \Pr[1/2 \leq p_A \leq 1 \,\&\, EH_1]} \\
&\quad = \int_{1/2}^{1} x^m (1-x)^n \, dx \Big/ \int_0^1 x^m (1-x)^n \, dx \ ,
\end{aligned}$$

and similarly

$$\Pr[0 \leq p_A \leq 1/2 \mid EH_1] = \int_0^{1/2} x^m (1-x)^n \, dx \Big/ \int_0^1 x^m (1-x)^n \, dx$$

(in each case the left-hand side is given by Condorcet as an *unconditional* probability). This completes the solution of the first question.

Proceeding to the second question we see that

$$\Pr[A \mid 0 \leq p_A \leq 1/2 \,\&\, EH_1] = \int_0^{1/2} x^{m+1} (1-x)^n \, dx \Big/ \int_0^{1/2} x^m (1-x)^n \, dx$$

$$\Pr[A \mid 1/2 \leq p_A \leq 1 \,\&\, EH_1] = \int_{1/2}^{1} x^{m+1} (1-x)^n \, dx \Big/ \int_{1/2}^{1} x^m (1-x)^n \, dx \ .$$

Thus

$$\begin{aligned}
\Pr[A \mid EH_1] &= \Pr[A \mid 0 \leq p_A \leq 1/2 \,\&\, EH_1] \Pr[0 \leq p_A \leq 1/2 \mid EH_1] \\
&\quad + \Pr[A \mid 1/2 \leq p_A \leq 1 \,\&\, EH_1] \Pr[1/2 \leq p_A \leq 1 \mid EH_1] \\
&= \int_0^1 x^{m+1} (1-x)^n \, dx \Big/ \int_0^1 x^m (1-x)^n \, dx \ ,
\end{aligned}$$

and similarly

$$\Pr[N \mid EH_1] = \int_0^1 x^m (1-x)^{n+1} \, dx \Big/ \int_0^1 x^m (1-x)^n \, dx \ .$$

Condorcet's solution to the third question runs as follows:

$$
\begin{aligned}
&\Pr\left[(A \;\&\; 1/2 \le p_A \le 1) \vee (N \;\&\; 1/2 \le p_N \le 1) \mid EH_1\right] \\
&\quad = \Pr\left[A \;\&\; 1/2 \le p_A \le 1 \mid EH_1\right] + \Pr\left[N \;\&\; 1/2 \le p_N \le 1 \mid EH_1\right] \\
&\quad = \Pr\left[A \mid 1/2 \le p_A \le 1 \;\&\; EH_1\right] \Pr\left[1/2 \le p_A \le 1 \mid EH_1\right] \\
&\qquad + \Pr\left[N \mid 1/2 \le p_N \le 1 \;\&\; EH_1\right] \Pr\left[1/2 \le p_N \le 1 \mid EH_1\right] \\
&\quad = \left[\frac{\int_{1/2}^{1} x^{m+1}(1-x)^n\,dx}{\int_{1/2}^{1} x^m(1-x)^n\,dx}\right]\left[\frac{\int_{1/2}^{1} x^m(1-x)^n\,dx}{\int_{0}^{1} x^m(1-x)^n\,dx}\right] \\
&\qquad + \left[\frac{\int_{1/2}^{1} x^{n+1}(1-x)^m\,dx}{\int_{1/2}^{1} x^n(1-x)^m\,dx}\right]\left[\frac{\int_{1/2}^{1} x^n(1-x)^m\,dx}{\int_{0}^{1} x^n(1-x)^m\,dx}\right] \\
&\quad = \int_{1/2}^{1} \left[x^{m+1}(1-x)^n + x^{n+1}(1-x)^m\right] dx \Big/ \int_0^1 x^m(1-x)^n\,dx \;.
\end{aligned}
$$

Condorcet next considers the same three questions under the assumption that [34]

> la probabilité changeante à chaque évènement, mais étant toujours pour le même, ou depuis 1 jusqu'à $\frac{1}{2}$, ou depuis 0 jusqu'à $\frac{1}{2}$. [p.182]

The solution presented by Condorcet is most confusing: the following is an attempt at interpretation.

We have firstly

$$
\Pr\left[E \;\&\; 0 \le p_A \le 1/2 \mid H_2\right] = \binom{m+n}{n} \frac{\left(\int_0^{1/2} x\,dx\right)^m \left(\int_0^{1/2} (1-x)\,dx\right)^n}{\left(\int_0^{1} x\,dx\right)^m \left(\int_0^{1} (1-x)\,dx\right)^n}
$$

$$\Pr[E \ \& \ 1/2 \leq p_A \leq 1 \mid H_2] = \binom{m+n}{n} \frac{\left(\int\limits_{1/2}^{1} x\,dx\right)^m \left(\int\limits_{1/2}^{1} (1-x)\,dx\right)^n}{\left(\int\limits_{0}^{1} x\,dx\right)^m \left(\int\limits_{0}^{1} (1-x)\,dx\right)^n}$$

where H_2 denotes the hypothesis of changing probability. The numerators in these two ratios are given (correctly) by Condorcet as

$$\binom{m+n}{n} 3^n \Big/ 8^{m+n} \quad \text{and} \quad \binom{m+n}{n} 3^m \Big/ 8^{m+n}$$

respectively. Hence

$$\begin{aligned}\Pr[1/2 \leq p_A \leq 1 \mid EH_2] &= \Pr[E \ \& \ 1/2 \leq p_A \leq 1 \mid H_2] / \Pr[E \mid H_2] \\ &= 3^m / (3^m + 3^n) \ ,\end{aligned}$$

a similar expression holding for $\Pr[0 \leq p_A \leq 1/2 \mid EH_2]$.

Condorcet now goes on to give

> la probabilité d'avoir une fois de plus l'évènement A, si la probabilité de A est depuis 1 jusqu'à $\frac{1}{2}$. [p.182]

This probability is found as follows:

$$\Pr[A \mid 1/2 \leq p_A \leq 1 \ \& \ EH_2] = \frac{\Pr[AE \ \& \ 1/2 \leq p_A \leq 1 \mid H_2]}{\Pr[E \ \& \ 1/2 \leq p_A \leq 1 \mid H_2]}$$

$$= \frac{\binom{m+n}{n}\left(\int\limits_{1/2}^{1} x\,dx\right)^{m+1}\left(\int\limits_{1/2}^{1} (1-x)\,dx\right)^{n} \Big/ \left(\int\limits_{0}^{1} x\,dx\right)^{m+1}\left(\int\limits_{0}^{1} (1-x)\,dx\right)^{n}}{\binom{m+n}{n}\left(\int\limits_{1/2}^{1} x\,dx\right)^{m}\left(\int\limits_{1/2}^{1} (1-x)\,dx\right)^{n} \Big/ \left(\int\limits_{0}^{1} x\,dx\right)^{m}\left(\int\limits_{0}^{1} (1-x)\,dx\right)^{n}}$$

$$= 3/4 \ ,$$

and similarly $\Pr[A \mid 0 \leq p_A \leq 1/2 \ \& \ EH_2] = 1/4$.

It then follows that

$$\begin{aligned}\Pr[A \mid EH_2] &= \Pr[A \mid 0 \leq p_A \leq 1/2 \ \& \ EH_2] \Pr[0 \leq p_A \leq 1/2 \mid EH_2] \\ &\quad + \Pr[A \mid 1/2 \leq p_A \leq 1 \ \& \ EH_2] \Pr[1/2 \leq p_A \leq 1 \mid EH_2] \\ &= \left[3^m(\tfrac{3}{4}) + 3^n(\tfrac{1}{4})\right] / (3^m + 3^n) \ ,\end{aligned}$$

and similarly

$$\Pr[N \mid EH_2] = \left[3^m(\tfrac{1}{4}) + 3^n(\tfrac{3}{4})\right] / (3^m + 3^n) \quad ,$$

these being Condorcet's solutions.

To answer the third question notice that

$$\Pr[A \;\&\; 1/2 \leq p_A \leq 1 \mid EH_2] = 3^m(\tfrac{3}{4}) / (3^m + 3^n)$$

$$\Pr[N \;\&\; 1/2 \leq p_N \leq 1 \mid EH_2] = 3^n(\tfrac{3}{4}) / (3^m + 3^n) \quad .$$

Thus

$$\Pr[(A \;\&\; 1/2 \leq p_A \leq 1) \vee (N \;\&\; 1/2 \leq p_N \leq 1) \mid EH_1] = \tfrac{3}{4} \quad .$$

As the final case Condorcet considers the answering of these three questions under the assumptions of Problem 3. Under the two hypotheses H_1 and H_2, the respective probabilities of E are as $\int_0^1 x^m(1-x)^n\,dx$ to $(3^m + 3^n)/4^{m+n}$, since

$$\Pr[E \mid H_1] = \binom{m+n}{n} \int_0^1 x^m(1-x)^n\,dx$$

and

$$\Pr[E \mid H_2] = \binom{m+n}{n} (3^m + 3^n)/4^{m+n} \quad .$$

It then follows, under the assumption that $\Pr[H_1] = \Pr[H_2]$, that

$$\begin{aligned}
&\Pr[1/2 \leq p_A \leq 1 \mid E] \\
&\quad = \frac{\Pr[E \;\&\; 1/2 \leq p_A \leq 1 \mid H_1] + \Pr[E \;\&\; 1/2 \leq p_A \leq 1 \mid H_2]}{\Pr[E \mid H_1] + \Pr[E \mid H_2]} \\
&\quad = \frac{\int_{1/2}^1 x^m(1-x)^n\,dx + 3^m/4^{m+n}}{\int_0^1 x^m(1-x)^n\,dx + (3^m + 3^n)/4^{m+n}} \quad .
\end{aligned}$$

Similarly the probability of obtaining one more A is

$$\begin{aligned}
&\Pr[A \;\&\; 1/2 \leq p_A \leq 1 \mid H_2] \\
&\quad = \frac{\int_{1/2}^1 x^{m+1}(1-x)^n\,dx + (3^{m+1} + 3^n)/4^{m+n+1}}{\int_0^1 x^m(1-x)^n\,dx + (3^m + 3^n)/4^{m+n}} \quad ,
\end{aligned}$$

while the probability of getting an event (either A or N) with probability between $\frac{1}{2}$ and 1 is

$\Pr[(A \ \& \ 1/2 \le p_A \le 1) \vee (N \ \& \ 1/2 \le p_N \le 1) \mid E]$

$$= \frac{\int_{1/2}^{1} \left[x^{m+1}(1-x)^n + x^{n+1}(1-x)^m\right] dx + \left(3^{m+1} + 3^{n+1}\right) / \, 4^{m+n+1}}{\int_0^1 x^m(1-x)^n \, dx + (3^m + 3^n) / \, 4^{m+n}} \; .$$

Problem 5

> Conservant les mêmes hypothèses, on demande quelle est, dans le cas du problème premier, la probabilité, 1^0. que celle de l'évènement A n'est pas au-dessous d'une quantité donnée; 2^0. qu'elle ne diffère de la valeur moyenne $m/(m+n)$ que d'une quantité a; 3^0. que la probabilité d'amener A, n'est point au-dessous d'une limite a; 4^0. qu'elle ne diffère de la probabilité moyenne $(m+1)/(m+n+2)$ que d'une quantité moindre que a. On demande aussi, ces probabilités étant données, quelle est la limite a pour laquelle elles ont lieu. [pp.183-184]

The solution presented to 1^0 runs as follows: since

$$\Pr[E] = \binom{m+n}{n} \int_0^1 x^m(1-x)^n \, dx$$

and

$$\Pr[EH] = \binom{m+n}{n} \int_a^1 x^m(1-x)^n \, dx$$

(where H is the proposition $a \le p_A \le 1$), it follows that

$$\begin{aligned} M &\equiv \Pr[H \mid E] = \Pr[EH] / \Pr[E] \\ &= \int_a^1 x^m(1-x)^n \, dx \Big/ \int_0^1 x^m(1-x)^n \, dx \\ &= 1 - \sum_{k=0}^{n} \frac{m! \, (n)_k}{(m+k+1)!} a^{m+k+1}(1-a)^{n-k} \Big/ \frac{m! \, n!}{(m+n+1)!} \end{aligned}$$

where $(n)_k = n(n-1)\ldots(n-k+1)$. This result is more elegantly given in terms of the incomplete beta-function as

$$M = 1 - I_a(m+1, n+1) \; .$$

Proceeding to the second question, Condorcet states that

$$\Pr[\alpha \le p_A \le 1 \mid E] = \int_\alpha^1 x^m(1-x)^n\,dx \Big/ \int_0^1 x^m(1-x)^n\,dx$$

$$\Pr[\beta \le p_A \le 1 \mid E] = \int_\beta^1 x^m(1-x)^n\,dx \Big/ \int_0^1 x^m(1-x)^n\,dx$$

where $\alpha = m/(m+n)+a$, $\beta = m/(m+n)-a$. Subtraction of the first of these formulae from the second then gives $\Pr[\beta \le p_A \le \alpha \mid E]$. Condorcet evaluates this probability, obtaining an expression analogous to M in the preceding question — in fact

$$I_\beta(m+1,n+1) - I_\alpha(m+1,n+1) \ .$$

The solution to the third question is given, if a is always the limit of the probability of A, by M in 1^0.

> On aura donc une probabilité égale que celle d'amener l'évènement A n'est pas au-dessous de a. [p.185]

A similar expression to that in question 2^0 is given for

$$\Pr[(m+1)/(m+n+2) - a \le p_A \le (m+1)/(m+n+2) + a] \ .$$

As a final remark Condorcet points out that the formulae given here serve equally to determine M in terms of a or a in terms of M, but that this latter value will be impossible to obtain rigorously. A general expression for M is given.

Problem 6

> En conservant les mêmes données, on propose les mêmes questions pour le cas où la probabilité n'est pas constante. [p.186]

As was the case in Problem 4, the treatment presented here by Condorcet is difficult to follow. The solution offered below is consistent with those of earlier problems, and results in the answer obtained by Condorcet.

In answer to the first question we note that

$$\Pr[E \ \& \ a \le p_A \le 1 \mid H_2]$$

$$= \binom{m+n}{n}\left(\int_a^1 x\,dx\right)^m\left(\int_a^1 (1-x)\,dx\right)^n \Big/ \left(\int_0^1 x\,dx\right)^m\left(\int_0^1 (1-x)\,dx\right)^n \qquad (6)$$

Thus

$$\Pr[a \le p_A \le 1 \mid EH_2] = \frac{\Pr[E \ \&\ a{\le}p_A{\le}1|H_2]}{\Pr[E \ \&\ a{\le}p_A{\le}1|H_2]+\Pr[E \ \&\ 0{\le}p_A{\le}a|H_2]}$$

$$= \frac{\left(1-a^2\right)^m\left(1-2a+a^2\right)^n}{(1-a^2)^m(1-2a+a^2)^n+a^{2m}(2a-a^2)^n} \ . \qquad (7)$$

Proceeding to the second question, Condorcet finds, exactly as above, $\Pr[E \ \&\ (b-a) \le p_A \le (b+a) \mid H_2]$ and the corresponding result analogous to (7).

As regards question 3, we have, from (6) with $m = 1$ and $n = 0$,

$$\Pr[A \ \&\ a \le p_A \le 1 \mid H_2] = \left((1/2) - a^2/2\right)/\,(1/2) = 1 - a^2$$

while, in answer to the fourth question,

$$\Pr[A \ \&\ (b-a) \le p_A \le (b+a) \mid H_2] = (b+a)^2 - (b-a)^2 \ .$$

As a remark following this problem Condorcet points out that the case resulting from a combination of the previous two can readily be solved by using Problems 3, 5 and 6.

Problem 7

> Supposant qu'un évènement A est arrivé m fois, & qu'un évènement N est arrivé n fois, on demande la probabilité que l'évènement A dans q fois arrivera $q - q'$ fois, & l'évènement N, q' fois. [pp.187-188]

Denoting by x and $1 - x$ the probabilities of A and N respectively, Condorcet shows in the usual way that[35]

$$\Pr\left[\left(q-q'\right) A\text{'s} \ \&\ q' \ \ N\text{'s} \mid E\right] = \frac{\binom{m+n}{n}\binom{q}{q'}\int_0^1 x^{m+q-q'}(1-x)^{n+q'}\,dx}{\binom{m+n}{n}\int_0^1 x^m(1-x)^n\,dx}$$

$$= \binom{q}{q'}\frac{(n+1)\ldots(n+q')(m+1)\ldots(m+q-q')}{(m+n+2)\ldots(m+n+q+1)}$$

$$= \binom{q}{q'}\frac{B(m+q-q'+1, n+q'+1)}{B(m+1, n+1)} \ .$$

Condorcet follows this with a remark in which he gives the probabilities of the events

$$q \ A\text{'s}; \ (q-1)A\text{'s} \ \&\ 1 \ \ N; \ldots; 1 \ A \ \&\ (q-1)N\text{'s}; \ q \ N\text{'s}$$

and he notes that the sum of these probabilities, irrespective of the values m, n and q, must of necessity be 1.

Problem 8

> On demande dans la même hypothèse, 1^0. le nombre des évènemens futurs étant $2q+1$, la probabilité que le nombre des évènemens N ne surpassera pas de $2q^{'}+1$ le nombre des évènemens A; 2^0. la probabilité que le nombre des évènemens A surpassera de $2q^{'}+1$ le nombre des évènemens N. [p.189]

The solutions are easily found on applying the result of the preceding problem: most of Condorcet's five and a half page solution is concerned with manipulations of the initial expressions.

Three remarks follow: in the first of these, Condorcet points out that the analogy between the formulae developed in this problem and those of the first part of the *Essai* shows that the latter may be used when m and n are large. In the second remark he finds the probability that the event A rather than N has happened, if one knows merely that one event has happened $2q^{'}+1$ times more than the other. Again this result is related to the corresponding one in Part 1. In the final remark various ratios of m to n are considered.

Problem 9

> Nous supposerons ici seulement que le nombre des Votans est $2q$, & la pluralité $2q^{'}$, & qu'on demande V & $V^{'}$ comme dans le Problème précédent. [p.197]

(Here V and $V^{'}$ are the probabilities desired in 1^0 and 2^0 respectively in the previous problem.) The solution is followed by a remark analogous to the second remark following the preceding problem: neither the present solution nor the remark contributes anything new to our discussion.

Problem 10

> On demande, tout le reste étant le même, la probabilité que sur $3q$ évènemens, 1^0. N n'arrivera pas plus souvent que A un nombre q de fois, 2^0. que A arrivera plus souvent que N un nombre q de fois. [p.199]

The method of solution parallels that of Problem 8, and will not be discussed here. Two remarks follow.

Problem 11

> La probabilité étant supposée n'être pas constante comme dans le Problème second, on demande 1^0. la probabilité d'avoir sur

> q évènemens, $q-q'$ évènemens A, & q' évènemens N; 2^0. la probabilité que sur $2q+1$ évènemens, N n'arrivera pas un nombre $2q'+1$ de fois plus souvent que A; 3^0. la probabilité que A arrivera un nombre $2q'+1$ de fois plus souvent que N. [pp.204-205]

Proceeding in the usual way we find that

$$\Pr\left[\left(q-q'\right) A\text{'s} \ \& \ q' \ N\text{'s} \mid E\right]$$

$$= \binom{q}{q'} \frac{\left(\int_0^1 x\,dx\right)^{m+q-q'} \left(\int_0^1 (1-x)\,dx\right)^{n+q'}}{\left(\int_0^1 x\,dx\right)^{m} \left(\int_0^1 (1-x)\,dx\right)^{n}}$$

$$= \binom{q}{q'} 2^{-q} \ .$$

This is the solution to the first question: the remaining two are special cases of certain results given in the first part of the *Essai.* In a remark Condorcet points out that when one is ignorant as to which of the two hypotheses holds, one should proceed as in Problem 3.

Problem 12

> On suppose que la probabilité d'un des évènemens est depuis 1 jusqu'à $\frac{1}{2}$, & celle de l'autre depuis $\frac{1}{2}$ jusqu'à zéro, & on demande dans cette hypothèse;
> 1^0. La probabilité que A arrivera $q-q'$ fois dans q évènemens, & N, q' fois; ou que l'évènement dont la probabilité est depuis 1 jusqu'à $\frac{1}{2}$, arrivera $q-q'$ fois, & celui dont la probabilité est depuis $\frac{1}{2}$ jusqu'à zéro, q' fois.
> 2^0. La probabilité que sur $2q+1$ évènemens, N n'arrivera point $2q'+1$ fois plus souvent que A; ou que l'évènement dont la probabilité est depuis $\frac{1}{2}$ jusqu'à zéro, n'arrivera pas $2q'+1$ fois plus souvent que l'évènement dont la probabilité est depuis 1 jusqu'à $\frac{1}{2}$.
> 3^0. La probabilité que sur $2q+1$ évènemens, l'évènement A arrivera $2q'+1$ fois plus que N; ou que l'évènement dont la probabilité est depuis 1 jusqu'à $\frac{1}{2}$, arrivera $2q'+1$ fois plus souvent que celui dont la probabilité est depuis $\frac{1}{2}$ jusqu'à zéro. [pp.205-206]

The solution to the first question is as follows (cf. Problem 4):

$$\Pr\left[\left(q-q'\right) A\text{'s} \;\&\; q' \;\; N\text{'s} \mid (1/2) \leq p_A \leq 1 \;\&\; EH_1\right]$$

$$= \binom{q}{q'} \int_{1/2}^{1} x^{m+q-q'}(1-x)^{n+q'}\,dx \Big/ \int_{1/2}^{1} x^m(1-x)^n\,dx \;.$$

Similarly[36]

$$\Pr\left[\left(q-q'\right) A\text{'s} \;\&\; q' \;\; N\text{'s} \mid 0 \leq p_A \leq (1/2) \;\&\; EH_1\right]$$

$$= \binom{q}{q'} \int_{0}^{1/2} x^{m+q-q'}(1-x)^{n+q'}\,dx \Big/ \int_{0}^{1/2} x^m(1-x)^n\,dx \;.$$

Now

$$\Pr\left[1/2 \leq p_A \leq 1 \mid EH_1\right] = \int_{1/2}^{1} x^m(1-x)^n\,dx \Big/ \int_{0}^{1} x^m(1-x)^n\,dx$$

and

$$\Pr\left[0 \leq p_A \leq 1/2 \mid EH_1\right] = \int_{0}^{1/2} x^m(1-x)^n\,dx \Big/ \int_{0}^{1} x^m(1-x)^n\,dx \;.$$

Thus, as in Problem 4,

$$\Pr\left[\left(q-q'\right) A\text{'s} \;\&\; q' \;\; N\text{'s} \mid EH_1\right]$$

$$= \binom{q}{q'} \int_{0}^{1} x^{m+q-q'}(1-x)^{n+q'}\,dx \Big/ \int_{0}^{1} x^m(1-x)^n\,dx \;.$$

By a procedure similar to that adopted in the solution to Problem 4, one finds that

$$\Pr\left[\left(\left(q-q'\right) A\text{'s} \;\&\; 1/2 \leq p_A \leq 1\right) \vee \left(\left(q-q'\right) N\text{'s} \;\&\; /2 \leq p_N \leq 1\right) \mid EH_1\right]$$

$$= \binom{q}{q'} \int_{1/2}^{1} \left[x^{m+q-q'}(1-x)^{n+q'} + x^{n+q-q'}(1-x)^{m+q'}\right] dx \Big/ \int_{0}^{1} x^m(1-x)^n\,dx \,.$$

The solutions to the first parts of articles 2 and 3 follow as in Problem 8. The answer to the second part of the second article is given as

$$(1/D)\left\{\int_{1/2}^{1}\left[x^{m+2q+1}(1-x)^{n}+x^{n+2q+1}(1-x)^{m}\right]dx\right.$$

$$+(2q+1)\int_{1/2}^{1}\left[x^{m+2q}(1-x)^{n+1}+x^{n+2q}(1-x)^{m+1}\right]dx+\cdots$$

$$\left.+\binom{2q+1}{q-q'+1}\int_{1/2}^{1}\left[x^{m+q-q'+1}(1-x)^{n+q+q'}+x^{n+q-q'+1}(1-x)^{m+q+q'}\right]dx\right\}$$

where $D=\int_0^1 x^m(1-x)^n\,dx$.

The solution to the second part of article 3 follows on using formulae from the first part of the *Essai.*

Condorcet points out in a remark that solutions to similar problems may now be obtained "sans peine".

Problem 13

> On suppose que la probabilité n'est pas constante, &, les autres hypothèses restant les mêmes que dans le Problème précédent, on propose les mêmes questions. [p.211]

Proceeding as in the solution to Problem 4, we note firstly that

$$\Pr\left[\left(q-q'\right)A\text{'s}\ \&\ q'N\text{'s}\mid 1/2\le p_A\le 1\ \&\ EH_2\right]$$

$$=\binom{q}{q'}\frac{\left(\int_{1/2}^{1}x\,dx\right)^{m+q-q'}\left(\int_{1/2}^{1}(1-x)\,dx\right)^{n+q'}}{\left(\int_0^1 x\,dx\right)^{q}\left(\int_{1/2}^{1}x\,dx\right)^{m}\left(\int_{1/2}^{1}(1-x)\,dx\right)^{n}}\ . \tag{8}$$

Similarly

$$\Pr\left[\left(q-q'\right)A\text{'s}\ \&\ q'N\text{'s}\mid 0\le p_A\le 1/2\ \&\ EH_2\right]$$

$$=\binom{q}{q'}\frac{\left(\int_0^{1/2}x\,dx\right)^{m+q-q'}\left(\int_0^{1/2}(1-x)\,dx\right)^{n+q'}}{\left(\int_0^1 x\,dx\right)^{q}\left(\int_0^{1/2}x\,dx\right)^{m}\left(\int_0^{1/2}(1-x)\,dx\right)^{n}}\ . \tag{9}$$

Thus on multiplying (8) and (9) respectively by the probabilities

$$\Pr\left[1/2\le p_A\le 1\mid EH_2\right]\ \text{and}\ \Pr\left[0\le p_A\le 1/2\mid EH_2\right]$$

(these being found as in Problem 3), and on setting

$$
\begin{aligned}
I_1 &= \left(\int_{1/2}^{1} x\,dx\right)^{m+q-q'} \left(\int_{1/2}^{1} (1-x)\,dx\right)^{n+q'} \\
I_2 &= \left(\int_{1/2}^{1} (1-x)\,dx\right)^{m+q-q'} \left(\int_{1/2}^{1} x\,dx\right)^{n+q'} \\
I_3 &= \left(\int_{0}^{1} x\,dx\right)^{q} \left(\int_{1/2}^{1} x\,dx\right)^{m} \left(\int_{1/2}^{1} (1-x)\,dx\right)^{n} \\
I_4 &= \left(\int_{0}^{1} x\,dx\right)^{q} \left(\int_{1/2}^{1} x\,dx\right)^{n} \left(\int_{1/2}^{1} (1-x)\,dx\right)^{m} ,
\end{aligned}
$$

we eventually find that

$$
\begin{aligned}
\Pr\left[\left(q-q'\right) A\text{'s} \,\&\, q' N\text{'s} \mid EH_2\right] &= \binom{q}{q'} (I_1+I_2)\,/\,(I_3+I_4) \\
&= \binom{q}{q'} \left(3^{m+q-q'} + 3^{n+q'}\right) \Big/ \, 4^q \left(3^m + 3^n\right) \;.
\end{aligned}
$$

In a similar fashion one can show that

$$
\begin{aligned}
&\Pr\left[\left(\left(q-q'\right) A\text{'s} \,\&\, 1/2 \le p_A \le 1\right) \vee \left(\left(q-q'\right) N\text{'s} \,\&\, 1/2 \le p_N \le 1\right) \mid EH_2\right] \\
&\quad = \binom{q}{q'} \, 3^{q-q'} \Big/ \, 4^q \;.
\end{aligned}
$$

The solutions to parts 2 and 3 are found in a manner analogous to that used in the corresponding parts of the previous problem.

Condorcet now suspends his examination of such matters and goes on to apply the preceding principles. The first question considered is concerned with the finding[37]

> des moyens de déterminer, d'après l'observation, la valeur de la probabilité de la voix d'un des Votans d'un Tribunal & celle de la décision d'un Tribunal donné. [p.213]

Two methods of solution are presented: the first does not concern us, and we shall comment but briefly on the second. In the latter, three hypotheses are considered:

(i) in each decision the vote of each voter has a constant probability;

(ii) the probability varies in each decision and for each voter;

(iii) both (i) and (ii) may be admitted together, by multiplying the probability which results from each by the probability that this hypothesis arises.

Condorcet advises against considering (i) on its own, finding the desired probability to be purely mathematical. The second hypothesis leads to the results of Problems 4 and 13, and so only (iii) need be considered, and under this hypothesis the results of Problems 4, 12 and 13 are applicable.

The remainder of this part of the *Essai* is devoted to the determination of the probabilities of decisions under certain conditions, and does not contribute anything to our study.

In the introduction to the *Essai* Condorcet describes the scope of the fourth part as follows:

> on donnera le moyen de faire entrer dans le calcul l'influence d'un des Votans sur les autres, la mauvais foi qu'on peut leur supposer, l'inégalité de lumières entre les Votans & les autres circonstances auxquelles il est nécessaire d'avoir égard pour rendre la théorie applicable & utile. [p.2]

Much use is made of the results of the third part: the integrals in the present part are not derived in as much detail as in the previous part, but no new results are to be found here[38].

In the fifth part various applications of the preceding theory are given: once again nothing pertinent is to be found.

The *Essai* concludes with the following words:

> la difficulté d'avoir des données assez sûres pour y appliquer le calcul, nous a forcés de nous borner à des apercus généraux & à des résultats hypothètiques: mais il nous suffit d'avoir pu, en établissant quelques principes, & en montrant la manière de les appliquer, indiquer la route qu'il faut suivre, soit pour traiter ces questions, soit pour faire un usage utile de la théorie. [p.304]

What we have discussed here provides ample evidence of Condorcet's ability — not only in handling abstruse probabilistic concepts, but also in rendering *obscurum per obscurius*. It is thus a bit severe of Cajori [1919a] to dismiss the work with the words

> [Condorcet's] general conclusions are not of great importance; they are that voters must be enlightened men in order to ensure our confidence in their decisions. [p.244]

5.6 Discours sur l'astronomie et le calcul des probabilités

This article[39], containing little to our purpose, was read at the *Lycée* in 1787. In the second half of the paper we once again find a reference to Pascal, de Méré and Fermat as the originators of the probability calculus, and this is followed by a passage in which Pearson [1978, p.503] finds Bayes's Theorem used. The pertinent extract runs as follows:

> Nous prouverons que le motif de croire à ces vérités réelles, auxquelles conduit le calcul des probabilités, ne diffère de celui qui nos détermine dans tous nos jugements, dans toutes nos actions, que parce que le calcul nous a donné la mesure de ce motif, et que nous cédons, par l'assentiment éclairé de la raison, à une force dont nous avons calculé le pouvoir, au lieu de céder machinalement à une force inconnue. [p.499]

I think an abundance of charity is needed to see any application of Bayes's work here, and there is nothing else even remotely relevant in the paper.

5.7 Eléméns du calcul des probabilités

This work, the full title of which is *Eléméns du calcul des probabilités et son application aux jeux de hasard, a la loterie, et aux jugemens des hommes. Avec un discours sur les avantages des mathématiques sociales*, was published posthumously in An XIII — 1805, together with an anonymous "notice sur M. de Condorcet". It is not discussed by Todhunter.

Intended as the fourth volume of Condorcet's annotated edition of Euler's *Lettres a une princesse d'Allemagne sur quelques sujets de physique et de philosophe* (an edition with which Lacroix was associated), this treatise contains the following general comment in the introductory note:

> On a justement reproché à tous les ouvrages mathématiques de Condorcet, d'ailleurs remplis de découvertes profondes dans l'analyse, d'être pénibles a lire et difficiles à entendre. Souvent même les méthodes qu'il emploie sont tellement généralisées, qu'elles échappent aux cas particuliers. Qu'il est loin de la clarté transparente de l'analyse d'Euler, ou de la simplicité élégante de celle de la Grange! [pp.vi-vij]

This book consists of seven articles[40], followed by a *Tableau Général de la Science*[41]. The first two articles[42] contain nothing relevant to the present study: we thus turn our attention immediately to the third, “Des principes fondamentaux du calcul des probabilités” [pp.56-79].

Speaking of equally possible events, Condorcet writes

> On cherche d’abord à déterminer le nombre de tous les évènemens également possibles, et il est absolument nécessaire de remonter à ceux auxquels il est permis de supposer cette égale possibilité, sans quoi le calcul deviendroit absolument hypothétique. On cherche ensuite, dans ce nombre d’évènemens également possibles, quel est le nombre de ceux qui remplissent une certaine condition, et on dit que la probabilité d’avoir un évènement qui remplisse cette condition, est exprimé par le second de ces nombres divisé par le premier. [p.56]

He then goes on to point out that

> Il n’est donc pas nécessaire, pour avoir la probabilité, de connaître le nombre total des évènemens, mais seulement le rapport du nombre de ceux qui l’on veut considére avec ce nombre total. [p.57]

The addition formula for mutually exclusive events is phrased as

> la probabilité d’avoir l’un ou l’autre des évènemens qui remplissent des conditions différentes, est égale à la somme des probabilités qu’on a pour les évènemens qui remplissent chacune de ces conditions. [p.59]

Condorcet next considers the question of sampling with replacement from an urn containing four balls (say) (white or black). If four draws result in three white balls and one black (event E, say), one might be interested in the probabilities of the various possible compositions of the urn. After some calculations, he passes on to consider the probability of getting a white ball on the next draw, all possible initial compositions of the urn being regarded as equally possible. This assumption

> est ici légitime, puisque, d’après la nature de la question, je suis dans une ignorance absolue sur ce rapport; et la seule donnée que j’aie pour evaluer la probabilité qu’il soit plutôt exprimé par un nombre que par un autre, dépend de l’observation des tirages successifs. [p.68]

Denoting by x the probability of drawing a white ball, one finds that

$$\Pr[\text{white ball } \& \, E] = \int_0^1 4\, x^4(1-x)\, dx \ .$$

Having shown that

$$\int_0^1 x^m(1-x)^n\,dx = m!\,n!/(m+n+1)! \ ,$$

Condorcet next shows that the probability that x has a specified value (say $\frac{1}{2}$) is nought, while the probability that x is $\frac{2}{3}$ rather than $\frac{1}{2}$ is given as $2^3/\,3^4 : 1/2^4$ (this being the ratio of $x^3(1-x)$ at $\frac{2}{3}$ to the same thing at $x=\frac{1}{2}$). He next evaluates

$$\Pr\left[(x>1/2)\ \&\ E\right] = \int_{1/2}^1 4\,x^3(1-x)\,dx$$

and

$$\Pr\left[(x<1/2)\ \&\ E\right] = \int_0^{1/2} 4\,x^3(1-x)\,dx \ .$$

The factor "4" is missing from both these expressions, which is not too serious an omission since one is really concerned with finding "s'il est plus probable que x est au dessus de $\frac{1}{2}$ qu'au dessous" [p.75]. More serious is the fact that Condorcet evaluates these integrals (without the "4") as $\left(1-1/2^5\right)/4.5$ and $\left(1/2^5\right)/4.5$ respectively.

Condorcet next shows that the probability of drawing a white ball after n white and m black balls have been drawn is $(n+1)/(m+n+1)$, and, more generally, that the probability of drawing a further p white and q black balls in $(p+q)$ draws is

$$\binom{m+p}{p}\binom{n+q}{q}\Big/\binom{m+n+p+q+1}{p+q} \ .$$

In the fourth article, "De la mesure des vérités auxquelles peut conduire le calcul des probabilités" [pp.79-100], we find a brief discussion of what Condorcet accepts as grounds for considering events to be equally possible, viz.

> l'égale possibilité des évènemens n'a été pour nous que l'ignorance absolue des causes qui peuvent déterminer un évènement plutôt qu'un autre. Enfin cette définition a supposè encore l'ignorance de l'évènement que l'on considère, soit que cette ignorance naisse de l'impossibilité où nous sommes de connaître les évènemens futurs, soit que l'évènement étant actuel ou passé nous soit inconnu par d'autres causes. [p.80]

Condorcet also ties up probability with belief by noting that the greater the probability of an event, the greater our reason for believing ("motif de croire") in its occurrence should be[43].

In the fifth article, "Sur la manière de comparer entre eux des évènemens de probabilités différentes, et de trouver une valeur moyenne qui puisse représenter les valeurs différentes entre elles d'évènemens inégalement probables" [pp.100-120], Condorcet attributes the invention of the probability calculus to Pascal and Fermat[44], and then, in a moment of perhaps justifiable pride, says

> cette remarque n'est pas inutile; elle peut servir à réfuter ceux qui se plaisent à répéter que la nature a refusé le don de l'invention, et n'accorde que celui de perfectionner aux hommes qui naissent entre Perpignan et Dunkerque. [p.100]

Nothing else from this monograph seems pertinent[45].

5.8 Appendix 5.1

I can find no trace of a work entitled "Sur les évènements futurs" [1803] attributed by Keynes [1921] to Condorcet. Keynes may have taken the reference from the bibliography in Laurent [1873].

Chapter 6

Laplace

Looke within; within is the fountaine of all good. Such a fountaine, where springing waters can never fail, so that thou digge still deeper and deeper.

Marcus Aurelius Antoninus.

6.1 Introduction

Pierre Simon, Marquis de Laplace[1] (1749-1827) was a prolific writer on a wide range of scientific and mathematical topics. The analytic table in the *Œuvres complètes de Laplace* covers 56 pages, and Stigler [1978, p.235] has indicated that there are in fact some writings by Laplace not included in this collection. I have not, of course, read all of Laplace's works (a feat beside which even the labours of Hercules would seem like child's play) but it is hoped that the present coverage is fairly complete.

Some dozen memoirs[2] have been identified as being pertinent to the present work, ranging from two early papers published in 1774 to the third edition of 1820 of the magnum opus *Théorie analytique des probabilités.* Of course, much of the early material is reprinted in the latter classic, yet it is, I think, of interest to examine the memoirs in chronological order, that some idea might be gained of the passage of Laplace's thought on Bayesian inference and methods. From each memoir we shall consider, in the main, only those parts specific to our topic.

6.2 Sur les suites récurro-récurrentes

This paper, fully entitled "Mémoire sur les suites récurro-récurrentes et sur leurs usages dans la théorie des hasards", was published in the *Mémoires de l'Académie royale des Sciences de Paris (Savants étrangers)*, Vol. VI [1774], pp.353-371, and contains, strictly speaking, nothing pertinent. The only point worth noting (in the context of the present work) is the appearance of an early "definition" of probability[3] (framed by Laplace as a "Principe"), i.e.

> La probabilité d'un événement est égale à la somme des produits de chaque cas favorable par sa probabilité divisée par la somme des produits de chaque cas possible par sa probabilité, et si chaque cas est également probable, la probabilité de l'événement est égale au nombre des cas favorables divisé par le nombre de tous les cas possibles. [pp.10-11]

(Page numbers refer to the *Œuvres complètes* edition of Laplace's works unless otherwise stated.)

We shall not enter into a discussion of equipossibility (or equiprobability) (an assumption to which Laplace was habituated (see Gillispie [1972, p.7])) here: suffice to say that, while Laplace is often viewed as the originator of this term, Hacking [1975, p.122] traces it back to Leibniz in 1678 (op. cit., pp.125, 127). Notice too that this principle is framed initially for cases which are not postulated to be equiprobable: this latter idea is only introduced in the second clause. (One might perhaps see in the first part of the principle the framing of the probability of an event in terms of the probabilities of *elementary* events.)

6.3 Sur la probabilité des causes

This "Mémoire sur la probabilité des causes par les événements", the first paper[4] in which Laplace discussed the probabilities of causes, was published[5] in 1774 in the sixth volume of the *Mémoires de l'Académie royale des Sciences de Paris (Savants étrangers)*. The memoir is in seven sections: since many of them contain relevant material, and since "scarcely any of the present memoir is reproduced by Laplace in his *Théorie ... des Prob.*" (Todhunter [1865, art.880]), we choose to give it rather more attention than it perhaps merits in the corpus of Laplace's works.

The essay opens with the following well-known words:

> La théorie des hasards est une des parties les plus curieuses et les plus délicates de l'Analyse, par la finesse des combinaisons

> qu'elle exige et par la difficulté de les soummettre au calcul. [p.27]

After mentioning certain other of his memoirs, Laplace explains the purpose of the present one as follows:

> je me propose de déterminer la probabilité des causes par les événements, matière neuve à bien des égards et qui mérite d'autant plus d'être cultivée que c'est principalement sous ce point de vue que la science des hasards peut être utile dans la vie civile. [p.28]

The importance of (parts of) this memoir to our present theme cannot be overstressed: indeed Todhunter says:

> This memoir is remarkable in the history of the subject, as being the first which distinctly enunciated the principle for estimating the probabilities of the causes by which an observed event may have been produced. [1865, art.868]

However, he goes on to say (loc. cit.) "Bayes must have had a notion of the principle ...", an assertion the reason for which is by no means clear[6]. Bayes does not explicitly refer to the "probability of causes", and, as we shall see later, there is room for doubt as to the exact connexion between Bayes's and Laplace's results (there is no mention of Bayes in the memoir)[7].

After an introductory Article, Laplace begins the second section of this memoir with a careful distinction between those cases in which the event (of interest) is uncertain, although the cause on which the probability of its occurrence depends is known, and those in which the event is known and the cause is unknown [p.29], that is, a distinction between direct and indirect (or inverse) probability. Stating that all problems in "la théorie des hasards" may be brought into one or other of these classes, Laplace declares his intent to restrict his attention only to those in the second class, to the furtherance of which end he asserts[8] the following fundamental principle[9]:

> *Principe.* — Si un événement peut être produit par un nombre n de causes différentes, les probabilités de l'existence de ces causes prises de l'événement sont entre elles comme les probabilités de l'événement prises de ces causes, et la probabilité de l'existence de chacune d'elles est égale à la probabilité de l'événement prise de cette cause, divisée par la somme de toutes les probabilités de l'événement prises de chacune de ces causes. [p.29]

In modern notation, this principle states the following two "facts":

(i) $$\frac{\Pr[A_i \mid E]}{\Pr[A_j \mid E]} = \frac{\Pr[E \mid A_i]}{\Pr[E \mid A_j]} \quad , \quad i, j \in \{1, 2, \ldots, n\}, \quad i \neq j$$

$$\text{(ii)}\quad \Pr[A_i \mid E] = \Pr[E \mid A_i] \Big/ \sum_1^n \Pr[E \mid A_j] \quad , \quad i \in \{1, 2, \ldots, n\} \ .$$

It is here perhaps that we have the first occurrence of the *so-called*[10] "Bayes's Theorem" with a uniform prior, a result which can be stated more generally as follows:

> Let E be an event (of positive probability) which can occur in conjunction with one of the mutually exclusive and exhaustive events $H_1, H_2, \ldots, H_n$, each of positive probability. Then, for each $i \in \{1, 2, \ldots, n\}$,
>
> $$\Pr[H_i \mid E] = \Pr[E \mid H_i] \Pr[H_i] \Big/ \sum_1^n \Pr[E \mid H_j] \Pr[H_j] \quad .$$

Several points are worthy of note in connexion with this principle: firstly, it is tacitly assumed that the prior probabilities of the causes are equal, and secondly, Laplace refers to "n" causes and uses the word "somme" — though the applications he indulges in are in fact not discrete. That the present nice distinction between $\sum$ and $\int$ was not observed during Laplace's time is of course well known: the point is clearly illustrated in the *Théorie analytique des probabilités*, Book II, art.23, where we find the sentence

> la somme des erreurs à craindre, abstraction faite du signe, multipliées par leur probabilité, est donc pour toutes les valeurs de x', moindres que l, $\int (l - x')y'\, dx'$. [p.339]

One might see then, in this fundamental principle, a continuous analogue of the above result, viz.,

$$f(x \mid y) = f(y \mid x) \Big/ \int f(y \mid x)\, dx \quad .$$

After applying this principle to a simple urn problem, Laplace proceeds, in his third article, to a problem[11] nearer to our investigation, viz.

> Si une urne renferme une infinité de billets blancs et noirs dans un rapport inconnu, et que l'on en tire $p+q$ billets dont p soient blancs et q soient noirs; on demande la probabilité qu'en tirant un nouveau billet de cette urne il sera blanc. [p.30]

In his solution of this problem, Laplace explains his choice of a (discrete) uniform prior in the following way:

> Le rapport du nombre des billets blancs au nombre total des billets contenus dans l'urne peut être un quelconque des nombres fractionnaires compris depuis 0 jusqu'à 1. [p.30]

(At least, as Edwards [1978] has observed[12], Bayes gave an *argument* for his assumptions!) Representing this unknown ratio by x, Laplace then says (correctly) that the probability of drawing p white (or blank) (lottery–) tickets and q black is $x^p(1-x)^q$. Consequently, by the principle of his preceding Article (and no additional argument is presented) the probability that x is the true ratio of the number of white to the total number of tickets is

$$x^p(1-x)^q\,dx \Big/ \int_0^1 x^p(1-x)^q\,dx \quad . \tag{1}$$

We might notice, in passing, that the expected binomial coefficients which would be here were the order in which the tickets were drawn not of importance, will in fact cancel out in this latter expression. Moreover, although x is rational, we may assume that the integrand is appropriately extended to the whole of [0,1] so that the denominator of this expression is well defined[13].

Using essentially the result (expressed in a modern notation)

$$\Pr[A \mid B] = \sum_i \Pr[A \mid B \,\&\, C_i]\Pr[C_i \mid B] \quad ,$$

Laplace deduces from (1) that the required probability is

$$\int_0^1 x^{p+1}(1-x)^q\,dx \Big/ \int_0^1 x^p(1-x)^q\,dx \quad ,$$

an expression which is shown (by repeated integrations by parts) to reduce to $(p+1)/(p+q+2)$. This result is immediately extended to obtain the probability of drawing m white and n black tickets, viz.

$$\int_0^1 x^{p+m}(1-x)^{q+n}\,dx \Big/ \int_0^1 x^p(1-x)^q\,dx$$

$$= \frac{(q+1)(q+2)\dots(q+n)(p+1)(p+2)\dots(p+q+1)}{(p+m+1)(p+m+2)\dots(p+q+m+n+1)} \quad . \tag{2}$$

(Once again, if no account is taken of the order in which the $(m+n)$ subsequent tickets are drawn, this expression should be multiplied by $\binom{m+n}{m}$.) For ease of future reference, let us denote the ratio (2) by $Q(p,q;m,n)$.

Supposing p and q to be very large, and m and n very small in comparison with p and q, Laplace shows that this latter probability is approximately

$$p^m q^n / (p+q)^{m+n} \quad .$$

He then goes on to point out the inadequacy of this approximation for larger values of m and n; indeed, if $m = p$ and $n = q$, the probability should be approximated by

$$\sqrt{\frac{1}{2}} p^m q^n \,/(p+q)^{m+n} \quad .$$

Laplace next points out that the solution of this problem provides a direct method of determining the probability of future events after ("d'après") those which have already occurred, but proposes to limit himself to a proof of the following theorem:

> On peut supposer les nombres p et q tellement grands, qu'il devienne aussi approchant que l'on voudra de la certitude que le rapport du nombre de billets blancs au nombre total des billets renfermés dans l'urne est compris entre les deux limites $p/(p+q) - \omega$ et $p/(p+q) + \omega$, ω pouvant être supposé moindre qu'aucune grandeur donnée. [p.33]

Using the preceding results, Laplace concludes almost immediately that the probability of the desired ratio's lying between the specified limits is

$$\int x^p (1-x)^q \, dx \Big/ \int_0^1 x^p (1-x)^q \, dx \quad ,$$

the integral in the numerator being taken over the region bounded by the limits $p/(p+q) - \omega$ and $p/(p+q) + \omega$. By what Todhunter [1865, art.871] calls "a rude process of approximation", Laplace shows that, for p and q infinitely large, and ω infinitely less than $(p+q)^{-1/3}$ and infinitely greater than $(p+q)^{-1/2}$, this probability becomes, approximately[14],

$$E \equiv \frac{(p+q)^{3/2}}{\sqrt{2\pi pq}} \int_0^\omega 2\, e^{-(p+q)^3 z^2/2pq} \, dz \quad , \tag{3}$$

which he goes on to say is approximately 1:

> on voit donc qu'en négligeant les quantités infiniment petites, nous pouvons regarder comme certain que le rapport du nombre des billets blancs au nombre total des billets est compris entre les limits $p/(p+q) + \omega$ et $p/(p+q) - \omega$, ω étant égal à $(p+q)^{-1/n}$, n étant plus grand que 2 et moindre que 3, et à plus forte raison n étant plus grand que 3; partant ω peut être supposé moindre qu'aucune grandeur donnée. [p.36]

He then discusses the error incurred in setting $E = 1$, concluding in fact [p.39] that

$$E = 1 - \frac{-\sqrt{pq}}{\omega\sqrt{2\pi}(p+q)^{3/2}}\left[\left(1+\frac{p+q}{p}\omega\right)^p\left(1-\frac{p+q}{q}\omega\right)^q\right.$$
$$\left.+\left(1-\frac{p+q}{p}\omega\right)^p\left(1+\frac{p+q}{q}\omega\right)^q\right] . \qquad (4)$$

In his fourth article Laplace applies his general principle to what Todhunter [1865, art.872] calls "the Problem of Points", i.e. two players, A and B, of unknown skills, play a game (e.g. piquet) under the condition that the first to win n points or matches ("parties") will win a sum a, laid down at the outset of the game. Suppose now that the players are forced to abandon the game at a stage at which A needs f matches and B needs h matches to win: how should the amount a be divided between the two players[15]?

To solve this, Laplace first states that, were the respective skills of A and B known, and in the ratio of p to q respectively (where $p+q=1$), the amount which B should receive is

$$aq^{f+h-1}\left[1+\frac{p}{q}(f+h-1)+\frac{p^2}{q^2}\frac{(f+h-1)(f+h-2)}{1.2}+\cdots\right.$$
$$\left.+\frac{p^{f-1}}{q^{f-1}}\frac{(f+h-1)\ldots(h+1)}{1.2.3\ldots(f-1)}\right] .$$

(This result is stated to have been proved "dans plusiers Ouvrages", including one of his own earlier memoirs of 1773.) Following Todhunter [1865, art.873], let us denote this amount by $\varphi(p,f,h)$.

Once again Laplace cavalierly concludes that ignorance (this time of the players' skills) should be reflected in the choice of a uniform distribution, his exact words being

> puisque la probabilité de A pour gagner une partie est inconnue, nous pouvons la supposer un des nombres quelconques, compris depuis 0 jusqu'à 1. [p.40]

Let us represent this unknown probability by x; then the probability that, in $2n-f-h$ matches, A and B will win $n-f$ and $n-h$ respectively is

$$x^{n-f}(1-x)^{n-h} .$$

Hence, by his fundamental principle, "la probabilité de la supposition que nous avons faite pour x" is

$$x^{n-f}(1-x)^{n-h}\,dx \Big/ \int_0^1 x^{n-f}(1-x)^{n-h}\,dx .$$

Now, x being the probability that A wins a match, the amount B ought to receive is $\varphi(x, f, h)$, and hence the amount B ought to receive is

$$\int_0^1 x^{n-f}(1-x)^{n-h}\varphi(x,f,h)\,dx \Big/ \int_0^1 x^{n-f}(1-x)^{n-h}\,dx \quad .$$

This expression is then evaluated.

In the fifth article Laplace applies his preceding results to the theory of errors: this is the first of Laplace's works on this important topic[16], the problem posed here being the following:

> Problème III — Déterminer le milieu que l'on doit prendre entre trois observations données d'un même phénomène. [p.42]

As a consequence of this restriction to three values, Todhunter [1865, art.875] somewhat harshly concludes "Thus the investigation cannot be said to have any practical value": however, when one appreciates the complexity of the solution, one cannot but admire Laplace.

Laplace takes as the density of the errors of observations the function $y = \varphi(x)$, a function which he supposes, firstly, to be even, to decrease asymptotically to zero as $x \to +\infty$ or $x \to -\infty$, and to have unit area. Let a, b, c be points on the line segment AB (see Figure 6.1) representing the instants at which a certain astronomical event has been recorded. Let p and q be the time (in seconds) between a and b and b and c respectively. Then

> on demande à quel point V de la droite AB on doit fixer le mileu que l'on doit prendre entre les trois observations a, b et c. [p.42]

If v is "le véritable instant du phénomène", at a distance x from a, the probability of realizing the given sequence of observations is[17]

$$y = f(x) = \varphi(x)\,\varphi(p-x)\,\varphi(p+q-x) \quad , \tag{5}$$

with a similar result (with x' replacing x) for any other v'. By the first part of the fundamental principle, the probabilities of the two hypotheses are in the ratio

$$\varphi(x)\,\varphi(p-x)\,\varphi(p+q-x) : \varphi(x')\,\varphi(p-x')\,\varphi(p+q-x') \ .$$

(The more modern approach would be to take $\varphi(x-v)$ as the density of deviations from v, to replace the above ratio (for the observations x_1, x_2 and x_3) by

$$\varphi(x_1-v)\,\varphi(x_2-v)\,\varphi(x_3-v) : \varphi(x_1-v')\,\varphi(x_2-v')\,\varphi(x_3-v') \ ,$$

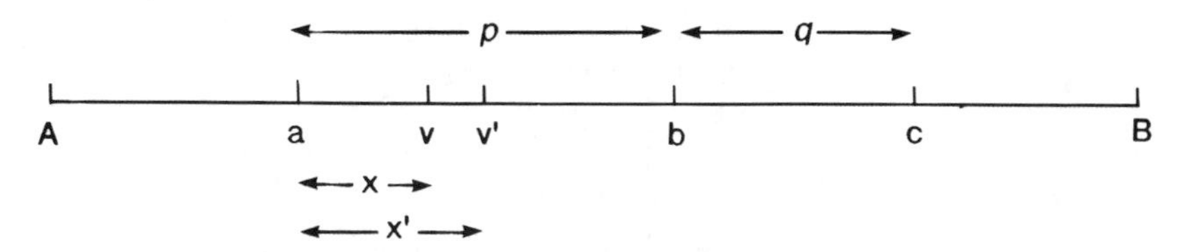

Figure 6.1: Instants at which an astronomical event is recorded.

and to use maximum likelihood estimation. For further discussion of this point see Sheynin [1977, p.3] and Stigler [1986a, pp.105-109].)

Wishing to find the mean, Laplace points out that one may intend one of two things[18]:

> La première est l'instant tel qu'il soit également probable que le véritable instant du phénomène tombe avant ou après: on pourrait appeler cet instant *milieu de probabilité.*
> La seconde est l'instant tel qu'en le prenant pour milieu, la somme des erreurs à craindre, multipliées par leur probabilité, soit un *minimum*; on pourrait l'appeler *milieu d'erreur* ou *milieu astronomique.* [p.44]

He next shows that the results obtained under these two conditions are equivalent. The first choice clearly leads to the median of (the posterior) $f(x)$ as defined above (a uniform prior is tacitly assumed). To find the second mean, it is necessary to choose a point v (see Figure 6.2) such that

$$\int |x - v| f(x)\, dx = \text{ minimum.}$$

This choice in fact yields exactly the same choice of mean as the first, a fact which can be observed on the following wise. Let $u - v = dx$, and let Q (respectively P) be the centre of gravity of the mass ("partie") uoL (respectively vOH), with abscissa a distance z (respectively z') from Ov (or ou). Let M and N be the respective masses.

The sum of the ordinates multiplied by their distances from the chosen point v is then

$$Mz + Nz' + (y\, dx)\left(\tfrac{1}{2}\, dx\right) \quad .$$

Similarly, taking u for this mean, we obtain

$$M(z - dx) + N\left(z' + dx\right) + (y\, dx)\left(\tfrac{1}{2} dx\right) \quad .$$

The second criterion implying that the difference between these two expressions should be zero, we find that $M = N$ — that is, Ov divides the area

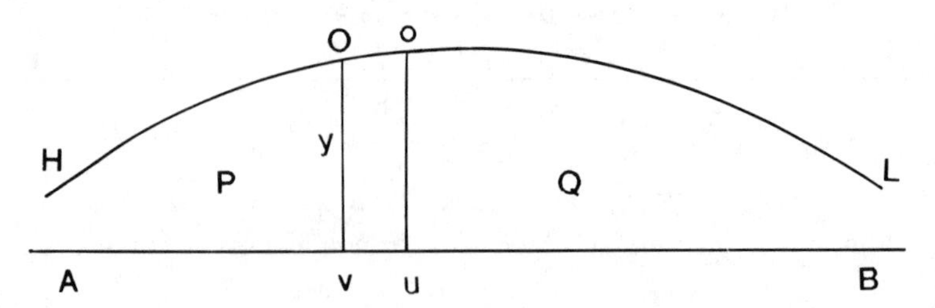

Figure 6.2: Illustration showing that the posterior median minimizes the posterior expected error.

under the curve into two equal parts.

Sheynin [1977, p.4] in fact states that the requirement of the second criterion is that

$$F(v) = \int_{-\infty}^{\infty} |x - v| f(x)\, dx$$

should be minimized. It follows from this that, for any α, β such that $-\infty < \alpha < v < \beta < \infty$,

$$0 = F'(v) = \left[\int_{-\infty}^{\alpha} |x - v| f(x)\, dx + \int_{\alpha}^{\beta} |x - v| f(x)\, dx + \int_{\beta}^{\infty} |x - v| f(x)\, dx \right]$$

whence

$$\int_{-\infty}^{v} f(x)\, dx = \int_{v}^{\infty} f(x)\, dx \ ,$$

or v is the median of $f(\cdot)$:

> On voit donc que le milieu *astronomique* ne diffère point de celui de *probabilité*, et que l'un et l'autre se déterminent par l'ordonnée OV qui divise l'aire de la courbe HOL en deux parties égales. [p.45]

(Laplace's V is our v.)

Fixing his attention on the line Ov, which divides the area under the curve into halves, Laplace points out that the finding of this ordinate requires knowledge of $\varphi(x)$, and adduces arguments which lead to[19]

$$\varphi(x) = \frac{m}{2} e^{-mx} \ . \tag{6}$$

(This, of course, is really derived under the assumption that $x > 0$: in general the form

$$\varphi(x) = \frac{m}{2} e^{-m|x|} \ , \quad -\infty < x < \infty \tag{7}$$

would result.) Using this (further details may be found in Sheynin [1977]), Laplace shows that the area S under the curve is given by

$$S = \frac{m^2}{8} e^{-m(p+q)} \left(1 - \frac{1}{3}e^{-mp} - \frac{1}{3}e^{-mq}\right)$$

(though in fact this seems to give only *half* the total area), and hence x, the abscissa of v, is found to be

$$x = p + \frac{1}{m} \ln \left(1 + \frac{1}{3}e^{-mp} - \frac{1}{3}e^{-mq}\right) .$$

For small values of $m, x \approx (2p+q)/3$ (i.e. the arithmetic mean). Further discussion of this point may be found in Sheynin (op. cit.) and Stigler [1986a, p.112], and we need say nothing more about it here.

What is, however, more germane to our present investigation is the case in which the parameter m is unknown. In this connexion Laplace writes [pp.48-49]

> D'après le principe fondamental de l'Article II, les probabilités des différentes valeurs de m sont entre elles comme les probabilités que, ces valeurs ayant lieu, les trois observations auront les distances respectives qu'elles ont entre elles. Or les probabilités que les trois observations a, b et $c\ldots$ s'éloigneront les unes des autres aux distances p et q sont entre elles comme les aires des courbes HOL, correspondantes aux différentes valeurs de m, comme il est facile de s'en assurer. D'où il résulte, par le principe de l'Article II, que la probabilité de m est proportionnelle à
>
> $$m^2 e^{-m(p+q)} \left(1 - \frac{1}{3}e^{-mp} - \frac{1}{3}e^{-mq}\right) dm .$$

To prove this assertion it is necessary firstly to recall expression (5), viz.

$$y = f(x) = \varphi(x)\,\varphi(p-x)\,\varphi(p+q-x) .$$

Sheynin [1977] chooses to interpret f as the conditional probability density function $f(x, m \mid p, q)$ where, using (5) and (7),

$$f(x, m \mid p, q) = \frac{m^3}{8} e^{-m(|x|+|p-x|+|p+q-x|)}, \quad -\infty < x < \infty .$$

It then follows from the formula of total probability (for the continuous case) that, in Sheynin's terminology,

$$\Pr[m] = c \int_{-\infty}^{\infty} f(x, m \mid p, q)\, dx$$

and, as Laplace noted, $\Pr[m = 0] = 0$.

The argument in Stigler [1986a, pp.112-113] runs as follows: interpreting f in (5) as $f(x, p, q \mid m)$, one has

$$f(p, q \mid m) = \int_{-\infty}^{\infty} f(x, p, q \mid m)\, dx \ .$$

Thus, by the Principle,

$$f(m \mid p, q) \propto \int_{-\infty}^{\infty} f(x, p, q \mid m)\, dx \ .$$

Notice that this latter integral can be written as

$$\frac{f(p, q)}{f(m)} \int_{-\infty}^{\infty} f(x, m \mid p, q)\, dx \ ,$$

and compare this expression with that given by Sheynin.

Still assuming m to be unknown, Laplace now turns his attention to the determination of the "best" x:

> si l'on nomme y la probabilité, correspondante à m, que le véritable instant du phénomène tombe à la distance x du point a, la probabilité entière que cet instant tombera à cette distance sera proportionnelle à
>
> $$\int y m^2 e^{-m(p+q)} \left(1 - \frac{1}{3} e^{-mp} - \frac{1}{3} e^{-mq}\right) dm \ ,$$
>
> l'intégrale étant prise de manière qu'elle commence lorsque $m = 0$, et finisse lorsque $m = \infty$; si donc on construit sur l'axe AB une nouvelle courbe $H'KL'$ dont les ordonnées soient proportionnelles à cette quantité, l'ordonée KQ qui divisera l'aire de cette courbe en deux parties égales coupera l'axe au point que l'on doit prendre pour milieu entre les trois observations. [p.49]

Laplace's y seems to be $f(x \mid p, q, m)$, and the integral in the above quotation is then

$$\begin{aligned} \int_0^{\infty} f(x \mid p, q, m) f(m \mid p, q)\, dm &= \int_0^{\infty} f(x, m \mid p, q)\, dm \\ &= f(x \mid p, q) \ . \end{aligned} \tag{8}$$

It thus follows, according to Laplace, that the posterior median (μ, say) may be found by solving

$$\int_{-\infty}^{\mu}\int_{0}^{\infty} f(x \mid p,q,m) f(m \mid p,q)\, dm\, dx = \frac{1}{2}\int_{-\infty}^{\infty}\int_{0}^{\infty} f(x \mid p,q,m) f(m \mid p,q)\, dm\, dx \,.$$

Using (8), this becomes

$$\int_{-\infty}^{\mu} f(x \mid p,q)\, dx = \frac{1}{2}\int_{-\infty}^{\infty} f(x \mid p,q)\, dx \quad ,$$

which is indeed true.

However, Laplace goes on to say

> L'aire de cette nouvelle courbe sera évidemment proportionnelle à l'intégrale du produit de l'aire de la courbe HOL par
>
> $$m^2 e^{-m(p+q)}\left(1 - \frac{1}{3}e^{-mp} - \frac{1}{3}e^{-mq}\right) dm \;.$$
>
> Donc, puisque, pour determiner x dans une supposition particulière pour m, on a
>
> $$m^2 e^{-m(2p+q-x)} = m^2 e^{-m(p+q)}\left(1 + \frac{1}{3}e^{-mp} - \frac{1}{3}e^{-mq}\right)$$
>
> on aura
>
> $$\int m^4 e^{-m(3p+2q-x)}\left(1 - \frac{1}{3}e^{-mp} - \frac{1}{3}e^{-mq}\right) dm$$
>
> $$= \int m^4 e^{-m(2p+2q)}\left(1 + \frac{1}{3}e^{-mp} - \frac{1}{3}e^{-mq}\right)\left(1 - \frac{1}{3}e^{-mp} - \frac{1}{3}e^{-mq}\right) dm \quad ,$$
>
> en intégrant de manière que les intégrales commencent lorsque $m = 0$, et finissent lorsque $m = \infty$. [p.49]

(See Figure 6.3.) The argument now seems to be that[20]

$$\int_{-\infty}^{\infty}\int_{0}^{\infty} f(x \mid p,q,m) f(m \mid p,q)\, dm\, dx$$

is proportional to

$$\int_{0}^{\infty}\int_{-\infty}^{\infty} f(x,p,q \mid m) f(m \mid p,q)\, dx\, dm \;.$$

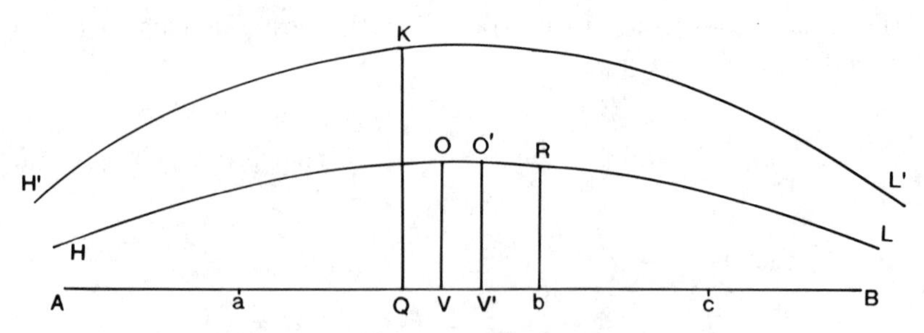

Figure 6.3: A posterior distribution.

Since $f(x \mid p, q, m) = f(x, p, q \mid m)/f(p, q \mid m)$, the first of these double integrals in fact becomes

$$\int_{-\infty}^{\infty} \int_{0}^{\infty} f(x, p, q \mid m) f(m \mid p, q)/f(p, q \mid m)\, dm\, dx \ ,$$

and it is immediately clear that Laplace's proportionality "constant" is in fact a function of m. Thus the statement at the start of the preceding quotation is false, and so therefore is the following statement, viz. since, when m is known, μ is given by solving

$$\int_{-\infty}^{\mu} f(x, p, q \mid m)\, dx = \frac{1}{2} \int_{-\infty}^{\infty} f(x, p, q \mid m)\, dx \ ,$$

it follows that, in this case,

$$\int_{-\infty}^{\mu} \int_{0}^{\infty} f(x, p, q \mid m) f(m \mid p, q)\, dm\, dx = \frac{1}{2} \int_{-\infty}^{\infty} \int_{0}^{\infty} f(x, p, q \mid m) f(m \mid p, q)\, dm\, dx \ .$$

We shall find later that the confusion engendered by Laplace's cavalier treatment of conditional probability is not limited to this memoir. Indeed, his lack of a precise notion of conditional probability contributes largely to the difficulty of reading much of his work.

In the present case Laplace obtains an equation of fifteenth degree for μ: he shows further that this equation has exactly one root in the open interval $(0, p)$, and discusses an iterative method for finding it. Stigler [1986a, p.116], by considering the corrected equation[21]

$$\int_{-\infty}^{\mu} \int_{-\infty}^{\infty} f(x, p, q \mid m)\, dm\, dx = \frac{1}{2} \int_{-\infty}^{\infty} \int_{0}^{\infty} f(x, p, q \mid m)\, dm\, dx$$

or

$$\int_0^\infty m^2 e^{-m(2p+q-\mu)}\, dm = \int_0^\infty m^2 e^{-m(p+q)} \left(1 + \frac{1}{3} e^{-mp} - \frac{1}{3} e^{-mq}\right) dm \ ,$$

obtains a cubic equation, whose roots in fact turn out to be even further from the corrections giving the arithmetic mean than do Laplace's.

Further comment on this problem may be found in Barnard [1988]. Here it is supposed that the time μ of a given event is to be estimated from three observations x_1, x_2 and x_3. Writing the errors of observation as

$$p_i = (x_i - \mu)/\sigma \ ,$$

Barnard transforms Laplace's joint density of the p_i to

$$\varphi(\mathbf{p}) = \frac{1}{8} e^{-(|p_1|+|p_2|+|p_3|)}$$

(cf. our earlier $f(x, m \mid p, q)$).

Turning to Laplace's problem of finding that function $g(x_1, x_2, x_3)$ which is such that the true value μ is as likely to fall short of g as to exceed it, Barnard notes that Laplace essentially assumes the joint prior

$$\varphi(\mathbf{p}) d\mathbf{p}\, d\mu\, d\sigma \ ,$$

that is, a uniform prior density element for μ and σ. If one wishes to allow an arbitrary prior for these parameters, one should rather consider

$$\varphi(\mathbf{p}) \pi(\mu, \sigma)\, d\mathbf{p}\, d\mu\, d\sigma \ .$$

The value of g obtained by Laplace is seen to be found in this case by taking $\pi(\mu, \sigma) \propto 1/\sigma$ — i.e. the Jeffreys non-informative prior — rather than using the uniform prior adopted by Laplace.

At the start of the sixth article Laplace poses the following problem[22]:

> je suppose que A joue avec B à croix ou pile, à ces conditions: savoir que, si A amène croix au premier coup, B lui donnera deux écus; qu'il lui en donnera quatre s'il ne l'amène qu'au second, huit s'il ne l'amène qu'au troisième, et ainsi de suite jusqu'au nombre x de coups. [pp.53-54]

In solving this problem Laplace supposes initially that the probability of a cross (i.e. a "head") is $(1 + \omega)/2$. Then A's expectation is

$$(1{+}\omega)\left[1 + (1-\omega) + (1-\omega)^2 + \cdots + (1-\omega)^{x-1}\right] = (1{+}\omega)\left[1 - (1-\omega)^x\right]/\omega \ .$$

A similar expression, *mutatis mutandis*, is given for the case in which the probability of a cross is $(1-\omega)/2$. Now, says Laplace, as the probability $(1+\omega)/2$ is as naturally attributed to cross as to pile (i.e. a "tail"), the expectation E of A is to be taken as

$$E = 1 + \left(1-\omega^2\right)\left[(1+\omega)^{x-1} - (1-\omega)^{x-1}\right]/2\omega \ , \tag{9}$$

which reduces, for ω so small that powers of ω higher than ω^2 may be neglected, to

$$E = x + \left[\frac{(x-1)(x-2)(x-3)}{1.2.3} - (x-1)\right]\omega^2 \ .$$

If one supposes that ω may take on equally any one of the values in the interval $(0, 1/q)$, one finds A's total expectation by multiplying (9) by q and integrating.

The remainder of the memoir is irrelevant to our purposes. However, before finishing off this discussion, let us note Laplace's remarks on the choice of a uniform prior[23]: he writes

> On suppose dans la théorie [i.e. des probabilités] que les différents cas qui amènent un événement sont également probables, ou, s'ils ne le sont pas, que leur probabilité est dans un rapport donné. Quand on veut ensuite faire usage de cette théorie, on regarde deux événements comme également probables, lorsqu'on ne voit aucune raison qui rende l'un plus probable que l'autre, parce que, quand bien même il y aurait une inégale possibilité entre eux, comme nous ignorons de quel côté est la plus grande, cette incertitude nous fait regarder l'un comme aussi probable que l'autre.
> Lorsqu'il n'est question que de probabilités simples, il paraît que cette inégalité de probabilités ne nuit en rien à la justesse de l'application du calcul aux objets physiques ... mais, lorsqu'il s'agit de probabilité composée, il me semble que l'application que l'on fait de la théorie aux événements physiques demande à être modifiée. [p.61]

6.4 Sur l'intégration des équations différentielles

The title of this memoir[24], viz. "Recherches sur l'intégration des équations différentielles aux différences finies et sur leur usage dans la théorie des hasards", is just right, and the actual contents do not concern us here.

It is, however, of interest to note the general remarks in the twenty-fifth article (the first section of the memoir in which probabilistic matters are broached), for it is here that we find a clear exposition of the distinction Laplace makes between "hasard" and "probabilité" (as well as a discussion of moral *vs* mathematical expectation)[25]:

> Nous regardons une chose comme l'effet du hasard, lorsqu'elle n'offre à nos yeux rien de régulier, ou qui annonce un dessein, et que nous ignorans d'ailleurs les causes qui l'ont produite. Le hasard n'a donc aucune réalité en lui-même; ce n'est qu'un terme propre à désigner notre ignorance sur la manière dont les différentes parties d'un phénomène se coordonnent entre elles et avec le reste de la Nature.
> La notion de probabilité tient à cette ignorance. Si nous sommes assurés que, sur deux événements qui ne peuvent exister ensemble, l'un ou l'autre doit nécessairement arriver, et que nous ne voyons aucune raison pour laquelle l'un arriverait plutôt que l'autre, l'existence et la non-existence de chacun d'eux est également probable. [p.145]

This is followed by an extension to three events.

A clear statement follows of the conditions under which probability is to be defined as the ratio of the number of favourable cases to the number of possible cases, viz.

> la probabilité de l'existence d'un événement n'est ainsi que le rapport du nombre des cas favorables à celui de tous les cas possibles, lorsque nous ne voyons d'ailleurs aucune raison pour laquelle l'un de ces cas arriverait plutôt que l'autre. Elle peut être conséquemment représentée par une fraction dont le numérateur est le nombre des cas favorables, et le dénominateur celui de tous les cas possibles. [p.146]

As Hacking [1975, p.131] has noted, the word "possibilité" does not occur in this definition: it is, however, used on p.149 with almost the sense of a physical probability.

Laplace next gives a precise definition of the purpose of the theory of chances, i.e.

> la théorie des hasards a pour objet de déterminer ces fractions [i.e. fractions de la certitude], et l'on voit par là que c'est le supplément le plus heureux que l'on puisse imaginer à l'incertitude de nos connaissances. [p.146]

As in the previous memoir, Laplace here draws a distinction between instances in which the causes are known but the events are to be determined, and those in which the events are known but the causes are unknown. The latter instances formed the subject of the previous memoir: the probabilistic parts of the present one are devoted to the former, their discussion being in terms of the finite difference methods introduced in the first 24 articles of the memoir.

6.5 Sur les probabilités

This, the second of Laplace's works which are particularly germane to our present study, appeared in the *Histoire de l'Académie royale des Sciences, Paris*, in the volume for 1778 — although it was submitted on the 19th July 1780 and was published[26] in 1781. Although no mention of Bayes or Price is made in this memoir, an anonymous abstract[27], in the same volume, of this article does in fact comment on their work. Because this summary is not reprinted in the *Œuvres complètes de Laplace*, and is perhaps therefore not readily accessible, I give the relevant part of it here:

> Toutes les questions du Calcul des Probabilités peuvent se réduire à une seule hypothèse, à celle d'une certaine quantité de boules de différentes couleurs mêlées ensemble, dont on suppose qu'on tire au hasard différentes boules dans un certain ordre ou dans certaines proportions. Si on suppose connu le nombre de boules de chaque espèce, on a le calcul ordinaire des probabilités tel que les Géomètres du dernier siècle l'ont considéré: mais si l'on suppose le nombre de boules de chaque espèce inconnu, & que par le nombre de boules de chaque espèce qu'on a tirées, on veuille juger ou de la proportion du nombre de ces boules, ou de la probabilité de les tirer dans la suite suivant certaines loix, on a une nouvelle classe de problèmes. Ces questions dont il paroît que M^rs^. Bernoulli & Moivre avoient eu l'idée, ont été examinées depuis par M^rs^. Bayes & Price; mais ils se sont bornés a exposer les principes qui peuvent servir à les résoudre. M. de la Place les a considérées avec plus d'étendue, & il y a appliqué l'analyse. [*Hist. Acad. r. des Sciences, Paris* 1778, pp.43-44]

The memoir begins with a clear statement of the scope of the study[28]:

> je me propose de traiter dans ce Mémoire deux points importants de l'analyse des hasards qui ne paraissent point avoir encore été suffisamment approfondis: le premier a pour objet

> la manière de calculer la probabilité des événements composés d'événements simples dont on ignore les possibilités respectives; l'objet du second est l'influence des événements passés sur la probabilité des événements futurs, et la loi suivant laquelle, en se développant, ils nous font connaître les causes qui les onts produits. [p.383]

These matters "forment une nouvelle branche de la théorie des probabilités" [p.383]. As regards the first point raised in the above quotation, Laplace proposes to give a general method for determining the probability of any event whatsoever, when only the law of possibility ("loi de possibilité") of the simple events is known, and, should that law be unknown, to determine what ought to be done. The second point leads him to the question of births. As a generalization of these investigations, he proposes a method which will lead to the determination not only of the possibilities * of simple events, but also of any future event whatever.

When one comes to consider the determination of the probability of events (following any law) compounded (or composed) of simple events of known possibilities, there are, claims Laplace in his second article, three ways of effecting this:

> 1. *a priori*, lorsque, par la nature même des événements, on voit qu'ils sont possibles dans un rapport donné; ... 2. *a posteriori*, en répétant un grand nombre de fois l'expérience qui peut amener l'événement dont il s'agit, et en examinant combien de fois il est arrivé; 3. enfin, par la considération des motifs qui peuvent nous déterminer à prononcer sur l'existence de cet événement. [pp.384-385]

The differences between these three methods are illustrated by an illuminating example: suppose that the respective skills of two players A and B are unknown. As one has no reason to suppose A more skilful than B, one may conclude that the probability of A's winning a match is $\frac{1}{2}$. The first of the above three methods gives the absolute possibility of the events; the second makes it approximately known, as will be seen in the sequel, and the third gives only their possibility relative to the state of our knowledge.

The relativeness of all probability to us (or to the state of our knowledge) is then emphasized, and Laplace stresses that this does not in fact blur the distinction between absolute and relative possibility.

He next returns to consideration of the problem of the two gamblers, already discussed in §6.3 above[29]. Assuming that $(1+\alpha)/2$ is the probability

*I shall thus translate Laplace's "possibilités".

that the more skilful player wins a game, and that there is no reason to suppose that A is more skilful than B, Laplace shows that the probability[30] that A will win the first n matches is

$$P = \frac{1}{2}\left[\left(\frac{1+\alpha}{2}\right)^n + \left(\frac{1-\alpha}{2}\right)^n\right] .$$

The next three articles are devoted to variations on, and generalizations of, this example[31]: we turn briefly to the continuation given in the sixth article. Supposing that $\alpha \in [0,q]$, and representing the probability of α by $\varphi(\alpha)$, we obtain

$$P = \int_0^q \left\{\left[(1+\alpha)^n + (1-\alpha)^n\right]/2^{n+1}\right\} \varphi(\alpha)\,d\alpha \quad .$$

If, for example, $\varphi(\alpha) = l$ (a constant), "en sorte que toutes les valeurs de α soient également possibles" [p.394], then $\int_0^q \varphi(\alpha)\,d\alpha = 1$ implies $l = 1/q$, and hence

$$P = \left[(1+q)^{n+1} - (1-q)^{n+1}\right]/\left[(n+1)q\,2^{n+1}\right] \quad ,$$

which, it should be noted, reduces, for $q = 1$, to $1/(n+1)$ — the same result as that obtained in the corollary to Bayes's eighth proposition[32].

Commenting, in the twelfth article, on the law of possibility of the skills of the players, Laplace points out that this law is able to be known only because of a long sequence ("suite") of observations, in the absence of which the most likely functions should be chosen — "l'analyse des hasards, qui n'est en elle-même que l'art d'apprécier les vraisemblances, doit donc nous guider dans ce choix" [p.409].

In his fourteenth article Laplace points out that, while one may very well have no reason initially to attribute more ability to any one of the players than to the others, new light is gained as to their respective skills as the matches continue, which skills would be exactly known were the number of games to become infinite. He proposes in this article to consider the effect past events exercise on future events[33].

Denoting by E the past event, by e the future event "dont on propose de calculer la probabilité P" [p.414] (though in fact a conditional rather than an absolute probability is found), and by $E + e$ "un événement composé de l'événement E arrivant le premier et de l'événement e arrivant ensuite" [pp.414-415], Laplace shows that

$$\Pr[e \mid E] \equiv P = \Pr[E+e]/\Pr[E] \quad .$$

From this it is but a short step to the idea of independence and the factorization $\Pr[E+e] = \Pr[E]\Pr[e]$, which in turn leads naturally to the question of the determination of the probability of causes as deduced from events[34].

So much really by way of introduction: it is only now that Laplace starts considering that which is actually our topic. For in his fifteenth article he turns his attention to the matter introduced at the end of the fourteenth (a matter already examined in his *Mémoire sur la probabilité des causes par les événements*). He supposes that an event E can occur in conjunction with one and only one of the n causes $A_1, A_2, \ldots, A_n$ (or $A, A', \ldots, A^{(n-1)}$ in his notation), and deduces the formula (here given in modern notation)

$$\Pr[A_i \mid E] = \Pr[E \mid A_i] \Big/ \sum_{j=1}^{n} \Pr[E \mid A_j], \quad i \in \{1, 2, \ldots, n\} \tag{10}$$

under the following assumptions: for each $i \in \{1, 2, \ldots, n\}$,

(i) $\Pr[A_i] = 1/n$, and

(ii) events $E_1, E_2, \ldots$ similar ("semblable") to the event E in question are conditionally independent with respect to each A_i.

One recognizes in (10) above, of course, a "discrete Bayes's Theorem" with uniform prior. Notice too that we have here a proof of the "Principe" discussed in §6.3 above. (A proof of (10) may be found in Appendix 6.2.) The article is concluded with a verbal expression of the final algebraic result.

In his sixteenth article Laplace proceeds to illustrate the results of the preceding article by what, on his own admission, is a very simple example. Let A and B be two players of unknown skills. It being exceedingly unlikely that these skills are perfectly equal, let $(1+\alpha)/2$ and $(1-\alpha)/2$ denote the greater and the lesser respectively. By Article II the probability that A will win the first two games is $P = \left(1+\alpha^2\right)/4$. If, however, one wants the probability that, B having won the first match, A will win the following two, it is clear that the preceding value of P is too large. In fact, if one considers each skill as a particular cause of the event, the probability that B's skill is $(1+\alpha)/2$ will be, by the preceding article, equal to the probability that B, having this skill, will win the first game, divided by the sum of the probabilities that he will win in having successively the skills $(1+\alpha)/2$ and $(1-\alpha)/2$, a probability that becomes

$$\frac{(1+\alpha)/2}{[(1+\alpha)/2 + (1-\alpha)/2]} = (1+\alpha)/2 \ .$$

In the notation of Article XIV, with E the winning of the first match by B and e the winning of the following two by A, we have $V \equiv \Pr[E] = (1+\alpha)/2$

or $(1-\alpha)/2$ according as the greater or lesser skill is B's. On taking the moiety of the sum of these values (and no further argument for this uniform assumption is given) we find that $V = \frac{1}{2}$. Similarly the probability v of $E+e$ is $[(1-\alpha)/2][(1+\alpha)/2]^2$ or $[(1+\alpha)/2][(1-\alpha)/2]^2$, and hence $v = \left(1-\alpha^2\right)/8$. Thus

$$P = v/V = \left(1-\alpha^2\right)/4 \ .$$

The preceding argument is then generalized to the case of finding the probability P that, B having won the first match, A will win the next n. This is

$$P = \left(1-\alpha^2\right)\left[(1+\alpha)^{n-1} + (1-\alpha)^{n-1}\right]/2^{n+1}$$

which, for small α, reduces approximately to

$$P = \frac{1}{2^n}\left\{1+\alpha^2\left[\frac{(n-1)(n-2)}{1.2} - 1\right]\right\} \ .$$

Thus far the use of the "discrete Bayes's rule". One should note the rôle played by the "equally likely" assumption (though this is often tacit).

Laplace next, in the seventeenth article, turns his attention to the probability of causes as deduced from events. Before doing so, however, he states quite clearly that absence of knowledge entails an equiprobable distribution of the possibilities: in his own words,

> lorsqu'on n'a aucune donnée *a priori* sur la possibilité d'un événement, il faut supposer toutes les possibilités, depuis zéro jusqu'à l'unité, également probables. [p.419]

Todhunter [1865, art.893] regards this as the same as the principle enunciated by Laplace in his *Mémoire sur la probabilité des causes par les événements*; but as we have already seen in our discussion of that work, the equiprobability assumption is at best tacit there, and the present memoir has the first clear statement of this assumption.

Laplace now applies this principle to the problem considered in Article III of his second memoir of 1774, stating it however in terms of births of boys and girls rather than the drawing of white and black tickets from an urn. Of $p+q$ children, p are boys and q girls: what is the probability P that $m+n$ future births will give rise to m boys and n girls? Now the probability that, in $p+q$ births, p will be boys and q girls (the event denoted by E in Article XIV) is

$$\lambda x^p (1-x)^q \ ,$$

where $\lambda = \binom{p+q}{p}$ and x is the probability of the birth of a boy. Similarly, the probability that, of the $p+q$ infants first born, p will be boys and q

girls, and that of the following $m+n$ births, m will be boys and n girls (the compound event denoted by $E+e$ in Article XIV) is

$$\gamma\lambda x^{p+m}(1-x)^{q+n}$$

where $\gamma = \binom{m+n}{m}$. (Note the use of independence and constant probability.)

Laplace now makes use of his equiprobability assumption. To leave no room for doubt, let us consider his exact words:

> maintenant, x étant susceptible de toutes les valeurs depuis $x = 0$ jusqu'à $x = 1$, et toutes ces valeurs étant *a priori* également probables, il faut, pour avoir la véritable probabilité de E, multiplier $\lambda x^p(1-x)^q$ par $a\,dx$, a étant constant, et prendre l'intégrale $\lambda \int ax^p(1-x)^q\,dx$ (depuis $x = 0$ jusqu'à $x = 1$) [p.420]

the value of a being determined from $\int_0^1 a\,dx = 1$, whence $a = 1$. Similarly, the probability of $E+e$ is

$$\lambda\gamma \int_0^1 x^{p+m}(1-x)^{q+n}\,dx \ ,$$

and thus the desired probability P, a probability which is of course conditional on the $p+q$ earlier births, is, by Article XIV,

$$P = \gamma \int_0^1 x^{p+m}(1-x)^{q+n}\,dx \Big/ \int_0^1 x^p(1-x)^q\,dx \ .$$

The remainder of this article is devoted to the evaluation of these integrals. Laplace shows firstly that

$$P = \gamma \frac{(q+1)(q+2)\dots(q+n)(p+1)(p+2)\dots(p+m)}{(p+q+2)(p+q+3)\dots(p+q+m+n+1)} \ . \tag{11}$$

Noting that

$$\log(1.2.3\dots u) = \tfrac{1}{2}\log 2\pi + \left(u+\tfrac{1}{2}\right)\log u - u + \frac{1}{12u} - \frac{1}{360u^2} + \cdots$$

Laplace suggests the use of the approximation[35]

$$1.2.3\dots u = \sqrt{2\pi}\, u^{u+\frac{1}{2}} e^{-u} \ . \tag{12}$$

If one supposes that p and q are "très grands nombres" [p.421] and that, approximately,

$$(p+q+1)/(p+q+m+n+1) = (p+q)/(p+q+m+n) \ ,$$

substitution of (12) in (11) yields

$$P = \gamma \frac{(q+n)^{q+n+\frac{1}{2}}(p+q)^{p+q+\frac{3}{2}}(p+m)^{p+m+\frac{1}{2}}}{p^{p+\frac{1}{2}}q^{q+\frac{1}{2}}(p+q+m+n)^{p+q+m+n+\frac{3}{2}}} \ .$$

Finally, if m and n are very small in comparison with p and q, the approximations

$$(p+m)^{p+m+\frac{1}{2}} \approx e^m p^{p+m+\frac{1}{2}}$$

$$(q+n)^{q+n+\frac{1}{2}} \approx e^n q^{q+n+\frac{1}{2}}$$

$$(p+q+m+n)^{p+q+m+n+\frac{3}{2}} \approx e^{m+n}(p+q)^{p+q+\frac{3}{2}}$$

enable us to write P as

$$P = \gamma \frac{p^m q^n}{(p+q)^{m+n}} \ ,$$

which may perhaps be more suggestively written in the form

$$P = \binom{m+n}{m} \left(\frac{p}{p+q}\right)^m \left(\frac{q}{p+q}\right)^n \ .$$

Laplace begins his eighteenth article by pointing out that the probability P obtained at the end of the preceding one is that which one would reach were one to suppose the possibilities of the births of boys and girls to be in the ratio of p to q, from which it is natural to conclude that these possibilities are (in fact) very nearly in that ratio, the true possibility of the birth of a boy thus being approximately $p/(p+q)$. (One sees here, indeed, the shadow of James Bernoulli, in the approximation of a possibility by an observed frequency.) This "approximately" is to be interpreted in a probabilistic sense — viz. that $p/(p+q)$ and neighbouring values are incomparably more probable than others — again *vide* Bernoulli.

This comment permits the reformulation of the preceding conclusion as follows:

> si l'on désigne par θ une quantité fort petite et par P la probabilité que la possibilité de la naissance d'un garçon est comprise dans les limites $p/(p+q) - \theta$ et $p/(p+q) + \theta$, la valeur de P différera d'autant moins de la certitude ou de l'unité que p et q seront de plus grands nombres, et l'on peut tellement faire croître p et q que la différence de P à l'unité soit moindre qu'aucune grandeur donnée, quelque petit que θ soit d'ailleurs [pp.422-423],

a result to the proof of which the present article is devoted[36].

Noting that this result is only true in the limit ("dans l'infini"), Laplace proposes to consider an approximation to P by a series that is rapidly convergent. From Article XV, the probability P that x (the possibility of the birth of a boy) lies between the two limits β_1 and β_2, say (where $\beta_1 < \beta_2$), is

$$P = \int_{\beta_1}^{\beta_2} x^p(1-x)^q\,dx \bigg/ \int_0^1 x^p(1-x)^q\,dx \quad .$$

The problem thus reduces to the evaluation of the incomplete beta-integral $\int_{\beta_1}^{\beta_2} x^p(1-x)^q\,dx$ when p and q are large. To this end, let $y = x^p(1-x)^q$: then

$$y\,dx = \frac{x(1-x)}{p-(p+q)x}\,dy \quad .$$

Letting $p = 1/\alpha$ and $q = \mu/\alpha$, where α is a very small fraction, we obtain

$$y\,dx = \alpha z\,dy \quad \text{or} \quad z = \frac{1}{\alpha}y\frac{dx}{dy} \quad ,$$

where $z = x(1-x)/[1-(1+\mu)x]$. Integration by parts yields

$$\int y\,dx = C + \alpha yz - \alpha^2 yz\frac{dz}{dx} + \alpha^2\int y\frac{d}{dx}\left(z\frac{dz}{dx}\right)dx + \cdots \tag{13}$$

where C is an arbitrary constant. (This expression is from Todhunter [1865, art.895]: Laplace gives it in the more suggestive form

$$\int y\,dx = C + \alpha yz\left\{1 - \alpha\frac{dz}{dx} + \alpha^2\frac{d(z\,dz)}{dx^2} - \alpha^3\frac{d[z\,d(z\,dz)]}{dx^3} + \cdots\right\} \quad .)$$

Laplace next shows that, for any $t < 1/(1+\mu)$,

$$\alpha yz\left(1 - \alpha\frac{dz}{dx}\right) \ll \int_0^t y\,dx < \alpha yz \quad .$$

> Cette remarque peut servir lorsque, sans chercher la valeur exacte de $\int y\,dx$, on veut s'assurer si elle est plus grande ou plus petite qu'un quantité donnée. [p.425]

Laplace next evaluates $\int y\,dx$ from $x = p/(p+q) - \theta$ to $x = p/(p+q) + \theta$ (or equivalently from $1/(1+\mu) - \theta$ to $1/(1+\mu) + \theta$), which, together with the Stirling-de Moivre approximation for $n!$, yields, on neglecting terms of order $\alpha^{5/2}$,

$$P = 1 - \frac{\sqrt{\alpha\mu}}{\theta\sqrt{2\pi(1+\mu)^3}} \left\{1 - \frac{[12\mu^2 + (1+\mu)^2(1+\mu+\mu^2)\theta^2]}{12\mu(1+\mu)^3\theta^2}\right\} \times$$

$$\left\{[1-(1+\mu)\theta]^{p+1}\left(1+\frac{1+\mu}{\mu}\theta\right)^{q+1} + [1+(1+\mu)\theta]^{p+1}\left(1-\frac{1+\mu}{\mu}\theta\right)^{q+1}\right\}$$

The factor in the last pair of braces being extremely small in the present question, Laplace concludes that

> il est visible que l'on peut tellement augmenter p et q, et, par conséquent, diminuer α, que cette différence de P à l'unité soit moindre qu'aucune grandeur donnée, ce qui est le théorème dont nous avons parlé au commencement de cet article. [p.429]

The results of this article are applied in the following one to the question[37] of the apparent excess of male over female births in Paris from 1745 to 1770. While Laplace states that his concern is to determine "combien il est probable que les naissances des garçons dans cette grande ville sont plus possibles que celles des filles" [p.429], it is in fact perhaps worth noting that what is actually found [p.430] is the probability[38] that the possibility of the birth of a boy is less than or equal to $\frac{1}{2}$. This is achieved by taking $\theta = (1-\mu)/2(1+\mu)$. With $p = 251,527$ and $q = 241,945$, the desired probability (i.e. that x exceeds $\frac{1}{2}$) is seen to differ from 1 by the fraction $1.1521/10^{42}$. Laplace's conclusion is

> on peut regarder comme aussi certain qu'aucune autre vérité morale, que la différence observée à Paris entre les naissances des garçons et celles des filles est due à une plus grande possibilité dans la naissance des garçons. [pp.431-432]

This is followed by some comments on births in London, it being noted that the ratio of births of boys to those of girls here is greater than the ratio in Paris[39].

In Article XX Laplace proposes, using the data of the preceding article, to determine the probability that, in any given year, the number of births of boys does not surpass that of girls. Supposing that of $2a$ births (the mean number in a given year) m are male, Laplace obtains from formula (11) above

$$P = \frac{(2a)!}{(p+q+2a+1)!} \; \frac{(p+q+1)!}{p!\,q!} \; \frac{(q+2a-m)!}{(2a-m)!} \; \frac{(p+m)!}{m!} \; ,$$

the "sum" of which, taken over all values of m, yields the desired result.

Denoting by y_m the expression $(q+2a-m)!\,(p+m)!\,/(2a-m)!\,m!$, Laplace shows that

$$y_m = \frac{(m+1)(q+2a-m)}{(2a-m)(p+m+1)}\, y_{m+1} \ .$$

More generally, he is led to consideration of the finite difference equation

$$y_m = z_m \,\Delta y_m \ ,$$

whence he deduces

$$\sum y_m = C + y_m z_{m-1} \{1 - \Delta z_{m-2} + \Delta\,(z_{m-2}\,\Delta z_{m-3})$$
$$-\Delta\,[z_{m-2}\,\Delta\,(z_{m-3}\,\Delta z_{m-4})] + \cdots\}$$

analogous to the expression (13) above. As in the discussion of that expression, two approximate bounds for the exact solution are obtained, and the results are applied to birth data from Paris (it being shown that the probability that the number of male births does not exceed that of female in one year is less than 1/259) and London, the probability of the event concerned being even smaller here. A similar, but more difficult, problem is treated in the *Théorie analytique des probabilités* (see Todhunter [1865, art.897]).

In the twenty-first article Laplace points out that the preceding theory required the knowledge of the number of times each simple event had happened. It is thus but a particular case of that part of the analysis of chance ("des hasards") which consists in going back from events to causes; and in subsequent articles he proposes to consider

> une méthode générale pour déterminer les possibilités des événements simples, quel que soit l'événement composé dont on a observé l'existence. [p.439]

The present article is really preparatory to the satisfaction of this avowed aim, and in it Laplace examines in more detail the problem, already discussed in Article III, of the determination of the skills of two gamblers.

In the twenty-second article Laplace considers a direct and general method for determining the possibilities of simple events, irrespective of the event observed. Denoting by x and $1-x$ the desired possibilities of simple events, he notes that the probability of the compound event in question will be a function of x multiplied by some coefficient. Calling this function y and denoting by a the value of x, positive and less than one, which maximizes it, Laplace notes that not only is this value the most probable, but it is also the limiting value of the true possibility x. This claim is illustrated by several examples [pp.441-442].

Laplace next points out that the integral $\int y\,dx$, taken over a very small interval about the maximum, is then very close to the same integral evaluated between 0 and 1:

> or le rapport de la première de ces intégrales à la seconde exprime la probabilité que la valeur de x est comprise dans cet intervalle. [p.442]

He concludes by mentioning (and adduces an example in support of this assertion) that compound events are not at all suitable for determining the possibilities of simple events.

The need for the evaluation of the ratio of the integrals of this article having been considered in Article XVIII (cf. also the start of Article XXV), Laplace proposes in Article XXIII to generalize these results and to extend them to all values of y. This leads to an equation of the form

$$y\,dx = \alpha z\,dy \quad ,$$

z being a function of x which contains no powers of order $1/\alpha$ at all.

While the methods of Article XVIII (aided by the "beau théorème de M. Stirling sur la valeur du produit $1.2.3\ldots u$, lorsque u est un très grand nombre" [p.445]) may be used, Laplace's search for a more direct method leads eventually to the evaluation[40] of $\int_0^\infty e^{-t^2}\,dt$. This he does by considering

$$\int_0^\infty \int_0^\infty e^{-s(1+u^2)}\,du\,ds \text{ and } \int_0^\infty \int_0^\infty e^{-s(1+u^2)}\,ds\,du$$

and equating the results. This leads, for very small α, to

$$\left(\int_0^1 y\,dx\right)^2 = 2\pi y^3 \Big/ -\frac{d^2y}{dx^2} \quad ,$$

the right-hand side being evaluated at $x = a$ (the value of x corresponding to the maximum of y). The problem is repeated in the *Théorie analytique des probabilités* (see Todhunter [1865, art.899]).

This question is further pursued in Article XXIV: certain errors in the formulae of this article are exposed in Todhunter [1865, art.900].

In Article XXV the methods of Articles XXIII & XIV are used to derive an approximate expression for $\int_0^1 x^p(1-x)^q\,dx$. Laplace's method ("si je ne me trompe" [p.456]) is stated to be "more direct" — independently of its generality — than those of Stirling and Euler, and this is illustrated by consideration of $y = x^p e^{-x}$, whence

$$p! = p^{p+\frac{1}{2}} e^{-p} \sqrt{2\pi} \left(1 + \frac{1}{12p} + \cdots \right) .$$

Laplace begins his next article[41] by reminding us that we saw, in Article XIX, that the ratio of births of boys to girls is sensibly greater in London than in Paris; an observation which seems to indicate a greater facility in London for the birth of boys. He asserts further that the preceding method, more easily than any other, will permit the ascertaining of how probable this difference is. Defining[42]

u —	the probability of the birth of a boy in Paris;
p —	the number of births of boys observed in that city;
q —	the number of births of girls observed in that city;
$u - x$ —	the possibility of the birth of a boy in London;
p' —	the number of births of boys observed there;
q' —	the number of births of girls observed there,

we find that the probability of the double event is proportional to

$$u^p (1-u)^q (u-x)^{p'} (1-u+x)^{q'} .$$

Thus the probability P that the birth of a boy is less possible in London than in Paris is given by

$$P = \frac{\int\int u^p (1-u)^q (u-x)^{p'} (1-u+x)^{q'} \, dx \, du}{\int\int u^p (1-u)^q (u-x)^{p'} (1-u+x)^{q'} \, dx \, du} .$$

The integration in the denominator is over all values of u and x, while that in the numerator is over $x = 0$ to u and $u = 0$ to 1 (Laplace's limits are wrong here). The rest of the article is devoted to the evaluation of the double integrals[43]: the approximate solution for the data gathered is $P = 1/410,458$, and the final conclusion[44] is

> ainsi l'on peut regarder comme une chose très probable qu'il existe, dans la première de ces deux villes, une cause de plus que dans la seconde, qui y facilite les naissances des garçons, et qui dépend soit du climat, soit de la nourriture et des mœurs. [p.466]

In the next article Laplace extends the theory of the preceding articles to a larger number of simple events, the theory being illustrated by an (infinite) urn problem (with balls of three different colours). Once again it is stated that the value of x which maximizes a certain integral is consequently the most probable value of x.

Consideration thus far has been limited to the case of a uniform prior: as Laplace writes at the beginning of his twenty-eighth article,

> jusqu'ici nous avons supposé la loi de possibilité des événements simples constante depuis zéro jusqu'à l'unité, et cette supposition est, comme nous l'avons observé dans l'article XVII, la seule que l'on doive adopter, lorsqu'on n'a aucune donnée relativement à ces possibilités. [p.469]

Here he proposes to consider the case in which the law (i.e. the prior) is known exactly. Limiting himself to the case of only two simple events of possibilities x and $1 - x$, Laplace deduces from Article XV that the probability P that the value of x lies between θ_1 and θ_2 (say) is

$$P = \int_{\theta_1}^{\theta_2} usy\, dx \Big/ \int_0^1 usy\, dx \quad ,$$

where u $(= u(x))$ denotes the facility of the possibility x of the first event, s denotes the facility[45] of the possibility $1 - x$ of the second event, and y is the probability of the observed event.

In his twenty-ninth article Laplace turns his attention to the question of the determination of a future event as determined by known events. Denoting by x and $1-x$ the possibilities of two simple events and by s and s' the facilities of x and $1-x$ respectively, one can calculate the probabilities, both of the observed event and of the future event, proceeding from these probabilities, a procedure which yields two functions of x, say y and u respectively. By Articles XIV & XV the desired probability P is then given by

$$P = \int_0^1 s\, s'\, uy\, dx \Big/ \int_0^1 s\, s'\, y\, dx \quad .$$

If the event is very complicated, the method of Article XXIII may be used to evaluate these integrals by a very rapidly convergent approximation.

Particular attention is paid to the case in which one has no information about the law of possibility of the two simple events, in which case[46] one must suppose $s = s' = 1$, and Laplace also points out that his approximation ceases to be exact if the future event concerned is itself very complicated.

The investigations of this article lead to the following "théorème assez remarquable":

> la probabilité d'un événement futur, pareil à celui que l'on a observé, est à cette même probabilité, déterminée en employant pour les possibilités des événements simples celles qui résultent de l'événement observé, comme 1 est à $\sqrt{2}$. [p.475]

As an illustration Laplace refers to the example already considered in Article XVII — i.e. given $p+q$ births (p boys), the probability P that $p+q$ future births will result in p boys and q girls is

$$P = \binom{p+q}{p} \left(\frac{p}{p+q}\right)^p \left(\frac{q}{p+q}\right)^q \frac{1}{\sqrt{2}} \; . \tag{14}$$

It is worth noting that the expression (14) given above for P is not obtainable from that of Article XVII simply by replacing m and n in that article by p and q respectively. Laplace's concern here seems to be with the repetition of the *compound* event of $p+q$ births. This observation is supported by his comment on the case of n repetitions, viz.

> si l'on cherche la probabilité P que l'événement observé sera suivi d'un nombre n d'événements pareils, on aura $u = y^n$, et l'on trouvera $P = v^n/\sqrt{n+1}$, v étant ce que devient y, lorsqu'on y substitue pour x la valeur a qui rend y un maximum, et cette équation a également lieu, n étant fractionnaire. [p.475]

But this technique is not to be regarded as universally applicable, and Laplace does in fact sound a warning:

> on s'exposerait donc alors à des erreurs considérables, en employant, dans le calcul de la probabilité des événements futurs, les possibilités des événements simples qui résultent de l'événement observé: en effet, il est visible que la petite erreur que l'on peut commettre, en faisant usage de ces possibilités, s'accumule en raison du nombre des événements simples qui entrent dans l'événement futur, et doit occasionner une erreur sensible lorsqu'ils y sont en très grand nombre. [p.475]

The rest of the memoir is devoted to a discussion of the theory of errors[47], most of which discussion is not of direct concern to us. However, since we have already looked at something on this topic in our examination of the 1774 *Mémoire sur la probabilité des causes par les événements*, and since we shall have occasion to consider something similar in the memoir to be discussed in §6.10 below, it seems wise to look at the topic now[48].

As one of the most useful problems in this part of the analysis of chance ("hasards"), which consists in going back from events to the causes which produce them, Laplace cites, in his Article XXX, the determination of the mean of the results of several observations. Having referred to his 1774 memoir and related work by Lagrange, Daniel Bernoulli and Euler, Laplace states that he proposes here to resume this matter and to present his results in such a way as to leave no doubt as to their precision.

In his memoir of 1774 Laplace had assumed that the errors were identically distributed: now, although still retaining the even distribution of the errors, he supposes that the facilities of the errors for the first, second, ... observer are $\varphi(x), \varphi'(x), \ldots$ respectively. Although, as Sheynin [1977, p.8] has pointed out, "the condition of asymptotic decline of the density is now omitted", this condition is reintroduced in Article XXXII. Supposing that the errors of the first observation ("celle qui fixe le plus tôt le phénomène" [p.476]), the second, third, ... are $x, p-x, p'-x, \ldots$, Laplace arrives at the density

$$y = \varphi(x)\,\varphi'(p-x)\,\varphi''(p'-x)\ldots$$

By Article XV the probabilities of the different values of x are to each other "comme les probabilités que, ces valeurs ayant lieu, les observations s'écarteront entre elles des quantités observées $p, p', p'', \ldots$" [p.477]. Thus the ordinates of y are proportional to the probabilities of the corresponding abscissae x, "et par cette raison nous la nommerons *courbe des probabilités*" [p.477].

Laplace next points out that by "milieu" or "résultat moyen" of any number of observations one may intend an infinity of different things, according as one subjects the result to some or other condition.

> Par exemple, on peut exiger que ce milieu soit tel que la somme des erreurs à craindre en *plus* soit égale à la somme des erreurs à craindre en *moins*; on peut exiger que la somme des erreurs à craindre en plus, multipliées par leurs probabilités respectives, soit égale à la somme des erreurs à craindre en moins, multipliées par leurs probabilités respectives. Ou peut encore assujettir ce milieu à être le point où il est le plus probable que doit tomber le véritable instant du phénomène, comme M. Daniel Bernoulli l'a fait. [p.477]

Following Sheynin [1977, pp.8-9] we formulate these conditions as follows:

$$\text{(i)}\ \int_{-N}^{0} y\,dx = \int_{0}^{N} y\,dx \left(\text{or} \sum_{\{x:x>N\}} (x-N) = \sum_{\{x:x<N\}} (N-x)\right);$$

$$\text{(ii)}\ \int_{-N}^{0} xy\,dx = \int_{0}^{N} xy\,dx \left(\text{or} \sum_{\{x:x>N\}} x\ \Pr[x] = \sum_{\{x:x<N\}} x\ \Pr[x]\right);$$

(iii) maximum likelihood,

where N is the maximum possible error.

In general, while one may impose an infinity of (other) similar conditions, each of which will give a different mean, they are not all arbitrary. There is one which obtains by the nature of the problem and which serves to fix the mean that it is necessary to choose between several observations:

> cette condition est que, en fixant à ce point l'instant du phénomène, l'erreur qui en résulte soit un minimum; or comme, dans la théorie ordinaire des hasards, on évalue l'avantage en faisant une somme des produits de chaque avantage à espérer, multiplié par la probabilité de l'obtenir, de même ici l'erreur doit s'estimer par la somme des produits de chaque erreur à craindre, multipliée par sa probabilité; le milieu qu'il faut choisir doit donc être tel que la somme de ces produits soit moindre que pour tout autre instant. [pp.477-478]

This may be represented symbolically as

(iv) $\int xy\,dx =$ minimum,

the integration being over all possible values of x.

Taking, then, the *courbe des probabilités* to be

$$y = \varphi(x)\,\varphi'(p-x)\,\varphi''(p'-x)\ldots$$

where $x \in [-f, c-f]$, Laplace first makes the substitution $x = z - f$, so that $z \in [0, c]$ — though, as Sheynin [1977, p.8] points out, he does not drop the assumption of evenness, so that in fact $\mid z \mid \leq c$. It is next noted that the probabilities of the different values of z are proportional to y or $\varphi(z-f)\,\varphi'(p-z+f)\ldots$, the proportionality constant being denoted by k. If h is "la valeur de z que l'on doit prendre pour le véritable instant du phénomène" [p.478], then the last condition mentioned above requires the minimization of

$$k\int_0^h (h-z)y\,dz + k\int_h^c (z-h)y\,dz \quad .$$

Differentiation with respect to h yields

$$\int_0^h y\,dz = \int_h^c y\,dz \quad .$$

The ordinate corresponding to this value of h, which determines the mean to be chosen, thus divides the area under the *courbe des probabilités*, between $z = 0$ and $z = c$, into two equal parts. The result, Laplace notes, is the "milieu de probabilité" [p.479] (see condition (i))[49].

In Article XXXI Laplace discusses the difference between the cases in which the laws of facility of the errors of observation are known and those in which they are unknown. In the former case it follows from the preceding article that the question of the determination of the mean of several observations reduces to the division of a given surface into two equal parts (a problem in pure Analysis). However, when the laws of facility are unknown, it is the calculus of probabilities that is needed to supply this ignorance. In this case we know from Article XIII that if $\pm a, \pm a^{'}, \pm a^{''}, \ldots$ are the limits of the error of the first, second, third, ... observations, one must suppose

$$\varphi(x) = \frac{1}{2a} \log \frac{a}{x}, \quad \varphi^{'}(x) = \frac{1}{2a^{'}} \log \frac{a^{'}}{x}, \ldots$$

(In his thirteenth article Laplace explains this choice in the following way:

> il est naturel de penser que les mêmes erreurs, en plus et en moins, sont également probables et que leur facilité est d'autant moindre qu'elles sont plus grandes; si l'on n'a aucune autre donnée, relativement à leur facilité, on retombe évidemment dans le cas du problème précédent: il faut donc supposer alors la possibilité, tant de l'erreur positive x, que de l'erreur négative $-x$, égale à $(1/2a) \log (a/x)$; et c'est cette loi de possibilité dont il faut partir, dans la recherche du milieu que l'on doit choisir entre les résultats de plusiers observations. [p.413])

Once again, only the inevitable difficulties of Analysis remain, though one must admit, says Laplace, that they make the preceding method very difficult to use.

His object here, Laplace states, has been rather to make known what light the analysis of chance ("hasards") can shed on this matter, than to present to observers a method both practical and easy to use — a method which, however, can be used on very delicate occasions.

Laplace starts his thirty-second article by pointing out "comme il est facile de s'en assurer" [p.480], that the ordinary rule for the arithmetic mean arises from this method when $a = a^{'} = a^{''} = \cdots = \infty$. He proposes here, however, to give a much more general theorem, showing that this rule always results under the following assumptions:

> 1° que la loi de facilité des erreurs est la même pour toutes les observations;

> 2° que les mêmes erreurs, soit en *plus*, soit en *moins*, sont également possibles;
> 3° qu'elles peuvent être infinies et que la fonction qui exprime leurs facilités ne décroît d'une quantité finie que lorsque x est infini, mais qu'alors elle va toujours en diminuant jusqu'au point de devenir nulle. [p.480]

Denoting by $\varphi(\alpha x)$ the law of facility of the errors of observation, α being infinitely small, and by q the value of $\varphi(\alpha x)$ when $\alpha x = 0$ (and as a result, whenever x is finite), one sees that the ordinate of the *courbe des probabilités* from $-x = 0$ to $-x = \infty$ is

$$y = \varphi(\alpha x)\,\varphi(\alpha p + \alpha x)\,\varphi(\alpha p' + \alpha x)\ldots$$

(Note that Laplace is once again assuming that all priors are the same.) If we suppose that there are n observations and if we ignore terms of order α^2, this last expression becomes[50]

$$y = [\varphi(\alpha x)]^n + \alpha\left(p + p' + \cdots + p^{(n-1)}\right)[\varphi(\alpha x)]^{n-1}\frac{d}{d(\alpha x)}\varphi(\alpha x) \ . \quad (15)$$

Laplace's integration I find rather confusing: his answer, however, is correct, as the following argument shows.

Firstly, from (15), and since φ is even, we have

$$\begin{aligned}
\int_{-\infty}^{0} y\,dx &= \int_{-\infty}^{0}\left\{[\varphi(\alpha x)]^n + \alpha\left(p + p' + \cdots\right)[\varphi(\alpha x)]^{n-1}\frac{d}{d(\alpha x)}\varphi(\alpha x)\right\}dx \\
&= \int_{0}^{\infty}[\varphi(-\alpha x)]^n\,dx + \left(p + p' + \cdots\right)\int_{0}^{\infty}\alpha[\varphi(-\alpha x)]^{n-1}\frac{d}{d(-\alpha x)}\varphi(-\alpha x)\,dx \\
&= \int_{0}^{\infty}[\varphi(\alpha x)]^n\,dx + \left(p + p' + \cdots\right)\int_{0}^{\infty}\alpha[\varphi(-\alpha x)]^{n-1}\frac{d}{d(-\alpha x)}\varphi(-\alpha x)\,dx \\
&= A - \left(p + p' + \cdots\right)\left.\frac{[\varphi(-\alpha x)]^n}{n}\right|_{x=0}^{\infty} \\
&= A - \left(p + p' + \cdots\right)\left.\frac{[\varphi(\alpha x)]^n}{n}\right|_{x=0}^{\infty} \\
&= A - \left(p + p' + \cdots\right)q^n/n \ ,
\end{aligned}$$

since $\varphi(\alpha x) = q$ when $x = 0$ and $\varphi(\alpha x) = 0$ for $x = \infty$. Consider next the interval $\left[0, p^{(n-1)}\right]$. Recalling the definition of q, one sees that one may suppose here that $\varphi(\alpha x) = \varphi(\alpha p - \alpha x) = \cdots = q$. Thus the ordinate y is just q^n, and

$$\int_0^{p^{(n-1)}} y\,dx = p^{(n-1)}q^n \quad .$$

Finally, for $x \in \left[p^{(n-1)}, \infty\right)$, one has

$$\begin{aligned} y &= \varphi(\alpha x)\,\varphi(\alpha x - \alpha p)\,\varphi(\alpha x - \alpha p')\ldots \\ &\approx [\varphi(\alpha x)]^n - \alpha\left(p + p' + \cdots\right)[\varphi(\alpha x)]^{n-1}\frac{d}{d(\alpha x)}\varphi(\alpha x) \quad . \end{aligned}$$

Now

$$\begin{aligned} \int_{p^{(n-1)}}^{\infty} [\varphi(\alpha x)]^n\,dx &= \int_0^{\infty}[\varphi(\alpha x)]^n\,dx - \int_0^{p^{(n-1)}}[\varphi(\alpha x)]^n\,dx \\ &= A - p^{(n-1)}q^n \end{aligned}$$

and

$$\int_{p^{(n-1)}}^{\infty} \alpha[\varphi(\alpha x)]^{n-1}\frac{d}{d(\alpha x)}\varphi(\alpha x)\,dx = \left.\frac{[\varphi(\alpha x)]^n}{n}\right|_{x=p^{(n-1)}}^{\infty} = \frac{-1}{n}q^n \quad .$$

Thus

$$\int_{p^{(n-1)}}^{\infty} y\,dx = A - p^{(n-1)}q^n + \frac{1}{n}\left(p + p' + \cdots + p^{(n-1)}\right)q^n \quad .$$

Hence the entire area under the *courbe des probabilités* is

$$\begin{aligned} A - \frac{1}{n}\left(p + p' + \cdots\right)q^n + p^{(n-1)}q^n + A - p^{(n-1)}q^n + \frac{1}{n}\left(p + p' + \cdots\right)q^n \\ = 2A \quad . \end{aligned}$$

If we now denote by h the abscissa whose ordinate divides this area into two equal parts, the part of the area which is to the left of this ordinate is clearly

$$A - \frac{1}{n}\left(p + p' + \cdots\right)q^n + h\,q^n$$

(because $h \in \left[0, p^{(n-1)}\right]$), and on setting this equal to A, we get

$$h = \frac{1}{n}\left(p + p' + \cdots + p^{(n-1)}\right) \quad ,$$

which yields the same value for h as "la règle des milieux arithmétiques" [p.481].

> Les suppositions qui nous ont conduit à ce résultat étant hors de toute vraisemblance, on voit combien il est nécessaire, dans les occasions délicates, de faire usage de la méthode que nous avons proposée. [pp.481-482]

In the final article of this memoir, Laplace considers the following problem: suppose that in repeated checks of an instrument one has found n different errors $p, p', p'', \ldots$ which are repeated $i, j, k, \ldots$ times respectively and which have respective facilities $x_1, x_2, x_3, \ldots$. The probability of the system of facilities will then be

$$x_1^i x_2^j x_3^k \ldots dx_1\, dx_2\, dx_3 \ldots \bigg/ \int^n x_1^i x_2^j x_3^k \ldots dx_1\, dx_2\, dx_3 \ldots \ ,$$

the integral being taken over all possible values of $x_1, x_2, x_3, \ldots$. Repeated integration shows that the probability that the facility x_1 lies between given limits θ_1 and θ_2 (say) is

$$\int_{\theta_1}^{\theta_2} x_1^i (1 - x_1)^{j+k+\cdots}\, dx_1 \bigg/ \int_0^1 x_1^i (1 - x_1)^{j+k+\cdots}\, dx_1 \quad ,$$

an expression we have already seen in Article XVIII. A similar example is adduced (see p.483), from which a simple rule, based on this result, follows for the correction of the instrument [p.484].

This concludes the memoir, one which "deserves to be regarded as very important in the history of the subject" (Todhunter [1865, art.905]). While the methods of approximation of definite integrals derived here are certainly important, the memoir is perhaps more noteworthy from the viewpoint of our present work for its applications of Bayes's and related results.

6.6 Sur les approximations des formules (suite)

This "Mémoire sur les approximations des formules qui sont fonctions de très grands nombres (suite)", published in 1786 in the volume for 1783 of the *Mémoires de l'Académie royale des Sciences de Paris*, pp.423-467, is a continuation (Article IV, *Application de l'analyse précedente à la théorie des hasards*, in fact) of an earlier memoir of the same title published in 1785 in the volume of the *Mémoires* for 1782, pp.1-88: the numbering here is a continuation of that of this earlier memoir, a summary of which is presented in Appendix 6.3 to this chapter.

In Number XXXII, the first section of the memoir, Laplace repeats certain elementary definitions and probabilistic notions which he had already stated in earlier writings. Here he gives a precise definition of the term "chance", viz[51].

> le mot *hasard* n'exprime donc que notre ignorance sur les causes des phénomènes que nous voyons arriver et se succéder sans aucun ordre apparent. [p.296]

Once again he repeats that "la probabilité est relative en partie à cette ignorance, en partie à nos connaissances" [p.296], and iterates the scope of the theory of chances, viz.

> la théorie des hasards consiste donc à réduire tous les événements qui peuvent avoir lieu relativement à un objet, dans un certain nombre de cas également possibles, c'est-à-dire tels que nous soyons également indécis sur leur existence, et à déterminer le nombre des cas favorables à l'événement dont on cherche la probabilité. [p.296]

Of the number of favourable cases he says further that "le rapport de ce nombre à celui de tous les cas possibles est la mesure de cette probabilité" [p.296], though I doubt whether he in fact finds a distinction between probability and the *measure* of probability.

Much of this article is concerned with the influence of past events on the probability of future events (a topic introduced in this initial number) and in this respect the memoir is irrelevant to our study of "inverse inference" — apart, of course, from any pertinent detail on the rule of succession.

In a short second number (XXXIII) Laplace repeats a formula given in his *Mémoire sur les probabilités*, viz. $\Pr[e \mid E] = \Pr[e + E]/\Pr[E]$, a formula which, he stresses, is basic to the whole theory of the probability of causes and of future events.

In Number XXXIV, under the assumption that each of n causes $e, e^{(1)}, \ldots, e^{(n-1)}$ has (prior) probability $1/n$, and denoting by $a, a^{(1)}, \ldots, a^{(n-1)}$ the (posterior) probabilities of an event E given these causes, Laplace deduces from the formula of the preceding number that[52]

$$p^{(r)} \equiv \Pr\left[e^{(r)} \mid E\right] = a^{(r)} \Bigg/ \sum_{i=0}^{n-1} a^{(i)} \qquad \left(a^{(0)} \equiv a\right) ,$$

a result which we recognize as a discrete Bayes's formula.

This result is then applied to a "sampling with replacement" scheme, an application which is worth noting since in it Laplace seems to assign equal probabilities *a priori* to combinations rather than permutations. An urn contains three balls which are white or black: m drawings (with replacement) result in m whites (event E, say). Denoting by $e, e^{(1)}, e^{(2)}, e^{(3)}$ the following four hypotheses respectively

All three balls are white,
Two balls are white and one is black,
One ball is white and two are black,
All three balls are black,

Laplace says that the probabilities of E conditional on each of these hypotheses are $1, (2/3)^m, (1/3)^m$ and 0 respectively. Thus the posterior probabilities are

$$\frac{3^m}{3^m+2^m+1}, \frac{2^m}{3^m+2^m+1}, \frac{1}{3^m+2^m+1}, 0$$

respectively.

It seems that Laplace is here considering the ordered triples (W, W, B), (W, B, W), (B, W, W) as indistinguishable, a situation which we may view as analogous to one in which we are presented with four indistinguishable urns, one of each of the four possible compositions $\{W, W, W\}$, $\{W, W, B\}$, $\{W, B, B\}$, $\{B, B, B\}$, from which one is chosen (at random) for sampling, rather than to one in which three balls are drawn "at random" from a very large population of equal numbers of black and white balls, which chosen three are then placed in an urn for further sampling (in this latter case the probabilities of the four different compositions possible would be $\frac{1}{8}, \frac{3}{8}, \frac{3}{8}, \frac{1}{8}$). The attribution of equal probabilities to combinations (and also to permutations) was suggested by W.E. Johnson in 1924 and fruitfully exploited in his theory of eduction.

Laplace begins his Number XXXV with the following words:

> la possibilité de la plupart des événements simples est inconnue et, considérée *a priori*, elle nous paraît également susceptible de toutes les valeurs depuis zéro jusqu'à l'unité; mais, si l'on a observé un résultat composé de plusiers de ces événements, la manière dont ils y entrent rend quelques-unes de ces valeurs plus probables que les autres. [p.302]

Expanding on this latter point, he denotes by x the possibility of a simple event, and by y [$\equiv y(x)$] the probability (obtained from "la théorie connue des hasards" [p.302]) of an observed result. It then follows, he asserts, from Number XXXIV that the probability of x will be

> égale à une fraction dont le numérateur est y et dont le dénominateur est la somme de toutes les valeurs de y. [p.302]

Multiplication by dx and appropriate integration then show that the probability that x lies between θ and θ' is

$$\int_{\theta}^{\theta'} y\,dx \Bigg/ \int_0^1 y\,dx \quad . \tag{16}$$

The final paragraph of this Number is important in that it indicates Laplace's reason for concentrating on equally probable causes. Suppose that the different values of x ("considérées indépendamment du résultat observé" [p.303]) are not equally possible, but that their probability can be expressed by $z = z(x)$. Laplace suggests that one then replace y by yz in the preceding formula, which amounts to supposing all the values of x equally possible and to considering the result observed as being formed of two independent results of probabilities y and z.

> On peut donc ramener de cette manière tous les cas à celui où l'on suppose une égale possibilité aux différentes valeurs de x et, par cette raison, nous adopterons cette hypothèse dans les recherches suivantes. [p.303]

Note that in this case Laplace is still finding $\Pr[\theta < x < \theta']$ and not $\Pr[\theta < z(x) < \theta']$. Of course, the usual problem that arises with the non-uniform prior is that one does not know what it is, and this difficulty is not solved by Laplace's proposal.

In Number XXXVI Laplace considers the evaluation of (16) — or more specifically, $\Pr[x \leq \theta]$, where θ is any number less than a (the most probable value of x, or that which maximizes y). This evaluation is accomplished by series expansion, two different results being given depending on the proximity of θ to a.

In his next number Laplace continues the preceding investigation, finding[53] $\Pr[a - \theta < x < a + \theta']$. This probability is shown to be

$$\frac{1}{\sqrt{\pi}} \int_0^{\infty} e^{-t^2}\, dt$$

when θ and θ' are very small, and is given more generally by

$$1 - \frac{\alpha^{\lambda/2}}{\sqrt{\pi}} e^{-1/\alpha^{\lambda}} + \cdots$$

when $\log y$ is of order $1/\alpha$ and $0 < \lambda < 1$ — in fact, $\sqrt{\log Y - \log J} \approx \alpha^{-\lambda/2}$, where $Y = y(a)$ and $J = y(a - \theta) = y(a + \theta')$. This leads to the following theorem:

> la probabilité que la possibilité des événements simples est comprise entre des limites qui se resserrent de plus en plus approche sans cesse de l'unité, de manière que, dans la supposition d'un nombre infini d'événements simples, ces deux limites venant à se réunir, et la probabilité se confondant avec la certitude, la véritable possibilité des événements simples est exactement égale à celle qui rend le résultat observé le plus probable. [pp.307-308]

Laplace stresses the two approximations found here (one relative to the limits which contain the value of x and which contract, and the other relative to the probability that x is found between these limits, a probability which approaches unity or certainty) and points out that these approximations differ from the ordinary ones, "dans lesquelles on est toujours assuré que le résultat est compris dans les limites qu'on lui assigne" [p.308].

Number XXXVIII is devoted to what is essentially a generalization of the problem of Number XXXVI above, leading to a sort of double Bayes's integral. Thus considering the question of two events, each composed of a large number of simple events of the same type, occuring (independently) in two different places, Laplace denotes by

x — the possibility of the simple event in the first place;
y — the function of x expressing the probability of the observed result in that place;
a — the value of x corresponding to the maximum of y;
x' — the possibility of the simple event in the second place;
y' — the function of x' expressing the probability of the observed result in that place;
a' — the value of x' corresponding to the maximum of y'.

Denoting further by P the probability that the possibility of the simple event is greater in the first place than in the second, Laplace claims that analogously to the discussion of Number XXXV,

$$P = \int_0^1 \int_0^x y\, y'\, dx'\, dx \Big/ \int_0^1 \int_0^1 y\, y'\, dx'\, dx \quad .$$

One might well rewrite this as

$$\begin{aligned} P &\equiv \Pr\left[x > x'\right] \\ &= \sum_y \Pr\left[x' < y \mid x = y\right] \Pr\left[x = y\right] \quad \text{or} \quad \int \Pr\left[x' < y \mid x = y\right] f_Y(y)\, dy \quad . \end{aligned}$$

Compare also the ratio of the integrals given in Article XXVI of Laplace's *Mémoire sur les probabilités*. The rest of this Number is devoted to approximations to these integrals.

As an application of the preceding result, Laplace addresses himself, in Number XXXIX, once again to the question of births (cf. Article XXVI of his *Mémoire sur les probabilités*). He begins by deriving an expression for the probability that the possibility x of the birth of a boy does not exceed any given θ ($p+q$ births having been observed, with p much greater than q), viz.

$$\Pr[x \le \theta]$$

$$= \frac{\theta^{p+1}(1-\theta)^{q+1}(p+q)^{p+q+3/2}}{\sqrt{2\pi}\,[p-(p+q)\theta]\,p^{p+1/2}q^{q+1/2}}\left\{1-\frac{(p+q)\theta^2+p(1-2\theta)}{[p-(p+q)\theta]^2}+\cdots\right\}.$$

Putting $\theta = \frac{1}{2}$ we get the probability that the possibility of the birth of a boy is less than that of a girl, viz.

$$\frac{(p+q)^{p+q+3/2}}{(p-q)2^{p+q+3/2}p^{p+1/2}q^{q+1/2}\sqrt{\pi}}\left[1-\frac{p+q}{(p-q)^2}+\cdots\right].$$

As an example Laplace considers the births in London, Paris and the Kingdom of Naples (excluding Sicily), and he determines the respective probabilities numerically. Todhunter [1865, art.909] regards the present exposition as a "much better investigation" than that presented in the *Mémoire sur les probabilités*.

Having observed that the ratios of births of boys to girls in London and Paris are 19:18 and 26:25 respectively, Laplace proposes in Number XL to determine with what likelihood the observations indicate that the possibility of the birth of a boy in the former city is greater than in the latter: this is a particular case of the theory of Number XXXVIII, with y (in Paris) being given by

$$y = \binom{p+q}{p} x^p (1-x)^q$$

and y' (in London) by

$$y' = \binom{p'+q'}{p'} x'^{p'} (1-x')^{q'}.$$

He finds that there is a more than 400,000 to 1 chance that there is a cause in London besides that ("de plus qu'a") in Paris facilitating the births of boys. A similar comparison is effected between Paris and the Kingdom of Naples, the probability that the possibility of the birth of a boy in the former is greater than in the latter being about 1/100.

In his Number XLI Laplace turns his attention to the question of the probability of future events, estimated ("prise") from past events, supposing that, having observed a result composed of any number whatsoever of simple events, one wishes to determine the probability of a future result composed of the same events. Denoting by x the possibility of the simple events, by y the corresponding probability of the observed result, and by z that of the future result (y and z both being functions of x), Laplace deduces from Number XXXIV that the probability P of the future event, given ("prise du") the observed result, is

$$P = \int_0^1 yz\,dx \Big/ \int_0^1 y\,dx \quad .$$

As an illustration Laplace considers the case of an urn containing an infinite number of white and black balls, from which one white ball has been drawn. What is the probability P that the next ball drawn will also be white? If one denotes by x the ratio of white balls in the urn to the total number of balls, "il est clair que x sera la probabilité, tant de l'événement observé que de l'événement futur" [p.326], and one has

$$P = \int_0^1 x^2\,dx \Big/ \int_0^1 x\,dx = \tfrac{2}{3} \quad .$$

Next Laplace considers the case of drawing one white ball, followed (in the future) by a sequence of n black balls. In this case[54]

$$P = \int_0^1 x(1-x)^n\,dx \Big/ \int_0^1 x\,dx = 2/(n+1)(n+2) \quad .$$

If, however, white and black balls are [known to be][55] in equal numbers in the urn, $P = 1/2^n$, a value which is less than that just obtained for $n \geq 4$. From this follows the result that, although the first draw makes it probable that there are more white than black balls, the probability of getting four black balls in the following four draws is much greater than if one supposes equal numbers of white and black balls. The apparent paradox is due, says Laplace, to the fact that (in modern notation)

$$\Pr[B_1B_2B_3\ldots] = \Pr[B_1]\Pr[B_2 \mid B_1]\Pr[B_3 \mid B_1B_2]\ldots,$$

the "probabilités partielles" [p.327] being always increasing and tending to 1 as $n \to \infty$.

This discussion is continued in Number XLII, but now it is supposed that the observed result (as well as the future one) is composed of a very large number of simple events: various approximations are derived.

In Number XLIII Laplace returns to the problem of births, and defines

p — the number of births of boys (in Paris);
q — the number of births of girls;
$2n$ — the annual number of births;
x — the possibility of the birth of a boy.

Denoting further by z the sum of the first n terms of the expansion

$$x^{2n} + 2nx^{2n-1}(1-x) + \frac{2n(2n-1)}{1.2}x^{2n-2}(1-x)^2 + \cdots,$$

which sum represents the probability that the number of boys will, in each year, prevail over that of the girls, and z^i being the probability that this superiority will be maintained during i consecutive years, one finds that the true probability P that this will happen is, by Number XLI,

$$P = \int_0^1 x^p z^i (1-x)^q \, dx \Big/ \int_0^1 x^p (1-x)^q \, dx \ .$$

The rest of this number is taken up with approximations for these integrals, and a numerical example is adduced[56].

At the start of the final number of this article Laplace relates the present memoir to its predecessor in the following words:

> les recherches précédentes suffisent pour faire voir les avantages de l'analyse exposée au commencement de ce Mémoire, dans la partie de la théorie des hasards, où il s'agit de remonter des événements observés à leurs possibilités respectives et de déterminer la probabilité des événements futurs. Cette analyse n'est pas moins utile dans la solution des problèmes où l'on cherche la probabilité d'un résultat formé d'un grand nombre d'événements simples, dont les possibilités sont connues. [pp.334-335]

6.7 Sur les naissances

Laplace's memoir[57], "Sur les naissances, les mariages et les morts a Paris,depuis 1771 jusqu'en 1784, et dans toute l'étendue de la France, pendant les années 1781 et 1782", published on pages 693-702 of the same volume as that in which the memoir discussed in the previous section appears[58], is devoted to an examination of the subjects of its title from the point of view of a théorie nouvelle et encore peu connue, celle de la probabilité des événements futurs prise des événements observés. [p.37]

By this means Laplace proposes to consider the following problem: suppose that, on the basis of past censuses, the ratio of births to population size is known for a given period in a large number of parishes in all provinces of France, these parishes being chosen in such a way that the birth-death ratios there found are the same as that in the whole kingdom. If, in addition, one knows the number of births in a given period in the whole of France, how should one estimate the total population size, and what can one say about the error incurred in such estimation[59]?

To solve this problem, Laplace considers the case of an urn containing an infinite number of white and black balls in unknown ratio. A preliminary drawing from this urn results in p white and q black balls, while a second drawing yields q' black and an unknown number of white balls, a number which is most naturally estimated by pq'/q. Denoting the true unknown number by P', our aim is to find

$$\Pr\left[|P' - pq'/q| < pq'\omega/q \mid p, q, q'\right]$$

for given ω, p, q and q'.

Laplace's reasoning seems somewhat confused: the following is an attempt at understanding it. Let X be the unknown ratio of white to total number of balls originally obtaining, with P' and Q' being appropriate random variables. Then

$$\Pr\left[P' = p', Q' = q' \mid X = x\right] = \binom{p'+q'}{p'} x^{p'}(1-x)^{q'} \quad .$$

Since

$$\Pr\left[Q' = q' \mid X = x\right] = \sum_{p'=0}^{\infty} \binom{p'+q'}{p'} x^{p'}(1-x)^{q'} = (1-x)^{-1} \quad ,$$

it follows that

$$\Pr\left[P' = p' \mid Q' = q', X = x\right]$$
$$= \Pr\left[P' = p', Q' = q' \mid X = x\right] \Big/ \Pr\left[Q' = q' \mid X = x\right]$$
$$= \binom{p'+q'}{p'} x^{p'}(1-x)^{q'+1} \quad ,$$

and based on the first drawing we have

$$\Pr[x < X < x + dx \mid p, q] = x^p(1-x)^q\, dx \Big/ \int_0^1 x^p(1-x)^q\, dx \quad .$$

Now by the definition of conditional probability (and using a less cumbersome though I trust sufficiently precise notation)

$$\Pr[p', x \mid p, q, q'] = \Pr[x \mid p, q] \Pr[p' \mid q', x] \Pr[q' \mid x] \Big/ \Pr[q' \mid p, q]$$

provided one assumes that (P, Q) and (P', Q') are conditionally independent given x. Laplace in fact assumes that

$$\Pr[p', x \mid p, q, q'] = \Pr[x \mid p, q] \Pr[p' \mid q', x] \ ,$$

and here again we take note of something we have already noticed before (see §6.3), viz. his conception of conditional distributions as being defined only up to proportionality (it is unfortunate that the proportionality "constant" in fact depends upon x)[60].

Using this last "equation" one obtains, finally,

$$\Pr\left[p' \mid p, q, q'\right] = \binom{p'+q'}{p'} \int_0^1 x^{p+p'}(1-x)^{q+q'+1}\, dx \Big/ \int_0^1 x^p(1-x)^q\, dx \quad ,$$

from which it follows that

$$\Pr\left[0 \le P' \le s \mid p, q, q'\right]$$

$$= \int_0^1 x^p(1-x)^{q+q'+1} \sum_{p'=0}^{s} \binom{p'+q'}{p'} x^{p'}\, dx \Big/ \int_0^1 x^p(1-x)^q\, dx \quad . \qquad (17)$$

If q' and s are both very large numbers (a condition which seems sufficient but at this stage unnecessary), one has[61]

$$\sum_{p'=0}^{s} \binom{p'+q'}{p'} x^{p'}(1-x)^{q'+1} = I_{1-x}\left(q'+1, s+1\right)$$

$$= \int_x^1 y^s(1-y)^{q'}\, dy \Big/ \int_0^1 y^s(1-y)^{q'}\, dy \quad .$$

Substitution in (17) yields

$$\Pr\left[0 \le P' \le s \mid p, q, q'\right] = \frac{\int_0^1 \int_x^1 x^p(1-x)^q y^s(1-y)^{q'}\, dy\, dx}{\int_0^1 \int_0^1 x^p(1-x)^q y^s(1-y)^{q'}\, dy\, dx} \quad . \qquad (18)$$

On applying the results of the memoir we have considered in §6.6 Laplace concludes that if s is less than and very little different to pq'/q then (18) becomes approximately

$$\frac{1}{\sqrt{\pi}}\int_T^\infty e^{-t^2}\,dt \ ,$$

where

$$T^2 = \frac{\left(p/(p+q) - s/(s+q')\right)^2 (p+q)^3(s+q')^3}{2sq'(p+q)^3 + 2pq\,(s+q')^3} \ .$$

A similar result follows if s is greater than, but not very different to, pq'/q, and it follows that, approximately,

$$\Pr\left[s \le P' \le s'\right] = 1 - \frac{1}{\sqrt{\pi}}\int_T^\infty e^{-t^2}\,dt - \frac{1}{\sqrt{\pi}}\int_{T'}^\infty e^{-t^2}\,dt$$

where T is as defined above and T' is defined similarly with s replaced by s'. If one in fact sets

$$s = (1-\omega)pq'/q \ , \quad s' = (1+\omega)pq'/q$$

then, on our neglecting terms of order ω^3, T^2 and T'^2 take on the common value

$$V^2 = \frac{pqq'\omega^2}{2(p+q)(q+q')} \ ,$$

and hence

$$\Pr\left[(1-\omega)pq'/q < P' < (1+\omega)pq'/q\right] = 1 - \frac{2}{\sqrt{\pi}}\int_V^\infty e^{-t^2}\,dt \ .$$

Having noted the ease with which this result can be applied to the question of population size (an application to which we shall shortly return), Laplace now considers how p (the number of people in the original census) should be determined so as to obtain a large probability that the error in p' (the predicted population size) is small. To this end, he supposes that $p = iq$ and $\omega pq'/q = a$, so that the expression for V^2 given previously yields

$$p = \frac{2i^2(i+1)q'^2V^2}{a^2 - 2i(i+1)q'V^2} \ .$$

Thus p will be determined provided that i, a, q' and V are known. A numerical example is supplied.

Whether Laplace's urn model is assimilable to his initial population problem is doubtful. Pearson [1928, app.II], in addition to finding the "treatment obscure" [p.168], finds the interpretation of the population parameters in terms of those of the urn to be questionable: he in fact writes

> I can see no justification for Laplace's method of reducing the problem to an urn problem. I see no reason why an additional birth in the sample means one fewer member of the population. I see further no ground whatever for considering the first sample and France as a whole as independent samples from an indefinitely large population. [p.172]

Pearson (loc. cit.) presents an analysis of Laplace's problem from the point of view of marked members of a population, the problem being restated as follows:

> A population of unknown size N is known to contain q' affected or marked members. It is desired to ascertain — on the hypothesis of inverse probabilities — a measure of the error introduced by estimating N to be $n \times q'/q$, where q is the number of marked individuals in a sample of size n. [p.172]

His final summing-up of the problem is as follows:

> I venture to think, therefore, that while Laplace's Problem is most important, it does not cover the case to which he applies it, and that his solution of the problem itself is not really correct. [p.174]

6.8 Sur les probabilités

In the *Journal de l'École Polytechnique*, VIIe et VIIIe *Cahiers*, juin 1812, was published[62] *Leçons de Mathématiques donnees a l'École Normale en 1795*, the *dixième séance* being of the above title. This popular statement of Laplace's views was later expanded into an introduction to his *Théorie analytique des probabilités*, and we shall postpone its consideration until we discuss this latter work.

6.9 Sur les approximations des formules

The "Mémoire sur les approximations des formules qui sont fonctions de très grands nombres et sur leur application aux probabilités", published in 1810 in volume X (1809) of the *Mémoires de l'Académie des Sciences, Ire Série*, pp.353-415, is notable chiefly for its contribution to the theory of errors[63]. However a supplement to this memoir is pertinent to our present purpose, and it therefore seems not inadvisable to say something about the memoir itself at this stage[64].

After an introductory section, Laplace turns his attention in the first article to the following problem:

> on suppose toutes les inclinaisons à l'écliptique également possibles depuis zéro jusqu'à l'angle droit, et l'on demande la probabilité que l'inclinaison moyenne de n orbites sera comprise dans des limites données. [p.305]

The formulae obtained in the solution of this problem are applied in the second article to the inclinations of planetary orbits, the result obtained indicating

> avec une très grande probabilité l'existence d'une cause primitive qui a déterminé les orbites des planètes à se rapprocher du plan de l'écliptique ou, plus naturellement, du plan de l'equateur solaire ... Ainsi l'existence d'une cause commune qui a dirigé ces mouvements dans le sens de la rotation du Soleil est indiquée par les observations avec une probabilité extréme. [p.308]

Laplace next discusses whether or no this cause has influence on the movement of comets, the intractability of the expressions derived leading him to another resolution of the problem in Article III (it was for this alternative method that he developed characteristic functions). These new formulae are then applied, in Article IV, to observed cometary data, further approximations being considered in Article V.

In the sixth article Laplace returns to the problem considered in the first article. He points out that this problem,

> relativement aux inclinaisons, est la même que celui dans lequel on se propose de déterminer la probabilité que l'erreur moyenne d'un nombre n d'observations sera comprise dans des limites données, en supposant que les erreurs de chaque observation puissent également s'étendre dans l'intervalle h. Nous allons maintenant considérer le cas général dans lequel les facilités des erreurs suivent une loi quelconque. [p.322]

The rest of the memoir is taken up with various ramifications of earlier results.

6.10 Supplément: sur les approximations des formules

The "Supplément au mémoire sur les approximations des formules qui sont fonctions de très grands nombres" was published in 1810 in the same volume, pp.559-565, as the preceding memoir. Here Laplace returns to a discussion of the theory of errors, undertaking yet a further generalization of the work of his earlier memoirs of 1774 and 1778. He proposes to consider the problem of the combination of several means, each of which is formed from a large number of identically distributed and independent observations, and although he assumes, as in the Memoir, that the individual means are normally distributed, some general discussion is also given.

Suppose that $n, n', n'', \ldots$ observations yield means $A, A+q, A+q', \ldots$ respectively, the laws of facility of the different errors being distinct. If $A+x$ is the true value, the error of the mean result of the first set of n observations is $-x$, the probability of this error, by the Memoir, being

$$\frac{1}{\sqrt{\pi}}\sqrt{\frac{k}{2k'}}\,\frac{dr}{dx}e^{-kr^2/2k'} \tag{19}$$

where $k = \int_{-h/2}^{h/2} \varphi(x/h)\,dx$ ($\varphi(x/h)$ being the true probability of the error $\pm x$), $k' = \int_{-h/2}^{h/2} (x^2/h^2)\,\varphi(x/h)\,dx$, and the limits[65] between which the probability of the error is required to lie are $\pm rh/\sqrt{n}$. With $x = rh/\sqrt{n}$ and $a = \sqrt{k/2k'}/h$, (19) becomes

$$\frac{1}{\sqrt{\pi}}a\sqrt{n}\,e^{-na^2x^2} \quad .$$

If one designates by $\psi(-x), \psi'(q-x), \psi''(q'-x), \ldots$ these diverse probabilities, the probability that the error of the first result will be $-x$ and that the others will differ from the first by $q, q', \ldots$ respectively, will be equal to the product

$$y = \psi(-x)\,\psi'(q-x)\,\psi''(q'-x)\ldots$$

Once again Laplace suggests that if one constructs a curve whose ordinate y is equal to this product, the ordinates of this curve will be proportional to the probabilities of the abscissae, and for this reason "nous la nommerons *courbe des probabilités*" [p.351].

Proceeding[66] as in his memoirs of 1774 and 1778, Laplace takes as his estimate of the mean that value of l such that

$$\int_0^l y\,dz = \int_l^\infty y\,dz \quad ,$$

this result being obtained by minimizing $\int |l - z|\, dz$ with respect to l.

Commenting on earlier work Laplace writes[67]

> Daniel Bernoulli, ensuite Euler et M. Gauss ont prise pour cette ordonnée la plus grande de toutes. Leur résultat coïncide avec le précédent lorsque cette plus grande ordonnée divise l'aire de la courbe en deux parties égales, ce qui, comme on va le voir, a lieu dans la question présente; mais, dans le cas général, il me paraît que la manière dont je viens d'envisager la chose résulte de la théorie même des probabilités. [p.352]

Both Stigler [1975, p.506] and Sheynin [1977, p.16] suggest that Laplace came to know of Gauss's treatise *Theoria Motus Corporum Coelestium* of 1809 after he had written his memoir and that this treatise might well have provided the impetus for the supplement.

Laplace now returns to the case in which the means are normally distributed. Writing $A + x \equiv A + X + z$ he considers the likelihood

$$y = p\,p^{'}\,p^{''}\ldots e^{-p^2\pi(X+z)^2 - p^{'2}\pi(q-X-z)^2 - p^{''2}\pi(q^{'}-X-z)^2\ldots}$$

where $p = a\sqrt{n}/\sqrt{\pi}$ "et par conséquent exprimant la plus grande probabilité du résultat donné par les observations n" [p.352] ($p^{'}, p^{''}, \ldots$ being similarly defined). X is now chosen in such a way that the term in z in the above exponential vanishes, which has the effect that the ordinate y corresponding to $z = 0$ divides the area under the curve into two equal parts, and at the same time is the greatest ordinate. One has, in this case,

$$X = \left(p^{'2}q + p^{''2}q^{'} + \cdots\right) \Big/ \left(p^2 + p^{'2} + \cdots\right) \ , \tag{20}$$

and thus y has the form

$$y = p\,p^{'}p^{''}\ldots e^{-M-Nz^2} \ ,$$

from which the effect mentioned above is immediate. Thus $A + X$ is the desired mean between the quantities $A, A + q, A + q^{'}, \ldots$.

Laplace also notes that the value of X given in (20) is that which minimizes

$$[pX]^2 + \left[p^{'}(q - X)\right]^2 + \left[p^{''}\left(q^{'} - X\right)\right]^2 + \cdots$$

(or $[p\,|X|]^2 + \left[p^{'}\,|q - X|\right]^2 + \cdots$), a function which is described as

> la somme des carrés des erreurs de chaque résultat, multipliées respectivement par la plus grande ordonnée de la courbe de facilité de ses erreurs. [p.353]

It is this remark which I think Stigler [1975, p.506] considers "a Bayesian justification for least squares": not only, says Stigler, do "the least squares estimates ... maximize the likelihood function, considered as a posterior distribution, but [they] also minimize the expected posterior error" [1975, p.506].

This property is characterized by Laplace as follows:

> ainsi cette propriété, qui n'est qu'hypothétique lorsqu'on ne considère que des résultats donnés par une seule observation ou par un petit nombre d'observations, devient nécessaire lorsque les résultats entre lesquels on doit prendre un milieu sont donnés chacun par un très grand nombre d'observations, quelles que soient d'ailleurs les lois de facilité des erreurs de ces observations. C'est une raison pour l'employer dans tous les cas. [p.353]

He concludes by showing that

$$\Pr\left[-T/\sqrt{N} \leq A + X \leq T/\sqrt{N}\right] = \frac{2}{\sqrt{\pi}} \int_0^T e^{-t^2}\, dt \ ,$$

the value of N, by what precedes, being $\pi\left(p^2 + p'^2 + p''^2 + \cdots\right)$.

6.11 Sur les intégrales définies

Published in the *Mémoires de l'Académie des Sciences, Ire Série*, Tome XI (Ire Partie) for 1810 (published 1811), pp.279-347, the "Mémoire sur les intégrales définies et leur application aux probabilités, et spécialement a la recherche du milieu qu'il faut choisir entre les résultats des observations" has a touch of both retrospection (Laplace recalls his earlier work on generating functions) and prospection (two references are made to an impending work, viz. "... une théorie que je me propose de publier bientôt sur les probabilités" [p.360], and "un Ouvrage que je vais bientôt publier sur les probabilités" [p.411]. Most of the memoir is devoted to the evaluation of certain definite integrals, but there is some discussion of three probability problems which receive scant attention from Todhunter (he discusses this memoir in his Articles 919-922); Laplace regards the investigations of this memoir as "d'une grande utilité dans la théorie des probabilités" [p.361].

Speaking of his calculus of generating functions Laplace says

> par ce moyen, on peut déterminer avec facilité les limites de la probabilité des résultats et des causes, indiqués par les événements considérés en grand nombre, et les lois suivant lesquelles cette

> probabilité approche de ses limites, à mesure que les événements se multiplient. [pp.360-361]

This research, "la plus délicate de la théorie des hasards" [p.361], deserves, he says, the attention of both mathematicians and philosophers.

We note also the following definition given in the introductory section of the memoir:

> j'entends par *erreur moyenne* la somme des produits de chaque erreur par sa probabilité. [p.362]

The first three articles are impertinent: we shall consider the others *seriatim.*

In his fourth article, headed "Application de l'analyse précédente aux probabilités", Laplace presents the first of his probability problems — 'tis what Todhunter [1865, art.921] refers to as "the problem of the Duration of Play". We shall not discuss this article here: the main tools in the solution of the problem posed are generating functions and techniques for evaluating certain definite integrals.

In Article V Laplace considers the following urn problem: suppose that two urns A and B each contain the same number n of balls, and that of the $2n$ balls there are as many white as black. A ball is drawn from each urn simultaneously and replaced in the other, the contents of the urns being shuffled before the trial is repeated. What is the probability that, after r repetitions, there are x white balls in urn A? Laplace develops certain formulae to solve this problem, as an application of which he considers what is essentially an urn problem involving the hypergeometric distribution. More precisely, he supposes that an urn C contains a vast number m of white balls and the same number of black balls. The contents of C having been shuffled, n balls are drawn and placed in urn A. One then places in urn B as many white (black) balls as there are black (white) balls in A. Under the assumptions usual for the appropriateness of this distribution, the desired probability (i.e. that A contains x white balls) is found to be

$$\binom{m}{x}\binom{m}{n-x}\Big/\binom{2m}{n} \quad ,$$

and an approximation to this probability (for large values of m, n and x) is derived from $s! = s^{s+\frac{1}{2}}e^{-s}\sqrt{2\pi}$.

In Article VI, headed "Du milieu qu'il faut choisir entre les résultats des observations", Laplace returns to a problem already considered in his earlier memoirs, viz. the finding of a mean between the results of several observations. He suggests firstly that one should write the observation C

in the form $C = m + pz$, where z is the correction to the element already approximately known. If, however, C is susceptible of an error ϵ, let $C+\epsilon = m+pz$, or $\epsilon = pz-\varphi$ (where $\varphi = C-m$). Denoting the error in the $(i+1)$th observation by $\epsilon^{(i)} = p^{(i)}z - \varphi^{(i)}$, Laplace considers

$$\sum_{i=0}^{s-1} \epsilon^{(i)} = z\sum_{i=0}^{s-1} p^{(i)} - \sum_{i=0}^{s-1} \varphi^{(i)} ,$$

where s denotes the total number of observations; and he deduces that, if the sum of the errors is to be zero, one must have

$$z = \sum_{i=0}^{s-1} \varphi^{(i)} \Big/ \sum_{i=0}^{s-1} p^{(i)}$$

(the "*résultat moyen* des observations" [p.388]).

He goes on next to suppose that, rather than requiring $\sum_{i=0}^{s-1} \epsilon^{(i)}$ to be zero, one may well look at a linear combination of the errors

$$q\epsilon + q^{(1)}\epsilon^{(1)} + q^{(2)}\epsilon^{(2)} + \cdots + q^{(s-1)}\epsilon^{(s-1)} , \tag{21}$$

where $q, q^{(1)}, \ldots \in \mathbb{Z}$. Substituting $p^{(i)}z - \varphi^{(i)}$ for $\epsilon^{(i)}$, and equating the result to zero, we get

$$z = \sum_i q^{(i)}\varphi^{(i)} \Big/ \sum_i p^{(i)}q^{(i)} .$$

He then shows that the probability that (21) lies between the limits $\pm ar$ is

$$\frac{1}{\sqrt{(k'\pi/k)\sum q^{(i)2}}} \int \exp\left(-kr^2/4k'\sum q^{(i)^2}\right) dr$$

where $k = 2\int_0^\infty \psi(x')\,dx'$, $k' = \int_0^\infty x'^2\,\psi(x')\,dx'$, $x' = x/a$ and $\psi(x/a)$ is the (prior) probability of an error x in each observation.

Laplace next considers the "valeur moyenne de l'erreur à craindre" [p.392], a mean value which, he says, "est donc la somme des produits de chaque erreur, abstraction faite du signe, par sa probabilité" [p.393]. The value concerned is

$$2a\sqrt{\frac{k'}{k\pi}}\frac{\sqrt{\sum q^{(i)^2}}}{\sum p^{(i)}q^{(i)}} . \tag{22}$$

While the values of $p, p^{(1)}, \ldots$ are given, those of $q, q^{(1)}, \ldots$ are arbitrary and must be determined by minimizing (22), whence

$$\frac{q^{(i)}}{\sum q^{(i)^2}} = \frac{p^{(i)}}{\sum p^{(i)}q^{(i)}} .$$

He then argues that $q = \mu p, q^{(1)} = \mu p^{(1)}, \ldots, q^{(s-1)} = \mu p^{(s-1)}$, from which it follows that μ must be chosen so that all of the $q, q^{(1)}, \ldots$ are integers. Then (22) becomes

$$\sqrt{\frac{k\pi}{k'} \sum p^{(i)^2}} \tag{23}$$

and

$$z = \sum p^{(i)} \varphi^{(i)} \Big/ \sum p^{(i)^2} \quad .$$

(Expression (23) does not seem correct: it should be $2a\sqrt{\frac{k'}{k\pi}}\left(\sum p^{(i)^2}\right)^{-1}$.) This result, he next points out, "est celui que donne la méthode des moindres carrés des erreurs" [p.395], and is exactly what arises on minimizing, with respect to z,

$$(pz - \varphi)^2 + \left(p^{(1)}z - \varphi^{(1)}\right)^2 + \cdots + \left(p^{(s-1)}z - \varphi^{(s-1)}\right)^2 \quad .$$

> Cette méthode doit donc être employée de préférence, quelle que soit la loi de facilité des erreurs, loi dont dépend le rapport k/k'. [p.395]

He also states (and demonstrates) that, although this law is almost always unknown, one may suppose $k/k' > 6$. He once again stresses, as in his memoir of 1778, that, under the hypotheses presented there (and repeated on p.396 of the present memoir), the (prior) probability of the error $\pm x$ should be taken to be $(1/2a)\log(a/x)$.

In the seventh article Laplace discusses, perhaps more clearly than in the memoir of §6.10, the question of the mean of a number of sample means, each based on a large number of observations. He supposes that an element is given successively by the mean result of $s, s', \ldots$ observations, these means being $A, A+q, \ldots$ respectively. If $A+x$ is "l'élément vrai" [p.398] the error from the first s observations will be $-x$. If C is now equal to

$$\sqrt{\frac{k}{k'}}\,\frac{\sqrt{\sum p^{(i)^2}}}{2a}$$

if the method of least squares is used to determine the mean, or to

$$\sqrt{\frac{k}{k'}}\,\frac{\sum p^{(i)}}{2a\sqrt{s}}$$

if the ordinary method is used, it follows from the preceding article that, for large s, the probability of that error is

$$\frac{C}{\sqrt{\pi}} e^{-C^2x^2} \quad .$$

Repeating this for the other sets of data, one finds that the probability that the errors are $-x, q-x, q^{'}-x, \ldots$ will be

$$\frac{C}{\sqrt{\pi}}\frac{C^{'}}{\sqrt{\pi}}\frac{C^{''}}{\sqrt{\pi}}\cdots e^{-C^2x^2-C^{'2}(x-q)^2-C^{''2}(x-q^{'})^2\cdots} \quad , \tag{24}$$

whence it follows that

> en la multipliant par dx et prenant l'intégrale depuis $x=-\infty$ jusqu'à $x=\infty$, on aura la probabilité que les résultats moyens des observations $s, s^{'}, s^{''}, \ldots$ surpasseront respectivement de $q, q^{'}, \ldots$ le résultat moyen des observations s. [p.399]

If one integrates between the given limits one obtains the probability that, the preceding condition (that of (24) I suspect) being satisfied, the error in the first result will lie within those limits. On dividing this probability by that of the condition itself one obtains the probability that the error of the first result will lie within the given limits, whenever it is certain that the given condition has actually arisen. (It is a little hard to see what Laplace is driving at here: I suspect he is considering something like

$$\Pr\left[l_1 < \text{ error}_1 < l_2 \mid \text{ condition}\right] = \frac{\Pr\left[l_1 < \text{ error}_1 < l_2 \ \& \text{ condition}\right]}{\Pr[\text{condition}]} \quad .)$$

This probability is given as

$$\int e^{-C^2x^2-C^{'2}(x-q)^2\cdots}\,dx \Big/ \int e^{-C^2x^2-C^{'2}(x-q)^2\cdots}\,dx$$

"l'intégrale du numérateur étant prise dans les limites données et celle du dénominateur étant prise depuis $x=-\infty$ jusqu'à $x=\infty$" [p.399]. He also shows that this probability can be written as

$$\frac{1}{\sqrt{\pi}}\int e^{-t^2}\,dt \quad .$$

Finally he concludes that

> la loi du minimum des carrés des erreurs devient nécessaire lorsque l'on doit prendre un milieu entre des résultats donnés chacun par un grand nombre d'observations. [p.401]

Continuing in the style of this article Laplace writes in Article VIII

> la méthode des moindres carrés des erreurs des observations est celle qui donne sur la correction des éléments la plus petite erreur moyenne à craindre. [p.401]

The method of Article VI is extended here to the case in which there are two elements, with z being the correction of the first and z' that of the second: here the observation C is supposed given by

$$C = A + pz + qz' .$$

As in Article VI we are led to

$$\epsilon^{(i)} = p^{(i)}z + q^{(i)}z' - \alpha^{(i)}$$

where $\alpha = C - A$. Proceeding as in that article Laplace shows that the corrections to be applied to the two elements are

$$z = \frac{\sum q^{(i)^2} \sum p^{(i)}\alpha^{(i)} - \sum p^{(i)}q^{(i)} \sum q^{(i)}\alpha^{(i)}}{\sum p^{(i)2} \sum q^{(i)2} - \left(\sum p^{(i)}q^{(i)}\right)^2}$$

$$z' = \frac{\sum p^{(i)^2} \sum q^{(i)}\alpha^{(i)} - \sum p^{(i)}q^{(i)} \sum p^{(i)}\alpha^{(i)}}{\sum p^{(i)2} \sum q^{(i)2} - \left(\sum p^{(i)}q^{(i)}\right)^2} ,$$

and he goes on to say that these are the corrections given by the method of least squares for errors of observations, on our minimizing

$$\sum \left(p^{(i)}z + q^{(i)}z' - \alpha^{(i)}\right)^2 .$$

He points out that this method may be extended to any number of elements whatsoever. Various formulae and comments, similar to those found in Article VI, follow.

6.12 Sur les comètes

Published in *Connaissance des Temps* for 1816, dated 1813, pp.213-220, this memoir is notable for the use it makes of posterior probability in the solution of a problem in celestial mechanics[68]. The problem[69] (that of the probability of a comet's having a particular orbit — elliptic, parabolic or hyperbolic — on the basis of certain data) considered has been the subject of several papers[70] which have shown up certain inadequacies in Laplace's discussion[71], though these seem not to have been noticed by Todhunter [1865, art.925]. Following Fabry [1893-1895], who has in fact considered a more general problem than did Laplace, we choose firstly to present Laplace's solution (albeit in a slightly modified form) before undertaking any criticism.

After a discussion of Herschel's views on the origin of the comets,

> qui consiste à les regarder comme de petites nébuleuses formées par la condensation de la matière nébuleuse répandue avec tant de profusion dans l'univers. Les comètes seraient ainsi, relativement au système solaire, ce que les aérolithes sont par rapport à la Terre, à laquelle ils paraissent étrangers [p.88],

views with which he professed himself to be in agreement[72], Laplace comments in fairly broad terms on the nature of the orbits, something to which he had applied the probability calculus. As a result of these investigations

> J'ai trouvé qu'en effet il y a un grand nombre à parier contre l'unité qu'une nébuleuse qui pénètre dans la sphère d'activité solaire, de manière à pouvoir être observée, décrira ou une ellipse très allongée ou une hyperbole qui, par la grandeur de son axe, se confondra sensiblement avec une parabole dans la partie que l'on observe. Cette application de l'analyse des probabilités pouvant intéresser les géomètres et les astronomes, je vais l'exposer ici. [p.89]

Soon after this we find the statement of the problem which I believe Laplace is trying to solve:

> Il faut donc déterminer quel est, dans ces limites, le rapport des chances qui donnent une hyperbole sensible aux chances qui donnent un orbe que l'on puisse confondre avec une parabole. [pp.89-90]

Here the word "limites", as Laplace uses it, refers to the velocity at the moment of entry of the comet into the sphere of the sun's activity[73], the magnitude and direction of which velocity lie within narrow limits.

It seems from this passage that what we want to find is (in a modern and — I hope — sufficiently self-explanatory notation) $\Pr[H]/\Pr[H']$, where H and H' are two hypotheses[74]. This is indeed what is found in the last part of the paper: in the first part what is found is $\Pr[H \mid \text{data}]/\Pr[H' \mid \text{data}]$, a typical Bayesian *quaesitum*. Laplace follows up the preceding quotation with a sentence in which he points out the importance of the prior distribution, writing

> Il est clair que ce rapport dépend de la loi de possibilité des distances périhélies des comètes observables ... [p.90]

Examination of extant data shows that, beyond a certain distance equal to the radius of the earth, the possibilities of the perihelion distances decrease very rapidly as these distances increase. This should be reflected in the law of these possibilities; but this being generally unknown, one is only able to determine the limit of the ratio concerned, or its value in the case most favourable to visible hyperbolas.

There is then some summary discussion, with which we need not concern ourselves at the moment, of the results obtained in the paper, this discussion being followed by some observations on the comets of 1682 and 1770.

Laplace now introduces some notation:

V — the velocity of a comet at the instant at which it penetrates into the sphere of the sun's activity (i.e. "cette partie de l'espace où l'attraction du Soleil est prédominante" [p.88]);
r — the radius vector of the comet at the same instant;
a — the semi-major axis of the orbit which it proceeds to describe about the sun;
e — the eccentricity of the orbit;
D — the perihelion distance (of the orbit which it is going to describe about the sun).

Taking as the unit of mass the mass of the sun, and, as the unit of distance, its mean distance to the earth (and ignoring the masses of the comets and planets relative to that star) one obtains the well-known formulae

$$\frac{1}{a} = \frac{2}{r} - V^2$$

$$rV \sin\omega = \sqrt{a\,(1-e^2)} \tag{25}$$

$$D = a(1-e)$$

where ω denotes the angle that the direction of the velocity V makes with the radius vector r. Fabry [1893-1895, p.35] gives these formulae (notation slightly altered) as[75]

$$\frac{1}{a} = \frac{2}{r} - \frac{V^2}{f}$$

$$k = rV \sin\omega$$

$$D = a(1-e)$$

$$a\left(1-e^2\right) = \frac{k^2}{f} \; .$$

We shall deal most often with Laplace's formulae, though reference to Fabry's will be made from time to time.

Elimination of a and e yields

$$\sin^2\omega = \left(2D - 2D^2/r + D^2V^2\right)/r^2V^2$$

$$\left[\text{or } f\left(2D - 2D^2/r + D^2V^2/f\right)/r^2V^2\right] \ ,$$

whence

$$1 - \cos\omega = 1 - \frac{\sqrt{1-D/r}}{rV}\sqrt{r^2V^2(1+D/r) - 2D} \ .$$

Fabry [1893-1895, p.6] points out that this latter expression holds for values of ω between 0 and $\pi/2$: for values in $(\pi/2, \pi)$ it is necessary to replace the $-$ sign in the second term by $+$.

Laplace now introduces an equiprobability assumption in the following words:

> Maintenant, si l'on imagine une sphère dont le centre soit celui de la comète et dont le rayon soit égal à la vitesse V, cette vitesse pourra être également dirigée vers tous les points de la moitié de cette sphère comprise dans la sphère d'activité du Soleil. [p.92]

A simple argument (using what would today be regarded as fairly elementary calculus) then shows that

$$\Pr\left[0 < \text{direction of } V < \omega\right] = 1 - \cos\omega \ .$$

(Laplace of course does not use this notation, and choice of "$<$" or "$\leq$" seems a matter of personal preference in interpreting his results.)

Laplace obtains this result by considering the ratio

$$\int_0^\omega 2\pi \ \sin\omega \, d\omega/(2\pi) \ ,$$

while Fabry's argument is somewhat different. He supposes that the sphere at whose centre the comet is, has radius 1, and that the velocities of comets which are almost at the limit of the sphere of the sun's activity are uniformly directed over that sphere (rather than the hemisphere considered by Laplace). This leads Fabry to[76]

$$\Pr\left[\beta < \text{direction } V \text{ makes with radius vector } < \beta + d\beta\right]$$

$$= 2\pi \ \sin\beta \, d\beta/4\pi \ ,$$

whence

$$\begin{aligned}\Pr[0 < \text{direction of } V < \beta] &= \tfrac{1}{2}\int_0^\beta \sin\beta\, d\beta \\ &= \tfrac{1}{2}(1 - \cos\beta)\ .\end{aligned}$$

Thus far the paper is unexceptionable. At this point, however, things become a little more complicated, and I believe that certain obscurities intrude. We shall consider the remainder of Laplace's discussion firstly in the case in which the prior is uniform, and secondly in the case in which it is non-uniform.

Suppose firstly, then, that the prior on D is uniform. The limits of the perihelion distance corresponding to the limits 0 and ω of [the direction of] V (the phrase in brackets is missing from the original) being 0 and D respectively,

> en supposant donc toutes les valeurs de D également possibles, on a pour la probabilité que la distance périhélie sera comprise entre zéro et D
>
> $$1 - \frac{\sqrt{1 - D/r}}{rV}\sqrt{r^2V^2(1 + D/r) - 2D}$$
>
> [p.92]

The reasoning here is perhaps clearer in Fabry's paper[77]: for a fixed value of V, ω varies between 0 and $\pi/2$, and D increases constantly with ω. Thus, for fixed V, the probability that the comet has a perihelion distance less than some given value is the same as the probability that the direction of the velocity is between 0 and ω.

Changing notation slightly, and denoting (henceforth) random variables by majuscules and reserving minuscules for values, we can write this latter probability as

$$\Pr[0 < D < d \mid V = v] = 1 - \frac{\sqrt{1 - d/r}}{rv}\sqrt{r^2v^2(1 + d/r) - 2d} \qquad (26)$$

(or $(1/2)\left[1 - (\sqrt{1 - d/r}/rv)\sqrt{r^2v^2(1 + d/r) - 2df}\right]$ in Fabry), where D and V denote perihelion distance and velocity respectively[78]. The supposition that all values of D are equally possible seems to be irrelevant.

Laplace now suggests that

> Il faut multiplier cette valeur par dV; en l'intégrant ensuite dans des limites déterminées et divisant l'intégrale par la plus grande

> valeur de V, valeur que nous désignerons par U, on aura la probabilité que la valeur de V sera comprise dans ces limites. [p.92]

It is at this stage, I believe, that things start becoming a little awkward (though Fabry sees problems arising only later in the Memoir[79]). Fabry's discussion sheds some light on the matter, and we shall therefore pursue it here. Before doing so, however, it might be wise to see exactly what it is we are trying to find.

From the last quotation it appears that what is wanted is $\Pr[v_1 < V < v_2]$, say. Yet further on Laplace gives

> la probabilité que la distance périhélie d'un astre qui entre dans la sphère d'activité du Soleil sera comprise dans les limites zéro et D, la valeur de V^2 n'excédant pas i^2/r [p.94],

which one might well write as $\Pr\left[0 \leq D < d \mid V \leq i^2/r\right]$. Again, on page 94 we find the words

> la probabilité que la distance périhélie étant comprise entre zéro et D, l'orbite sera ou elliptique, ou parabolique, ou une hyperbole dont le demi-grand axe sera au moins égal à 100,

which seems to refer to $\Pr[\text{conic of the described form} \mid 0 < D < d]$.

Todhunter [1865, p.493], on the other hand, argues that the problem is really one in inverse probability, and that what we need to find is

$$\Pr[v_1 < V < v_2 \mid 0 < D < d] .$$

Fabry's approach is slightly different: he considers a ratio of two numbers, which then, by his definition of probability[80], gives the desired result. The final result, obtained by simplification of the integrands and subsequent integration, is described by Fabry as follows:

> Laplace donne cette expression comme représentant la probabilité que la distance périhélie soit inférieure à q et la vitesse initiale inférieure à $i/\sqrt{r}$. [p.12]

Fabry writes elsewhere that

> le nombre des comètes de vitesses compris entre v et $v+dv$ qui se trouvent à l'intérieur d'une unité de volume située dans la région considérée de l'espace, vers la limite de la sphère d'activité du Soleil, peut être représenté par $\varphi(v)\,dv$, $\varphi(v)$ étant une certaine fonction de v. [p.8]

Multiplication of (26) above by $\varphi(v)\,dv$ gives the number of *comètes visibles* having velocities between v and $v + dv$ which may be found in a unit of volume in the region of space considered. Thus the number of *comètes visibles* with initial velocities between v_1 and v_2 that will be found in a unit of volume is

$$\int_{v_1}^{v_2} \Pr[0 < D < d \mid V = v]\,\varphi(v)\,dv \ .$$

The definition of U also occasions some difficulty. Supposing all values of V between 0 and some value U to be equally probable, Fabry points out that his $\varphi(v)$ will then be constant for all values of V less than U and zero for all values of V greater than U. Thus the total number of comets in the interior of each unit of volume, in the region of space considered, will be

$$\int_0^U \varphi(v)\,dv = U\varphi \ ,$$

where φ denotes the constant value of $\varphi(v)$. One thus obtains the ratio of the number of *comètes visibles* with velocities between v_1 and v_2 contained in a unit of volume, to the total number of comets in that volume (φ being assumed constant and the same in all regions of space situated towards the limit of the sphere of activity of the sun) as

$$\frac{1}{2U}\int_{v_1}^{v_2}\left[1 - \sqrt{\frac{1-d/r}{rv}}\sqrt{r^2v^2(1+d/r) - 2df}\right] dv$$

(the factor $\frac{1}{2}$ being of course missing in Laplace's formulation). Fabry's result may thus be written as

$$\int_{v_1}^{v_2} \Pr[0 < D < d \mid V = v]\,\varphi(v)\,dv \bigg/ \int_0^U \varphi(v)\,dv \ .$$

Notice next that

$$\varphi(v)\,dv = \text{ number of comets with velocities in } (v, v+dv) \ .$$

Thus

$$\begin{aligned}\frac{\varphi(v)\,dv}{\int_{\mathcal{V}} \varphi(v)\,dv} &= \frac{\text{number of comets with velocities in } (v, v+dv)}{\text{total number of comets}} \\ &= \Pr[v < V < v + dv] \\ &= f_V(v)\,dv \ ,\end{aligned}$$

where f_V is a probability density function. Thus, as Fabry finds,

$$\int_{v_1}^{v_2} \Pr[0 < D < d \mid V = v]\, \varphi(v)\, dv \Big/ \int_0^U \varphi(v)\, dv$$

$$= \int_{v_1}^{v_2} \Pr[0 < D < d \mid V = v] \left\{ \varphi(v) \Big/ \int_0^U \varphi(v)\, dv \right\} dv$$

$$= \int_{v_1}^{v_2} \Pr[0 < D < d \mid V = v]\, f_V(v)\, dv$$

$$= \Pr[0 < D < d \;\&\; v_1 < V < v_2] \quad .$$

Thus, if $\varphi(v)$ is constant ($\equiv \varphi$),

$$f_V(v) = \varphi \Big/ \int_0^U \varphi\, dv = 1/U,$$

and

$$\Pr[0 < D < d \;\&\; v_1 < V < v_2] = \int_{v_1}^{v_2} \Pr[0 < D < d \mid V = v]\, dv/U \quad .$$

Moreover,

$$\begin{aligned} \Pr[0 < D < d] &= \int_0^U \Pr[0 < D < d \mid V = v]\, f_V(v)\, dv \\ &= \int_0^U \Pr[0 < D < d \mid V = v] \left\{ \varphi(v) \Big/ \int_0^U \varphi(v)\, dv \right\} dv \\ &= \int_0^U \Pr[0 < D < d \mid V = v]\, dv/U \quad , \end{aligned}$$

if $\varphi(v) \equiv \varphi$ is constant. Thus

$$\Pr[v_1 < V < v_2 \mid 0 < D < d]$$

$$= \Pr[v_1 < V < v_2 \;\&\; 0 < D < d] / \Pr[0 < D < d]$$

$$= \int_{v_1}^{v_2} \Pr[0 < D < d \mid V = v]\, dv \Big/ \int_0^U \Pr[0 < D < d \mid V = v]\, dv \quad .$$

This is all but Todhunter's result: the only difference is in the limits of integration in the denominator.

Returning to Laplace's result, we note, from the quotation following equation (26) above, that what is wanted is

$$\int_{v_1}^{v_2} \Pr[0 < D < d \mid V = v]\, dv/U \ ,$$

which, by our preceding discussion, with constant φ, is just

$$\Pr[0 < D < d \ \& \ v_1 < V < v_2] \ .$$

Now

$$\begin{aligned} &\Pr[0 < D < d \ \& \ v_1 < V < v_2] \\ &= \int_{v_1}^{v_2} \Pr[0 < D < d \mid V = v]\, f_V(v)\, dv \qquad (27) \\ &= \int_{v_1}^{v_2} \Pr[0 < D < d \mid V = v] \left\{ \varphi(v)\, dv \bigg/ \int_0^U \varphi(v) dv \right\} dv \ , \end{aligned}$$

where $f_V(v) = \varphi(v) \big/ \int_0^U \varphi(v)\, dv$. For constant φ this becomes

$$\begin{aligned} \Pr[0 < D < d \ \& \ v_1 < V < v_2] &= \int_{v_1}^{v_2} \Pr[0 < D < d \mid V = v]\, [\varphi/U\varphi]\, dv \\ &= \frac{1}{U} \int_{v_1}^{v_2} \Pr[0 < D < d \mid V = v]\, dv \ , \end{aligned}$$

which in fact follows from (27) on our supposing that V is uniformly distributed over $(0, U)$ — i.e. on our dividing the integral by the maximum value of V. This is just what Laplace advocates in the quotation following equation (26) above.

Now to the limits v_1 and v_2: for the lower limit Laplace takes that value of V which makes $\sqrt{r^2v^2(1 + d/r) - 2d}$ zero, viz. $\sqrt{2d/r(r + d)}$ (in Fabry's notation this becomes $\sqrt{2df/r(r + d)}$). Denoting this lowest value by v_0 and the upper limit by V_0, and defining z by

$$\sqrt{r^2v^2(1 + d/r) - 2df} = rv\sqrt{1 + d/r} - z \ , \qquad (28)$$

we have

$$\begin{aligned} I &\equiv \int \left[1 - \frac{\sqrt{1 - d/r}}{r} \sqrt{r^2v^2(1 + d/r) - 2df} \right] dv \\ &= v + (\sqrt{1 - d/r}/r) \left[z/2 - 2\sqrt{2df} \arctan\left(z/\sqrt{2df}\right) - df/z \right] + c \ . \end{aligned}$$

Now the integral being zero for $v = v_0$, or $z = \sqrt{2df}$, we have

$$c = -v_0 + 2\sqrt{2df}\left(\sqrt{1-d/r}\Big/ r\right)\pi/4 \ ,$$

and hence

$$I = v + \frac{\sqrt{1-d/r}}{r}\left[z/2 - 2\sqrt{2df}\ \arctan\left(z/\sqrt{2df}\right) - df/z\right] - \frac{\sqrt{2df}}{r\sqrt{1+d/r}} + 2\sqrt{2df}\frac{\sqrt{1-d/r}}{r}\frac{\pi}{4} \ . \tag{29}$$

The value of I between the limits v_0 and V_0 is then easily obtained.

As can be seen from the above manipulation, the upper limit for z is a complicated function of V_0. Laplace thus proposes a series solution, for which he lets $i = V_0\sqrt{r}$. The upper limit

$$z = i\sqrt{r}\ \sqrt{1+d/r}\left[1 - \sqrt{1 - 2df/i^2 r(1+d/r)}\right] \tag{30}$$

is then developed by Laplace (Fabry's notation) as

$$z = \frac{df}{i\sqrt{r}}\left[1 - \frac{d}{2r}\left(1 - \frac{f}{i^2}\right) + \cdots\right] \ ,$$

substitution of which in (29) yields

$$\frac{\sqrt{2df}}{r}\left(\frac{\pi}{2} - 1\right) - \frac{df}{ir\sqrt{r}} \ . \tag{31}$$

Fabry then shows that

$$\frac{1}{2U}\int_{v_0}^{V_0} \Pr\left[0 < D < d \mid V = v\right] dv = \frac{1}{2}\left[\frac{(\pi-2)\sqrt{2df}}{2Ur} - \frac{df}{iUr\sqrt{r}}\right] \ , \tag{32}$$

Laplace's value being twice this with $f = 1$.

Notice next that substitution of $i = v/r$ in $1/a = 2r - v^2/f$ yields

$$1/a = \left(2f - i^2\right)/fr \ .$$

The orbit is thus elliptic or parabolic according as $i^2 < 2f$ or $i^2 > 2f$. Supposing, for example, that $a = -100$, or $-100R$ in Fabry's notation, where R is the radius of the terrestrial orbit, we have

$$i^2 = (200R + r)f/100R \ .$$

At this stage Fabry's explanation errs a little: he starts to interpret probabilities as numbers — e.g. the expression (31) is referred to as

> le nombre des orbites dont la distance périhélie est inférieure à q [$\equiv d$] et qui sont elliptiques, paraboliques, ou hyperboliques avec un demi-grand axe supérieur à $100R$ en valeur absolue. [p.13]

It therefore seems better to follow Laplace's original here.

Denoting by A the event that the orbit is elliptic, parabolic or hyperbolic with semi-major axis at least equal to 100 (or $100R$ "en valeur absolue", Fabry [1893-1895, p.13]) we find, with i^2 as above,

$$\Pr[0 < D < d \ \& \ A] = \frac{1}{2}\left[\frac{(\pi-2)\sqrt{2df}}{2Ur} - \frac{10d\sqrt{f}}{rU\sqrt{r(200R+r)/R}}\right] .$$

Denoting by $i' \equiv U\sqrt{r}$ the value of i corresponding to the upper limit U of the velocity, we have, again from (31)

$$\Pr[0 < D < d] = \frac{1}{2}\left[\frac{(\pi-2)\sqrt{2df}}{2Ur} - \frac{df}{i'Ur\sqrt{r}}\right] .$$

Thus

$$\begin{aligned}\Pr\left[0 < D < d \ \& \ \overline{A}\right] &= \Pr[0 < D < d] - \Pr[0 < D < d \ \& \ A] \\ &= \frac{1}{2}\left[\frac{10d\sqrt{f}}{Ur\sqrt{r(200R+r)/R}} - \frac{df}{i'Ur\sqrt{r}}\right]\end{aligned}$$

where $\overline{A}$ denotes the event that the comets are "sensiblement hyperboliques" (Fabry [1893-1895, p.13]). Thus

$$\Pr[0 < D < d \,\&\, A] / \Pr\left[0 < D < d \,\&\, \overline{A}\right]$$

$$= \left[\frac{(\pi-2)\sqrt{2df}}{2} - \frac{10d\sqrt{f}}{\sqrt{r(200R+r)/R}}\right] \Bigg/ \left[\frac{10d\sqrt{f}}{\sqrt{r(200R+r)/R}} - \frac{df}{i'\sqrt{r}}\right] ,$$

an expression which of course depends on the value of U through i'. Letting i' (and therefore U) be infinite, we obtain

$$\frac{\Pr[0 < D < d \,\&\, A]}{\Pr\left[0 < D < d \,\&\, \overline{A}\right]} = \frac{(\pi-2)}{10}\sqrt{\frac{r}{2d}\left(200 + \frac{r}{R}\right)} - 1 , \tag{33}$$

or, in Laplace's words,

> ainsi la distance périhélie étant supposée comprise entre zéro et D, la probabilité que l'orbe sera ou une ellipse, ou une parabole, ou une hyperbole d'un demi-grand axe au moins égal à 100, est à la probabilité qu'il sera une hyperbole d'une demi-grand axe inférieur, comme
>
> $$\frac{(\pi-2)}{10}\sqrt{\frac{r}{2D}(r+200)}-1:1$$
>
> [p.95].

A numerical example now follows: taking $d = 2R$ ("la limite des distances périhélies des comètes que nous pouvons voir" — Fabry [1893-1895, p.14]) and $r = 10^5 R$, (33) yields the value 5712.668. As Laplace says,

> il y a donc à fort peu près cinquante-six à parier contre l'unité que, sur cent orbes cométaires observables, aucun ne doit être une hyperbole d'un demi-grand axe inférieur à 100. [p.95]

After noting that the preceding analysis supposes all values of D between 0 and 2 to be equally possible (for all comets which one can perceive), Laplace points out that comets with perihelion distance greater than one are much less numerous than those with this distance less than one. He next attempts to prove that the probability of sensibly hyperbolic comets is further diminished by this fact. Although the examination of this case is difficult, Todhunter [1865] dismisses it all with the words "he proceeds to consider how this will modify his result" [p.494]: we propose to be somewhat more explanatory here[81].

The introduction of $\varphi(D)$ — rather than 1 — as the prior on D is handled as in the 28th article of Laplace's *Mémoire sur les probabilités* of 1778. The method used here consists in differentiating both numerator and denominator with respect to D, multiplying each of the resulting expressions by $\varphi(D)$, and considering the ratio of the two resulting expressions[82].

Laplace thus proceeds as follows: note firstly that

$$\Pr[0<D<d]=\frac{(\pi-2)\sqrt{2d}}{2Ur}-\frac{d}{iUr\sqrt{r}}\rightarrow\frac{(\pi-2)\sqrt{2d}}{2Ur} \tag{34}$$

as $i\rightarrow\infty$, and also that

$$\Pr\left[0<D<d\ \ \&\ \ a>r/(2-i^2)\right]$$

$$=\Pr[0<D<d]-\Pr\left[0<D<d\ \ \&\ \ a<r/(2-i^2)\right]$$

$$= \Pr\left[0 < D < d\right] - \Pr\left[0 < D < d \;\&\; V^2 < i^2/r\right]$$

$$= \frac{(\pi-2)\sqrt{2d}}{2Ur} - \left[\frac{(\pi-2)\sqrt{2d}}{2Ur} - \frac{d}{iUr\sqrt{r}}\right]$$

$$= \frac{d}{iUr\sqrt{r}} \;. \tag{35}$$

The expressions (34) and (35) are then each differentiated with respect to d, each multiplied by $\varphi(d)$, and the products then integrated, their ratio giving

$$\Pr\left[a > r/(2-i^2) \mid 0 < D < d\right]$$

$$= \Pr\left[0 < D < d \;\&\; a > r/(2-i^2)\right] / \Pr\left[0 < D < d\right]$$

in the case in which $\varphi(d)$ is not necessarily identical to one. (This is in fact the procedure set out in Article 28 of Laplace's *Mémoire sur les probabilités.*)

Laplace checks that this ratio yields his previously derived result in the case in which $\varphi(d) \equiv 1$, and also considers the prior density $\varphi(d) = k\exp(-d^2)$, an assumption[83] which he finds supported by empirical data. In this case it transpires that

> il y a donc alors, à fort peu près, 8263 à parier contre l'unité qu'une nébuleuse qui pénètre dans la sphère d'activité du Soleil décrira un orbe dont le demi-grand axe sera au moins égal à 100. [p.97]

The final conclusion is

> ainsi l'on peut regarder la supposition de $\varphi(D)$ constant, et ne s'étendant que jusqu'à $D = 2$, comme la limite des suppositions favorables aux mouvements hyperboliques sensibles, en sorte qu'il y a au moins 56 à parier contre l'unité que, sur cent comètes observables, aucune n'aura un semblable mouvement. [p.97]

Comments on Laplace's results.

Students of celestial mechanics were not slow to realize that certain lacunae were evident in Laplace's exposition. Perhaps the first of those to point out the errors was Gauss, who devoted a major part of his review of 1815 of the appropriate volume of *Connaissance des temps*, to this memoir. Others who commented critically were Schiaparelli, Seeliger and Fabry[84].

There are essentially two points to which exception may be taken: the first of these is concerned with the lower limit v_1 (as given in the integral in equation (27)), while the other concerns the series expansion

$$z = \frac{df}{i\sqrt{r}}\left[1 - \frac{d}{2r}\left(1 - \frac{f}{i^2}\right) + \cdots\right] \tag{36}$$

(connected with this is the assumption that i' — or U — tends to infinity). We shall treat these matters *seriatim*, following Fabry in the main[85].

Firstly, in the integral

$$\int_{v_1}^{v_2}\left[1 - \frac{\sqrt{1-d/r}}{r}\sqrt{r^2v^2(1+d/r) - 2df}\right]dv$$

we have seen that Laplace takes as lower limit that value v_1 which makes $\sqrt{r^2v^2(1+d/r)-2df}$ zero, that is, $v_1 = \sqrt{2df/r(r+d)}$. Now there seems no reason for omitting smaller, positive values of V, though of course their inclusion will necessitate the addition of a term $\int_0^{v_1}\cdots dv$ (for an appropriate integrand, of which we shall say more anon) to the extant $\int_{v_1}^{v_2}$. Indeed, the perihelion distance is always less than d, no matter what the angle β may be, and these values of V should therefore not be left aside. This lapse, charitably described as an "Ueberheilung" by Gauss [1874, p.582], is not as serious as it might at first sight appear to be: the velocities omitted correspond always to elliptic orbits[86], and their inclusion thus serves but to strengthen Laplace's conclusion — indeed, Gauss shows that the odds of 56:1 found by Laplace for the observing, among 100 comets, of one which does not have a "sensibly hyperbolic orbit", are in fact raised to 157:1.

Secondly, let us pass on to the infinite series and the approximation used by Laplace. That something is indeed amiss is evident on our noting that, for any finite α,

$$\int_{\alpha}^{\infty}\left[1 - \frac{\sqrt{1-d/r}}{rv}\sqrt{r^2v^2(1+d/r) - 2df}\right]dv$$

is infinite[87], while Laplace's evaluation of this integral, obtained by a series expansion and a limiting process, yields a finite quantity.

In his discussion of this point Fabry points out that Laplace's development is correct when i is a quantity of moderate size, or more precisely when it is of the same order as $\sqrt{f}$. Under this assumption Fabry shows that the formulae (29) and (30) above lead to

$$z = \frac{df}{i\sqrt{r}}\left[1 - (d/2r)\left(1 - f/i^2\right)\right] \tag{37}$$

which is identical to (36), the terms neglected in the brackets being at least $(d/r)^2$. On substituting this value for z, and V, (or $i/\sqrt{r}$) for v in (29) Fabry obtains

$$\frac{\sqrt{2df}}{r}\left(\frac{\pi}{2}-1\right)-\frac{df}{ir\sqrt{r}} \tag{38}$$

as did Laplace. Fabry [1893-1895] emphasizes that the terms neglected here are at least of order $(d/r)^2$:

> la formule [(38)] peut donc bien remplacer la formule [(29)] dans le cas où i est une quantité finie de grandeur modérée (du même ordre que $\sqrt{f}$). [p.19]

The formula (33) above was derived under the assumption that $U \to \infty$ (and hence that $i' \to \infty$). In this case, however, certain of the terms neglected above become infinite, and these terms are no longer negligible with respect to the terms conserved which remain finite. If one continues the infinite series further than Laplace did one finds terms which become infinite for i infinite, and in this case (38) ceases to be equal to (29).

Moreover, as Gauss noted, the assumption that all values of the velocities are equally probable over $[0, \infty)$ is inadmissible[88], since this leaves an infinitely small probability for each finite velocity. This implies that orbits which are nearly parabolic will be infinitely less probable, and all the probability will on the contrary be placed on orbits which are indistinguishable from straight lines and will be traversed with infinite velocity[89].

Fabry notes that, even in the case in which U is infinite, the series expansion requires rectification: the velocities of celestial bodies being of the same order of magnitude as the velocity of the earth in its orbit (which value is $\sqrt{f/R}$), if U and V are magnitudes of this order, i is of order $\sqrt{fr/R}$. Fabry proposes then [p.20] to repeat his earlier development, keeping V in the calculation and carrying the expansions further. He shows that, on neglecting terms of order $1/r^3$ and higher [p.22], (29) becomes

$$\frac{\sqrt{2df}}{r}\left(\frac{\pi}{2}-1\right)\left(1-\frac{d}{2r}\right)+\frac{1}{2}\frac{d^2}{r^2}\left(V-\frac{2f}{Vd}\right) . \tag{39}$$

(He also verifies carefully [p.22] that the terms neglected in this latter expression are really negligible.) It is worth noting that if V becomes infinite so does (39), which is in accord with what we have already said.

Introducing these two corrections into Laplace's discussion[90] and expanding the series appropriately, Fabry shows that (29) becomes[91], on neglecting terms of order $1/r^3$,

$$\frac{\sqrt{2df}}{r}\frac{\pi}{2}\left(1-\frac{d}{2r}\right)+\frac{1}{2}\frac{d^2}{r^2}\left(V-\frac{2f}{Vd}\right) . \tag{40}$$

Thus the ratio of the number of comets which are not sensibly hyperbolic to the number of those which are is

$$\frac{\sqrt{2df}(\pi/2)(2r-d)+d^2\left[V-(2f/Vd)\right]}{d^2\left[U-V-(2f/d)\left(1/U-1/V\right)\right]}\ .$$

Here U is the (hypothesized) largest value of the velocities, and V is the velocity corresponding to the semi-major axis $-100R$ (or to whatever other semi-major axis we choose to separate the orbits of comets which are sensibly hyperbolic from those which are not).

Fabry now discusses a numerical example showing that comets with sensibly hyperbolic orbits ought to be exceedingly rare, as Laplace in fact asserted[92]. Much of Fabry's monograph is in fact devoted to the study of this question, taking into account complicating factors such as the movement of the sun in space and comets near to the sun: we shall not pursue the matter further here[93].

Let us conclude this discussion with two pertinent quotations. The first is from Gauss [1874]:

> wenn inzwischen die Wahrscheinlichkeitsrechnung auch gleich keinen entscheidenden Beweis *für* die Hypothese liefern kann, so entscheidet sie doch, eben wegen unsrer Unwissenheit über die Grenze U, auch durchaus nichts gegen die Hypothese. [p.583]

The second is from Schiaparelli [1874]:

> en 1813 les astronomes n'avaient pas beaucoup de confiance dans les spéculations de W. Herschel sur le mouvement propre du système solaire: on pouvait donc raisonnablement en exclure la considération. Cela n'est plus permis aujourd'hui. En reprenant donc le problème sous le point de vue de Laplace, mais avec la supposition que le système solaire se transporte dans l'espace avec la vitesse u comparable à celles des planètes, on trouvera non seulement un très grand excès de probabilité en faveur des orbites fortement hyperboliques, mais on verra de plus, que les hyperboles dont l'axe approche de la quantité $-1/u^2$ doivent être plus fréquentes que toutes les autres. Cela étant contraire à l'observation, il faut conclure que les comètes ne sont point des corps de nature stellaire. [p.80]

6.13 Two memoirs

In the *Connaissance des Temps* for 1818, printed in 1815, Laplace published two articles bearing on our subject, viz. *Sur l'application du calcul des prob-*

abilités a la philosophie naturelle and *Sur le calcul des probabilités appliqué a la philosophie naturelle.* The material of these papers being reproduced in the first Supplement to the *Théorie analytique des probabilités* (and part of the first paper being repeated in the introduction to that book), we shall postpone consideration of the contents until we discuss the latter work[94]. Indeed, there seems little in the two memoirs, as originally printed, which is pertinent.

6.14 Théorie analytique des probabilités

6.14.1 Introduction

The introduction to the *Théorie analytique des probabilités*[95], published separately under the title *Essai philosophique sur les probabilités*, is a much expanded version of a *Leçon* on probabilities delivered by Laplace at the *Écoles Normales* in 1795 under the title *Sur les probabilités*[96]. A sketch of certain passages in the *Essai* appeared in Laplace's *Notice sur les probabilités*[97] of 1810, and the *Essai* itself underwent drastic changes at Laplace's hands, from the first edition of 1814 to the fifth edition of 1825. It is to this last edition, the last to appear before Laplace's death, that attention will be paid here[98].

Seven general principles of probability are given in the section of the *Essai* entitled "Principes généraux du Calcul des Probabilités". The sixth and seventh of these are particularly pertinent to the present work, and are accordingly given below (all page references are to the Thom/Bru 1986 edition of the *Essai*).

> VI. Chacune des causes, auxquelles un événement observé peut être attribué, est indiquée avec d'autant plus de vraisemblance, qu'il est plus probable que, cette cause étant supposée exister, l'événement aura lieu; la probabilité de l'existence d'une quelconque de ces causes est donc une fraction dont le numérateur est la probabilité de l'événement, résultante de cette cause, et dont le dénominateur est la somme des probabilités semblables relatives à toutes les causes: si ces diverses causes considérées *a priori* sont inégalement probables, il faut au lieu de la probabilité de l'événement, résultante de chaque cause, employer le produit de cette probabilité, par la possibilité de la cause elle-même. C'est le principe fondamental de cette branche de l'Analyse des hasards, qui consiste à remonter des événements aux causes. [p.42]
>
> VII. La probabilité d'un événement futur est la somme des pro-

> duits de la probabilité de chaque cause, tirée de l'événement observé, par la probabilité que, cette cause existant, l'événement futur aura lieu. [p.44]

Symbolically the two parts of the sixth principle can be expressed as

$$\Pr[E \mid H_i] > \Pr[E \mid H_j] \Rightarrow \Pr[H_i \mid E] > \Pr[H_j \mid E]$$

$$\Pr[H_i \mid E] = \Pr[E \mid H_i] / \sum_j \Pr[E \mid H_j] \quad ,$$

both of which are true if $\Pr[H_i] = \Pr[H_j]$ for all i and j. More generally, of course,

$$\Pr[H_i \mid E] = \Pr[E \mid H_i] \Pr[H_i] / \sum_j \Pr[E \mid H_j] \Pr[H_j] \quad .$$

Laplace also notes that "Ce principe donne la raison pour laquelle on attribue les événements réguliers à une cause particulière" [p.42]. Pearson [1978, p.658] asserts that Laplace took this Principle without acknowledgement from Condorcet, who had in turn developed it from Bayes. While it is true that this result is not in Bayes's Essay, it is to be found in Laplace's *Mémoire sur la probabilité des causes par les événements* of 1774, a memoir which seems to antedate any pertinent writings published by Condorcet.

Principle VII may be symbolized as

$$\Pr[E_2 \mid E_1] = \sum_j \Pr[H_j \mid E_1] \Pr[E_2 \mid H_j] \quad ,$$

which is true if we assume the conditional independence of E_1 and E_2 with respect to $\{H_j\}$. (This formulation is supported by the example given by Laplace following its presentation.)

The seventh principle is followed by a discussion of the case in which the probability of the simple event is unknown. The suggestion in this case is to suppose all values from zero to one equally probable. The pertinent passage is perhaps a little confusing, and since it embodies a discrete form of Bayes's Theorem, we quote it here in full:

> Quand la probabilité d'un événement simple est inconnue, on peut lui supposer également toutes les valeurs depuis zéro jusqu'à l'unité. La probabilité de chacune de ces hypothèses, tirée de l'événement observé, est par le sixième principe une fraction dont le numérateur est la probabilité de l'événement dans cette hypothèse, et dont le dénominateur est la somme des probabilités semblables relatives à toutes les hypothèses. Ainsi,

> la probabilité que la possibilité de l'événement est comprise dans des limites données, est la somme des fractions comprises dans ces limites. Maintenant, si l'on multiplie chaque fraction par la probabilité de l'événement futur, déterminée dans l'hypothèse correspondante, la somme des produits relatifs à toutes les hypothèses sera par le septième principe la probabilité de l'événement futur, tirée de l'événement observé. [p.45]

In modern terms this may be written as follows: let E denote the initial single event, O the observed event, F the future event, and H_i the hypothesis $\Pr[E] = p_i$, where $p_i \in [0,1]$ and $i \in \{1,2,\ldots,n\}$. Then

$$\Pr[H_i \mid O] = \Pr[O \mid H_i] / \sum_j \Pr[O \mid H_j]$$

and

$$\begin{aligned}\Pr[x < p_i < x' \mid O] &= \sum{}' \Pr[H_i \mid O] \\ &= \sum{}' \Pr[O \mid H_i] / \sum_j \Pr[O \mid H_j] \quad ,\end{aligned}$$

where $\sum{}'$ indicates that the summation is taken over all $p \in (x, x')$. Furthermore,

$$\Pr[F \mid O] = \sum_j \Pr[F \mid H_j] \Pr[O \mid H_j] / \sum_j \Pr[O \mid H_j] \quad .$$

It is assumed here, of course, that $\Pr[H_i] = \Pr[H_j]$, and that F and O are independent with respect to $\{H_i\}$.

From this the rule of succession follows. Laplace applies this principle to the problem of the sun's rising, and points out that his solution is different from Buffon's[99], since

> la vraie manière de remonter des événements passés à la probabilité des causes et des événements futurs, était inconnue à cet illustre écrivain. [p.46]

(See §4.6 of the present work for a discussion of Buffon's result[100].)

There is nothing more that is pertinent to our present study until we reach the eleventh section, entitled "De la probabilité des témoignages", in which Laplace applies his earlier principle on the probability of causes elicited from observed events[101]. Here Principle VI is used to estimate the veracity of a witness, as shown in the following example: suppose that a number has been drawn [at random] from an urn containing 1000 numbers [assumed distinct]. A witness to the drawing announces that the number 79 was drawn: what is the probability that he tells the truth? Let E denote the event that it is announced that the number drawn is 79: then

$$\begin{aligned}\Pr[E] &= \Pr[E \mid \text{witness lies}]\,\Pr[\text{witness lies}] \\ &\quad + \Pr[E \mid \text{witness tells the truth}]\,\Pr[\text{witness tells the truth}]\ . \qquad (41)\end{aligned}$$

Now from prior experience it is known that

$$\Pr[\text{witness lies}] = \frac{1}{10} = 1 - \Pr[\text{witness tells the truth}]\ .$$

Furthermore,

$$\Pr[E \mid \text{witness tells the truth}] = \Pr[79 \text{ drawn}] = \frac{1}{1000}$$

and

$$\begin{aligned}\Pr[E \mid \text{witness lies}] &= \Pr[E \mid 79 \text{ not drawn}] \times \Pr[79 \text{ not drawn}] \\ &= \frac{1}{999} \times \frac{999}{1000} = \frac{1}{1000}\ .\end{aligned}$$

(These probabilities are stated to be determined *a priori*. Note the tacit equiprobability assumption.) Thus, finally,

$$\begin{aligned}\Pr[\text{witness tells the truth} \mid E] &= \Pr[E \mid \text{witness tells the truth}] \times \\ &\qquad \Pr[\text{witness tells the truth}] \,/\, \Pr[E] \\ &= \frac{9}{10000} \Big/ \left(\frac{9}{10000} + \frac{1}{10000}\right) \\ &= \frac{9}{10}\ .\end{aligned}$$

(Similarly for falsehood.) Laplace also considers the case in which the witness has an interest in the number drawn, and discusses how this will affect the final result.

> Le bon sens nous dicte que cet intérêt doit inspirer de la défiance, mais le calcul en apprécie l'influence. [p.120]

A further example, in similar vein, concerning an urn containing one white and 999 black balls shows that

> la probabilité de l'erreur ou du mensonge du témoin devient d'autant plus grande, que le fait attesté est plus extraordinaire [p.122],

while another example admits of the possibility of the witness's being mistaken: in this case the two hypotheses of the earlier example are replaced by the following four[102]:

(i) the witness neither lies nor is deceived;

(ii) the witness does not lie but is deceived;

(iii) the witness lies and is not deceived;

(iv) the witness lies and is deceived.

And a further example shows that

> une conséquence impossible est la limite des conséquences extraordinaires, comme l'erreur est la limite des invraisemblances; la valeur des témoignages, qui devient nulle dans le cas d'une conséquence impossible, doit donc être très affaiblie dans celui d'une conséquence extraordinaire. [p.123]

In the thirteenth section, "De la probabilité des jugements des tribunaux", Laplace uses a Bayes-type argument. It will be more convenient, however, to postpone discussion of this point until consideration of the First Supplement to the *Théorie analytique des probabilités.*

In the final section of the *Essai* we find the only reference to Bayes, viz.

> Bayes, dans les *Transactions Philosophiques* de l'année 1763, a cherché directement la probabilité que les possibilités indiquées par des expériences déjà faites sont comprises dans les limites données, et il y est parvenu d'une manière fine et très ingénieuse, quoiqu'un peu embarrassée. Cet objet se rattache à la théorie de la probabilité des causes et des événements futurs, conclue des événements observés, théorie dont j'exposai quelques années après les principes, avec la remarque de l'influence des inégalités qui peuvent exister entre les chances que l'on suppose égales. Quoique l'on ignore quels sont les événements simples que ces inégalites favorisent, cependant cette ignorance même accroît souvent la probabilité des événements composés. [pp.200-201]

Finally, as a summary of the *Essai*, let us note the following sentence from the last paragraph:

> on voit par cet Essai que la théorie des probabilités n'est au fond que le bon sens réduit au calcul: elle fait apprécier avec exactitude, ce que les esprits justes sentent par une sorte d'instinct, sans qu'ils puissent souvent s'en rendre compte. [p.206]

6.14.2 Livre 1: Calcul des fonctions génératrices

This, the first of the two Books into which the *Théorie analytique des probabilités*[103] is divided, does not contain anything pertinent to our present topic. It is devoted to a study of generating functions and (in the second part) certain approximations of functions of large numbers[104].

6.14.3 Livre 2: Théorie générale des probabilités

In the first chapter, entitled "Principes généraux de cette theorie", Laplace presents[105] in both words and mathematical symbols, those "principes généraux de l'Analyse des Probabilités" [p.190] of which he had already written in the *Essai.*

He begins by stating his usual definition of the probability of an event, viz.[106]

> la probabilité d'un événement est le rapport du nombre des cas qui lui sont favorables au nombre de tous les cas possibles, lorsque rien ne porte à croire que l'un de ces cas doit arriver plutôt que les autres, ce qui les rend, pour nous, également possibles [p.181],

and he goes on to say

> la juste appréciation de ces cas divers est un des points les plus délicats de l'Analyse des hasards. [p.181]

Cognisance is also taken of the situation in which the cases are not equally possible:

> si tous les cas ne sont pas également possibles, on déterminera leurs possibilités respectives, et alors la probabilité de l'événement sera la somme des probabilités de chaque cas favorable. [p.181]

These definitions are then expressed in mathematical symbols.

Laplace next passes to the question of independence, pointing out that if $\{E_i\}$ is a sequence of independent simple events, with $p_i = \Pr[E_i]$, then $\Pr[E_1E_2\ldots E_n] = \prod_1^n p_i$. Attention is then turned to dependence, and Laplace states that in the case of two simple events where the supposition of the occurrence of the first (E_1) affects the probability of the occurrence of the second (E_2), we have $\Pr[E_1E_2] = \Pr[E_2 \mid E_1]\Pr[E_1]$.

Laplace also provides an explanation of the *a priori* determination of probability: he says that the probability of an event is determined *a priori* "ou indépendamment de ce qui est déjà arrivé" [p.183]. He thus deduces what he describes as "ce nouveau principe" [p.183], that for any future event E_2 depending on an observed event E_1 (and E_1, E_2 need not necessarily be simple), $\Pr[E_2 \mid E_1] = \Pr[E_1E_2]/\Pr[E_1]$, where each term on the right-hand side is determined *a priori*.

One might call this a *prospective* use of conditional probability; however, immediately after stating this principle Laplace goes on to frame a *retrospective* definition[107]: he writes

> de là découle encore cet autre principe relatif à la probabilité des causes, tirée des événements observés [p.183],

and he then formulates the following expressions: for a sequence $\{H_i\}$ of causes and an event E,

(i) $\Pr[H_i] : \Pr[H_j] :: \Pr[E \mid H_i] : \Pr[E \mid H_j], \quad i \neq j$

(ii) $\Pr[H_i \mid E] = \Pr[E \mid H_i]/\sum_j \Pr[E \mid H_j]$.

No mention of *a priori* equipossibility is initially made, though it is explicitly stated when the translation into symbols is effected, and the corresponding formula for the non-equipossible case is also given [p.184].

Following an example, Laplace states the following principle:

> la probabilité d'un événement futur est la somme des produits de la probabilité de chaque cause, tirée de l'événement observé, par la probabilité que, cette cause existant, l'événement futur aura lieu [p.186],

or[108] $\Pr[E_2 \mid E_1] = \sum_j \Pr[H_j \mid E_1] \Pr[E_2 \mid H_j]$.

Consideration is given to the question of obtaining two heads (say) in two tosses of a coin known to be biased, though whether in favour of heads or tails is unknown. The first toss of this coin will yield heads still with probability $\frac{1}{2}$, though the probability of the result of the second toss is modified. This procedure is then extended to any events whatsoever [pp.188-189].

Finally, there is some discussion of moral and mathematical expectation, with the usual definition of the latter being given [p.189].

In Chapter II, "De la probabilité des événements composés d'événements simples dont les possibilités respectives sont données", Laplace contrasts [p.193] "l'espérance mathématique" with "crainte": thus the translation of the former as "mathematical hope" rather than "mathematical expectation" is not altogether something to be avoided.

Perhaps the only other point worth noting here is that Laplace shows [pp.278-279] that the appropriate law to describe a distribution of errors is

$y = a\exp(-2ax)$, under the assumptions that the law is initially unknown, that the probabilities tend to zero as the errors increase in absolute value, and that the probability of an error of $+\epsilon$ is equal to that of $-\epsilon$.

Chapter III, "Des lois de la probabilité qui résultent de la multiplication indéfinie des événements", is devoted to a proof of Bernoulli's Theorem and certain examples using this result — a result of which Laplace writes

> la détermination de ces accroissements et de ces limites est une des parties les plus intéressantes et les plus délicates de l'analyse des hasards. [p.280]

Laplace shows firstly that, if p and $1-p$ are the respective probabilities of two events A and B, then in a very large number of trials ("coups"),

(i) the most probable of all combinations which can arise is that in which each event is repeated proportionally to its probability, and

(ii) the probability that the difference between the ratio of the number of times that the event A can occur to the total number of trials, and the "facilité" p of that event, lies between the limits

$$\frac{x-np}{n} \pm \frac{t\sqrt{2xx'}}{n\sqrt{n}} \ ,$$

is

$$\frac{2}{\sqrt{\pi}}\int_0^t e^{-u^2}\,du + \frac{\sqrt{n}}{\sqrt{2\pi xx'}}e^{-t^2} \tag{42}$$

where n is the total number of trials, in which A and B occurred x and x' times respectively, $t = l\sqrt{n}\Big/\sqrt{2xx'}$, and l is that term in the expansion of $[p+(1-p)]^{x+x'}$ which contains $p^{x-l}(1-p)^{x'+l}$. (The approximation (42) is derived under the assumptions that terms of order $1/n$ are neglected, and that l^2 does not exceed n in order of magnitude.)

Although I do not intend to prove the above result here, it is perhaps not inadvisable briefly to outline some steps in Laplace's procedure. To this end, suppose that p and $q \equiv 1-p$ are the (initial) probabilities of the events A and B respectively. Denoting by X the random variable indicating the number of times A occurs in n trials, we require $\Pr[\alpha_n < X < \beta_n \mid n, p]$.

Now the probability that A and B occur x and $x' = n-x$ times respectively is

$$\binom{n}{x}p^x(1-p)^{x'} \ ,$$

the greatest value of which is achieved when $p : 1-p :: x : x'$. Laplace then shows that $\Pr[x - r < X < x + r \mid n, p]$ is approximately

$$\frac{2}{\sqrt{\pi}} \int_0^t e^{-u^2}\, du + \frac{\sqrt{n}}{\sqrt{2\pi x x'}} e^{-t^2} \quad , \tag{43}$$

that is, the approximate value of the sum of $(2r+1)$ terms of the expansion of $[p + (1-p)]^n$, the greatest term in this expansion being the middle term in the $(2r+1)$ terms. Writing $x = np + z$ and $t = r\sqrt{n} \Big/ \sqrt{2xx'}$, we see that (43) gives the probability

$$\Pr\left[np + z - t\sqrt{2xx'} \Big/ \sqrt{n} \leq X \leq np + z + t\sqrt{2xx'} \Big/ \sqrt{n}\right] \tag{44}$$

or

$$\Pr\left[z/n - t\sqrt{2xx'} \Big/ n\sqrt{n} \leq X/n - p \leq z/n + t\sqrt{2xx'} \Big/ n\sqrt{n}\right] \quad . \tag{45}$$

This is the result given in (ii) above, and is as far as Laplace takes the direct result, although some remarks not pertinent to our present investigations follow.

Both Keynes [1921, chap.30] and Todhunter [1865, art.993], however, carry the argument further. Supposing that for large n, z may be ignored in comparison with np, we have $xx' \approx n^2pq$. In this case the probability (45) becomes

$$\Pr\left[-t\sqrt{2pq} \Big/ \sqrt{n} \leq X/n - p \leq t\sqrt{2pq} \Big/ \sqrt{n}\right] \quad , \tag{46}$$

this being approximately given by

$$\frac{2}{\sqrt{\pi}} \int_0^t e^{-u^2}\, du + \frac{1}{\sqrt{2\pi npq}} e^{-t^2} \quad . \tag{47}$$

This latter expression is in fact the only one given by Keynes (loc. cit.), who makes no mention of the expressions (43) and (44), the only ones given by Laplace.

Laplace proceeds next to an inverse form of the theorem[109], writing

> si l'on connaît le nombre de fois que sur n coups l'événement a [$\equiv A$] est arrivé, le formule (o) [$\equiv$ (42)] donnera la probabilité que sa facilité p, supposée inconnue, sera compris dans des limites données. [p.286]

This is shown as follows: denoting by i the number of times that A occurs in the n trials, Laplace states that his preceding result gives the probability that $i/n - p$ will be contained within the limits

$$\frac{z}{n} \pm \frac{T\sqrt{2xx'}}{n\sqrt{n}} \tag{48}$$

where T is the limit of t. Since $T\sqrt{2xx'}\,/n\sqrt{n}$ is of order $1/\sqrt{n}$, and since terms of order $1/n$ are being neglected in deriving the approximations, one may substitute i for x and $n - i$ for x', with the result that the limits in (48) become

$$\frac{i}{n} \pm \frac{T\sqrt{2i(n-i)}}{n\sqrt{n}} \ . \tag{49}$$

It thus follows that the probability that the "facilité" p of A lies within these limits is given by

$$\frac{2}{\sqrt{\pi}}\int_0^T e^{-u^2}\,du + \frac{\sqrt{n}e^{-T^2}}{\sqrt{2\pi i(n-i)}} \ . \tag{50}$$

From this Laplace concludes that, as n increases, the interval of the limits contracts, and the probability that p falls within these limits approaches 1. "C'est ainsi que les événements, en se développant, font connaître leurs probabilités respectives" [p.287].

However, Laplace's discussion does not end here. He proposes an alternative method (for trenchant criticism of which see Keynes [1921, chap.30]) with the words

> on parvient directement à ces résultats, en considérant p comme une variable qui peut s'étendre depuis zéro jusqu'à l'unité, et en déterminant, d'après les événements observés, la probabilité de ses diverses valeurs, comme on le verra lorsque nous traiterons de la probabilité des causes déduite des événements observés. [p.287]

This alternative procedure is explored in Chapter VI: we shall postpone discussion of it for the nonce.

The rest of the present chapter contains applications of this result to various urn problems, and consideration is also given (i) to the result obtained when more than two events are considered, and (ii) to the case in which the probability p is replaced by some specific function $f(p)$.

Chapter IV, "De la probabilité des erreurs des résultats moyens d'un grand nombre d'observations et des résultats moyens les plus avantageux", contains the development of least squares theory. Writing of this chapter Todhunter [1865] calls it "the most important in Laplace's work, and perhaps the most difficult" [art.1001], and he goes on to say

> Laplace's processes in this Chapter are very peculiar, and it is scarcely possible to understand them or feel any confidence in their results without translating them into more usual mathematical language. [art.1001]

Happily only the last two articles (numbers 23 and 24) of this chapter are at all pertinent to the present work.

In Article 23 Laplace proposes to switch from the consideration of observations not yet made, to the consideration of the mean result of observations already made, and whose respective deviations ("écarts") are known. Consider s observations, with results $A, A+q, A+q^{(1)}, \ldots$, with the same law of errors (here $q, q^{(1)}, \ldots$ may, without loss of generality, be assumed positive and increasing). If $A+x$ is the true result, the errors of the first, second, third ... observations are then $-x, q-x, q^{(1)}-x, \ldots$. Denoting by $\varphi(z)$ the probability of the error z (the same for each observation), we see that the probability of the simultaneous existence of all the errors is

$$\varphi(-x)\,\varphi(q-x)\,\varphi(q^{(1)}-x)\ldots$$

In considering the infinity of values of which x is supposed susceptible as the causes of the observed event, we find, from Article 1, that "la probabilité de chacune d'elles sera" [p.339] or, as Todhunter [1865] has it, "the probability that the true value lies between x and $x+dx$" [art.1013] is[110]

$$\varphi(-x)\,\varphi(q-x)\ldots dx \Big/ \int \varphi(-x)\,\varphi(q-x)\ldots dx \ ,$$

the integral being taken over all values of x. Denote the denominator by $1/H$.

Now consider a curve with abscissa x and ordinate

$$y = H\varphi(-x)\,\varphi(q-x)\ldots$$

Laplace states quite baldly that

> la valeur qu'il faut choisir pour résultat moyen est celle qui rend l'erreur moyenne à craindre un minimum [p.339],

and he goes on to say that the mean value of the error to be apprehended ("l'erreur à craindre") is the sum of the products of each error, all regarded as positive, by its probability. To determine the abscissa necessary to be chosen to minimize this sum, Laplace places a new origin at the left-hand end ("la première extrémité") of the curve, with co-ordinates now denoted by $x^{'}$ and $y^{'}$. If l is the value to be chosen then (see the discussion in §6.3 above)

$$\int_0^l y^{'}\,dx^{'} = \int_l^{\max x^{'}} y^{'}\,dx^{'} \ .$$

> Il suit de là que l'abscisse qui rend l'erreur moyenne à craindre un minimum est celle dont l'ordonée divise l'aire de la courbe en deux parties égales. [p.340]

This number is called the *milieu de probabilité.* Contrasting his value with that given by earlier mathematicians, Laplace writes

> des géomètres célèbres ont pris pour le milieu qu'il faut choisir celui qui rend le résultat observé le plus probable, et par conséquent l'abscisse qui répond à la plus grande ordonnée de la courbe; mais le milieu que nous adoptons est évidemment indiqué par la théorie des probabilités. [p.340]

Supposing next that $\varphi(x) = \exp\big(-\psi\,(x^2)\big)$ (i.e. assuming only that positive and negative errors are equally likely), one has

$$y = H\exp\Big(-\psi(x^2) - \psi(x-q)^2 - \psi(x-q^{(1)})^2 - \ldots\Big) \ .$$

Laplace then shows that, taking the "*average* of the results furnished by observations as the *most probable* result" [Todhunter 1865, art.1014], y is necessarily given by

$$y = \sqrt{\frac{k}{\pi}}\, e^{-kx^2}$$

(where k is constant). Laplace also notes the equivalence of the above assumption with that of the method of least squares, with the words

> cette valeur [de x qu'il faut choisir pour résultat moyen des observations] est celle que donne la règle des milieux arithmétiques; la loi précédente des erreurs de chaque observation donne donc constamment les mêmes résultats que cette règle, et l'on a vu qu'elle est la seule loi qui jouisse de cette propriété [p.344]

and

> la loi précédente des erreurs de chaque observation conduit donc aux mêmes résultats que cette méthode [i.e. la méthode des moindres carrés des erreurs des observations]. [p.345]

Writing further of the method of least squares of errors, Laplace says that it

> devient nécessaire lorsqu'il s'agit de prendre un milieu entre plusiers résultats donnés, chacun, par l'ensemble d'un grand nombre d'observations de divers genres. [p.345]

A detailed discussion is given by Todhunter [1865, art.1015], an argument similar to that discussed earlier in the present article being employed. Some history of the method of least squares is given in Article 24.

Chapter V is entitled "Application du calcul des probabilités à la recherche des phénomènes et de leurs causes". It contains nothing pertinent to our study, and we accordingly pass on immediately to Chapter VI, "De la probabilité des causes et des événements futurs, tirée des événements observés". Here we find the alternative method of inverting Bernoulli's Theorem suggested by Laplace in his third chapter.

Since, Laplace argues, the probability of most simple events is unknown, in considering this probability *a priori*, "elle nous paraît susceptible de toutes les valeurs comprises entre zéro et l'unité" [p.370]. Calling the law followed by the true possibility of the simple event x, Laplace notes that the theory discussed in preceding chapters yields the probability of the observed result as a function y of x. By the third principle of his first article it follows that the probability of x (say p_x) is equal to

> une fraction dont le numérateur est y, et dont le dénominateur est la somme de toutes les valeurs de y. [p.370]

Laplace then multiplies both numerator and denominator of this fraction by dx, to get

$$p_x\,dx = y\,dx \Big/ \int_0^1 y\,dx \ .$$

Hence

$$\Pr\left[\theta < x < \theta'\right] = \int_\theta^{\theta'} y\,dx \Big/ \int_0^1 y\,dx \ .$$

Further, let a denote that value of x which maximizes y (the most probable value of x).

Laplace next notes that if the values of x are not equally possible (considered independently of the observed result), then on our denoting by z "la fonction de x qui exprime leur probabilité" [p.371], it follows from Chapter I that

$$\Pr\left[\theta < x < \theta'\right] = \int_\theta^{\theta'} yz\,dx \Big/ \int_0^1 yz\,dx \ .$$

Since this amounts to considering all values of x as equipossible, with the observed result as being formed from two independent results with probabilities y and z, Laplace proposes, in what follows, always to adopt the equipossible hypothesis.

The results of Book I of the *Théorie analytique des probabilités* on the evaluation of definite integrals by approximations, are to be used here to determine the law of probability of the values of x as they deviate from the most probable value a. It is perhaps worth noting that, since Laplace is usually concerned with data drawn from a large number of observations, most of the integrands occurring in this chapter are of the form $\exp(-kt^2)$.

Laplace shows here [pp.371-374] that

$$\Pr\left[a - t\sqrt{\alpha}/k < x < a + t\sqrt{\alpha}/k\right] = \frac{2}{\sqrt{\pi}}\int_0^\infty e^{-t^2}\,dt$$

where α is an extremely small fraction, $k = \left(-\frac{\alpha}{2Y}\frac{d^2Y}{dx^2}\right)^{1/2}$ and $Y = y(x)|_{x=a}$. Notice the following observation:

> il résulte de cette expression que la valeur de x la plus probable est a, ou celle qui rend l'événement observé le plus probable, et qu'en multipliant à l'infini les événements simples dont l'événement observé se compose, on peut à la fois resserrer les limites $a \pm t\sqrt{\alpha}/k$, et augmenter la probabilité que la valeur de x tombera entre ces limites; en sorte qu'à l'infini, cet intervalle devient nul, et la probabilité se confond avec la certitude. [p.374]

Attention is next focused on a double Bayes's integral: supposing that the observed event depends on simple events of two different types, of possibilities x and x' respectively, we find that

$$\Pr\left[\theta_1 < x < \theta_2, \theta_1' < x' < \theta_2'\right] = \int_{\theta_1'}^{\theta_2'}\int_{\theta_1}^{\theta_2} y\,dx\,dx' \bigg/ \int_0^1\int_0^1 y\,dx\,dx'$$

(notation altered), where y denotes the probability of the observed compound event. Once again Laplace passes almost immediately from this expression to one with integrand $\exp(-t^2 - u^2)$.

Then follows a brief comment to the effect that, in the drawing of a large number n of balls from an urn containing balls of many colours, p of which draws result in balls of the first colour, q of the second, r of the third etc., the probabilities $x, x', x'', \ldots$ which render the observed event most probable are the observed sample frequencies.

> Ainsi les valeurs les plus probables sont proportionnelles aux nombres des arrivées des couleurs, et lorsque le nombre n est un grand nombre, les probabilités respectives des couleurs sont à très peu près égales aux nombres de fois qu'elles sont arrivées divisés par le nombre des tirages. [p.376]

Some examples involving Laplace's method of approximation now follow: in the first of these two players A and B play a match subject to the condition that the first to win two out of three games wins the match. If in a large number n of matches A has won i, then the probability that x, the probability that A wins a game, lies between $a - r/\sqrt{n}$ and $a + r/\sqrt{n}$ (with a determined as before) turns out to be

$$\frac{6\sqrt{2}}{\sqrt{\pi(3-2a)(1+2a)}} \int_0^\infty \exp(-18r^2 /(3-2a)(1+2a))\, dr$$

(compare the method used in Chapter III). Various ramifications of this situation follow: we shall not explore them here.

The second example is concerned with births — the sort of question to which Laplace finds his preceding analysis chiefly applicable. Laplace proposes to find the probability that the possibility x of the birth of a boy in Paris exceeds $\frac{1}{2}$, based on data[111] for the 40-year period 1745-1784. If p and q denote the numbers of male and female births respectively, then

$$\Pr[0 < x < 1/2] = \int_0^{1/2} y\, dx \Big/ \int_0^1 y\, dx$$

where $y = x^p(1-x)^q$. The righthand side of this expression being approximately

$$\frac{(p+q)^{p+q+3/2}}{(p-q)\sqrt{\pi}\, 2^{p+q+3/2} p^{p+\frac{1}{2}} q^{q+\frac{1}{2}}} \left[1 - \frac{p+q}{(p-q)^2} - \frac{(p+q)^2 - 13pq}{12pq(p+q)} - \cdots\right] ,$$

we find that for the observed values $p = 393,386$ and $q = 377,555$,

$$\Pr[0 < x < 1/2] = \frac{1}{\mu}(1 - 0.0030761) ,$$

where μ, a complicated function of p and q, has $\ln \mu = 72.2511780$. Thus the probability that, in Paris, the possibility of the birth of a boy exceeds that of a girl is very close to 1, and hence

> l'on voit que l'on doit regarder cette probabilité comme étant égale, au moins, à celle dans faits historiques les plus avérés. [p.387]

The third problem, discussed in Section 29, is also concerned with births: more precisely, having noticed that the ratios of male to female births are 19:18 in London and 25:24 in Paris[112], Laplace proposes to determine the probability of the constant cause to which he attributes the difference in the ratios. Denoting by p and q the numbers of baptisms of boys and girls respectively in Paris, and by p' and q' the similar numbers in London, he shows that the probability that the possibility of the baptism of a boy is greater in London than in Paris is approximately given (for large values of p, q, p', q') by

$$\frac{k}{\sqrt{\pi}} \int_0^\infty e^{-k^2(t-h)^2}\, dt \ ,$$

where

$$k^2 = (p+q)^3(p'+q')^3 \Big/ \Big[2p'q'(p+q)^3 + 2pq(p'+q')^3\Big]$$

$$h = (p'q - pq') \Big/ (p+q)(p'+q') \ .$$

Substitution of the values

$$\begin{array}{ll} p = 393,386 & q = 377,555 \\ p' = 737,629 & q' = 698,958 \end{array}$$

shows that it is 328,268 to one that the possibility of the baptism of a boy is greater in London than in Paris. (Laplace notes further that the baptism of foundlings in Paris has a sensible effect on the ratio observed in that city.)

The examples discussed in Sections 30 and 31 refer to the probabilities of *future* events, and thus, as Todhunter [art.1029] has noted, they are more logically placed after Section 32. We shall accordingly postpone discussion of them for the moment, turning rather to the problems of Section 32, where we find discussed[113]

> la probabilité des événements futurs, tirée des événements observés, et supposons qu'ayant observé un événement composé d'un nombre quelconque d'événements simples, on cherche la probabilité d'un résultat futur, composé d'événements semblables. [p.401]

Denoting by x the probability of each simple event, by y the probability of the observed result, and by z that of the future result, Laplace states that P, "la probabilité entière de l'événement futur", is given by

$$P = \int_0^1 yz\, dx \Big/ \int_0^1 y\, dx \ .$$

No proof of this result is presented: we provide the following heuristic argument as corroboration.

Let $\varphi_m \equiv \varphi(E_{i_1}, E_{i_2}, \ldots, E_{i_m})$ be the observed result depending on the simple events $E_{i_1}, E_{i_2}, \ldots, E_{i_m}$, and $\psi_n \equiv \psi(E_{i_1}, E_{i_2}, \ldots, E_{i_n})$ be the future result. Let $H_i = [x_{i-1}, x_i)$, where $i \in \{1, 2, \ldots, N-1\}$, $H_N = [x_{N-1}, x_N]$, and $0 = x_0 < x_1 < \cdots < x_N = 1$: the H_i are mutually exclusive and exhaustive, with $\bigcup_1^N H_i = [0,1]$. Finally, let $\Pr[H_i] = x_i - x_{i-1} \equiv \Delta x$ (i.e. an assumption of equiprobability) and let $y_m^{(i)}(\xi)$ and $z_n^{(i)}(\xi)$ be given, for any $\xi \in H_i$, by

$$y_m^{(i)}(\xi) = \Pr[\varphi_m \mid H_i], \qquad z_n^{(i)}(\xi) = \Pr[\psi_n \mid H_i] \ .$$

Then

$$\begin{aligned}\Pr[\varphi_m] &= \sum_{i=1}^N \Pr[\varphi_m \mid H_i] \Pr[H_i] \\ &= \sum_{i=1}^N y_m^{(i)}(\xi)\, \Delta x \ ,\end{aligned}$$

and by the usual sort of limiting argument (refining the partition), we obtain

$$\Pr[\varphi_m] = \int_0^1 y_m(x)\, dx \ .$$

Similarly, assuming that $\Pr[\varphi_m \psi_n \mid H_i] = \Pr[\varphi_m \mid H_i] \Pr[\psi_n \mid H_i]$, we have

$$\Pr[\psi_n\, \varphi_m] = \int_0^1 y_m(x) z_n(x)\, dx \ .$$

Thus

$$\Pr[\psi_n \mid \varphi_m] = \int_0^1 y_m(x) z_n(x)\, dx \Big/ \int_0^1 y_m(x)\, dx \ . \tag{51}$$

As a first example illustrating the use of this result, Laplace supposes that an event has happened m times running. If the probability of the simple event is x, the probability that the event will occur the next n times is

$$\begin{aligned}P &= \int_0^1 x^{m+n}\, dx \Big/ \int_0^1 x^m\, dx \\ &= (m+1)/(m+n+1) \ ,\end{aligned}$$

a result which we recognize as the rule of succession. It follows (though Laplace does not give this extension here) that if

$$y(x) = x^m(1-x)^n, \quad z(x) = x^p(1-x)^q \ ,$$

then

$$P = \frac{\Gamma(m+p+1)\Gamma(n+q+1)\Gamma(m+n+2)}{\Gamma(m+n+p+q+2)\Gamma(m+1)\Gamma(n+1)} \quad . \tag{52}$$

Laplace next supposes that the observed event is composed of a vast number of simple events. If the future event is composed of relatively few simple events, then $P = Z$ (approximately), where Z is the value of $z(x)$ for $x = a$, the value of x which maximizes $y(x)$. (This can be seen in a special case by applying the Stirling-de Moivre approximation to the factorials in (52) above: Laplace's argument is more general.) If, next, the future result is a function of the observed result, so that $z = \varphi(y)$, say, then it turns out that

$$P = \varphi(Y) \Big/ \sqrt{1 + Y\varphi'(Y)/\varphi(Y)}$$

where $Y = y(a)$. In particular, if $\varphi(y) = y^n$, then

$$P = Y^n \Big/ \sqrt{1+n} \quad .$$

On the other hand, the probability of the future result, given $x = a$, is Y^n. Thus, notes Laplace,

> on voit ainsi que les petites erreurs qui résultent de cette supposition [i.e. that $x = a$] s'accumulent à raison des événements simples qui entrent dans le résultat futur, et deviennent très sensibles lorsque ces événements sont en grand nombre. [p.404]

In Section 30 Laplace turns his attention to the probabilities of results based on tables of mortality or assurance, constructed from a large number of observations. He supposes firstly that

> sur un nombre p d'individus d'un âge donné A, on ait observé qu'il en existe encore le nombre q à l'âge $A + a$; on demande la probabilité que, sur p' individus de l'âge A, il en existera $q' + z$ à l'âge $A + a$, la raison de p' et q' étant la même que celle de p à q. [p.392]

The solution to this problem is clearly given by formula (51), with

φ_m — of p people aged A, q survive to age $A+a$;

ψ_n — of p' people aged A, $q'+z$ survive to age $A+a$;

Y_m — Pr [q people survive to age $A+a \mid p$ survive to age A]

$$= \binom{p}{q} x^q (1-x)^{p-q};$$

z_n — Pr [$q'+z$ people survive to age $A+a \mid p'$ survive to age A]

$$= \binom{p'}{q'+z} x^{q'+z} (1-x)^{p'-q'-z};$$

where x denotes the probability that an individual of age A survives to age $A+a$. Thus, by (51),

$$P \equiv \Pr[\psi_n \mid \varphi_m]$$

$$= \binom{p'}{q'+z} \int_0^1 x^{q+q'+z}(1-x)^{p+p'-q-q'-z}\, dx \Big/ \int_0^1 x^q(1-x)^{p-q}\, dx \quad .$$

Laplace then considers approximations to this value.

He next supposes that, of p individuals aged A, q live to age $A+a$ and r to age $A+a+a'$: what is the probability that, of p' individuals of age A, $(qp'/p)+z$ and $(rp'/p)+z'$ will survive to ages $A+a$ and $A+a+a'$ respectively? This probability is found in the same manner as that just discussed, and, extending the procedure somewhat, Laplace shows that

> l'expression précédente de P est donc la probabilité que les erreurs de $q', r', s', \ldots$ sont comprises dans les limites zéro et z, zéro et z', zéro et z'', etc. [p.397]

In Section 31 Laplace applies his analysis to the question of the error incurred in the determination of the population of a large empire, based on the numbers of births[114]. He reduces this to an urn problem, supposing that from an urn containing an infinite number of white and black balls, p draws are made and q of these yield white balls. A second series of draws is then made, q' of these resulting in white balls. Assuming that the unknown ratio of white to black balls in the urn initially is $x:1$, we require the probability that the number of balls drawn in the second series lies within the limits $(pq'/q) \pm z$. Proceeding exactly as in the preceding section, Laplace shows that this probability is approximately given by

$$P = 1 - \frac{2}{\sqrt{2\pi\sigma^2}} \int_z^\infty e^{-t^2/2\sigma^2}\, dt \ ,$$

where $\sigma^2 = pq'(p-q)(q+q')/q^3$.

In Section 33 Laplace continues his study of births. It had been noted that in Paris, over a number of years, the number of registrations of baptisms of boys exceeded that of girls. Laplace proposes here to determine the probability that this superiority will be maintained for a given period (for example, a century). Denoting by $2n$ the number of annual baptisms, of which p are of boys and q of girls, and by x the probability that an infant about to be born and baptised will be a boy, one finds that the probability that in each year the number of baptisms of boys will exceed that of girls is the sum z of the first n terms of the series

$$x^{2n} + 2n\, x^{2n-1}(1-x) + \frac{2n(2n-1)}{1.2} x^{2n-2}(1-x)^2 + \cdots$$

Then z^i is the probability that this superiority will be maintained for i consecutive years, and the probability P, given the pertinent data, that this superiority is maintained for i years is, by the formula of Section 32,

$$P = \int_0^1 x^p(1-x)^q z^i\, dx \Big/ \int_0^1 x^p(1-x)^q\, dx \ .$$

Using the data obtained from the years 1745-1784, during which

$$p = 393,386 \qquad q = 377,555 \ ,$$

Laplace finds by appropriate approximation that $P = 0.782$.

We now turn to Chapter VII, entitled "De l'influence des inégalités inconnues qui peuvent exister entre des chances que l'on suppose parfaitement égales". The chief topic of concern is the question of the tossing of a coin known to be biased (though whether towards heads or tails is uncertain), a topic which Laplace had considered earlier[115].

In the final paragraphs of this chapter we find integrals reminiscent of those of the preceding chapter. These arise in the following way: let P denote the probability of a compound event composed of two simple events of probabilities p and $1-p$. Suppose further that p is susceptible of an unknown error z which can take on values in $[-\alpha, +\alpha]$, and let φ be the probability of $p+z$. Then, says Laplace, one will have "pour la vraie probabilité de l'événement compose" [p.415]

$$\int_{-\alpha}^{\alpha} P'\varphi\, dz \Big/ \int_{-\alpha}^{\alpha} \varphi\, dz \ , \tag{53}$$

where $P^{'}$ is what P becomes on substitution of $p+z$ for p.

The derivation of this result, as expounded by Laplace (and as just indicated), is by no means clear to me. It seems, though, that it can be deduced in the same fashion as the expression (51), with y_m and z_n replaced by φ and $P^{'}$ respectively.

Laplace next goes on to say that, if z is determined only by an observed event (formed of the same simple events) of probability Q, then the probability of the compound event will be

$$\int_{-p}^{1-p} P^{'} Q \, dz \Big/ \int_{-p}^{1-p} Q \, dz \quad . \tag{54}$$

He then concludes that "ce qui est conforme à ce que nous avons trouvé dans le Chapitre précédent" [p.415]. While the ratios given here are certainly of the same form as (51), it must be remembered that this latter expression refers to *future* events, which is not the case in (53) and (54) above; though if we let Q and $P^{'}$ in (53) and (54) correspond respectively to the probabilities of φ_m (the observed result) and of ψ_n (the future result) in the discussion leading to (51), we see that (53) and (54) do in fact agree with (51) (some extra investigation of the limits is perhaps called for).

In Section 35, the first section of Chapter VIII, "Des durées moyennes de la vie, des mariages et des associations quelconques", Laplace discusses the mean duration of life of n infants, where n is very large; and he finds the probability that the sum of the ages attained by n infants lies within given limits. In Section 36 Laplace continues his study of mortality, concerning himself now with the mean duration of life when one of the causes of mortality dies out.

In Section 37 we find a discussion which more nearly concerns us — a discussion of the mean duration of marriages[116]. Laplace's statement of the problem is as follows: suppose that a large number n of marriages are entered into between lads of age a and lasses of age $a^{'}$. Let us determine how many marriages are still going strong after x years[117].

Todhunter [1865, art.1036] finds Laplace's investigation of this problem "very obscure": nevertheless we shall try to present the latter's solution in as lucid a manner as possible, before discussing the alternatives presented by Todhunter. If φ and ψ denote respectively the probabilities that a boy and a girl who married at ages a and $a^{'}$ will reach ages $a+x$ and $a^{'}+x$, then the probability[118] that their marriage will last to the x-th year is $\varphi\psi$. Thus[119]

$$\Pr\left[i \text{ out of } n \text{ marriages will last } x \text{ years}\right] = H(\varphi\psi)^i(1-\varphi\psi)^{n-i} \quad .$$

The next problem is the estimation of the product $\varphi\psi$. To this end Laplace refers to his §16 [p.281], where he showed that, in such a binomial situation, the greatest term in the expansion is that in which the value of i is given by

$$i = [(n+1)\varphi\psi] \ ,$$

where $[x]$ here denotes the integral part of x. By the same article, Laplace says, it is extremely probable that the number of marriages that last differs only very slightly from that number — i.e. that this is the most probable number. Thus (again by §16) it follows that $i = n\varphi\psi$ approximately.

From mortality tables, suppose we can find $p^{'}$ (the number of men living at age a) and $q^{'}$ (the number surviving to age $a+x$): then, approximately,

$$n\varphi = nq^{'}/p^{'} \ .$$

Similarly $\psi = q^{''}/p^{''}$, and thus[120]

$$i = nq^{'}q^{''}/p^{'}p^{''} \ .$$

This, then, is the "best" estimate of i; and having found it, Laplace goes on to consider the problem of finding

> la probabilité que l'erreur de la valeur précédente de i sera comprise dans des limites données. [p.424]

I am forced to agree with Todhunter that this investigation is "very obscure", and fearing that too slavish an exposition of the original might but render *obscurum per obscurius*, I choose rather to present the argument as I see it.

Let us suppose, as Laplace initially did, for ease of calculation that $a = a^{'}$, $q^{''} = q^{'}$, $p^{''} = p^{'}$, $\varphi = \psi$. Then the value i of I (a random variable) found above becomes

$$i = n\left(q^{'}/p^{'}\right)^2 \ .$$

Now if of a large number p of individuals of age a, q are alive at age $a+x$, then, by Article 30, the probability that of $p^{'}$ other individuals of age a, Z will reach age $a+x$ is such that

$$\begin{aligned} f_Z(z)\,dz &= \Pr\left[\left(p^{'}q/p\right) + z < Z < \left(p^{'}q/p\right) + z + dz\right] \\ &= p^3 e^{-Q} dz \Big/ \sqrt{2\pi\, qp^{'}(p-q)(p+p^{'})} \end{aligned}$$

where $Q = p^3 z^2 \Big/ 2qp^{'}(p-q)(p+p^{'})$. If one supposes p and q very large[121], then $\varphi = q/p$, and hence

$$f_Z(z)\,dz = e^{-\Phi}\,dz \Big/ \sqrt{2\pi\, p^{'}\varphi(1-\varphi)}$$

where $\Phi = z^2/2p'\varphi(1-\varphi)$ (since $1+(p'/p)\approx 1$).

Suppose next that, conditional on the value of Z, $I \sim b\left(n,\varphi^2\right)$. Then, by Article 16,

$$f_{I|Z}(l \mid z)\,dl = \Pr\left[n\varphi^2 + l < I < n\varphi^2 + l + dl \mid Z = z\right]$$

$$= e^{-L}\,dl \Big/ \sqrt{2n\pi\,\varphi^2(1-\varphi^2)}$$

where $L = l^2/2n\,\varphi^2(1-\varphi^2)$. Recalling now that $\varphi = q/p$ and setting $q' = \left(p'q/p\right) + z$, we find that $\varphi = (q'-z)/p'$. Thus, on neglecting terms involving z^2, we have $n\varphi^2 = n\left(q'/p'\right)^2 - 2nq'z/p'^2$. If we now put $s = l - 2nq'z/p'^2$ then

$$f_{I|Z}(s \mid z)\,ds$$

$$= \Pr\left[n\left(q'/p'\right)^2 - 2nq'z/p'^2 + l < I < n\left(q'/p'\right)^2 - 2nq'z/p'^2 + l + dl \mid Z = z\right]$$

$$= \Pr\left[n\left(q'/p'\right)^2 + s < I < n\left(q'/p'\right)^2 + s + ds \mid Z = z\right]$$

$$= e^{-S}\,ds \Big/ \sqrt{2n\pi\,\varphi^2(1-\varphi^2)}$$

where $S = \left(s + 2nq'z/p'^2\right)^2 / 2n\varphi^2(1-\varphi^2)$. Thus

$$\Pr\left[n\left(q'/p'\right)^2 + s < I < n\left(q'/p'\right)^2 + s + ds,\right.$$

$$\left.\left(p'q/p\right) + z < Z < \left(p'q/p\right) + z + dz\right]$$

$$= f_{I|Z}(s \mid z) f_Z(z)\,ds\,dz\ ,$$

and hence

$$\Pr\left[s_0 < I - n\left(q'/p'\right)^2 < s_1\right] = \int_{s_0}^{s_1}\int_{-\infty}^{\infty} f_{I|Z}(s \mid z) f_Z(z)\,dz\,ds$$

$$= \left[2\pi\sqrt{np'\varphi^3(1-\varphi)^2(1+\varphi)}\right]^{-1}\int_{s_0}^{s_1}\int_{-\infty}^{\infty} e^{-(\Phi+S)}\,dz\,ds\ . \qquad (55)$$

On setting

$$k^2 = p' \Big/ \left\{2n\varphi^2(1-\varphi)\left[p' + \left(p' + 4n\right)\varphi\right]\right\}$$

we obtain

$$\Pr\left[s_0 < I - n\left(q'/p'\right)^2 < s_1\right] = \frac{k}{\sqrt{\pi}} \int_{s_0}^{s_1} e^{-k^2 s^2}\, ds \ , \qquad (56)$$

and hence

$$\Pr\left[\left|I - n\left(q'/p'\right)\right| < s_0\right] = \frac{2k}{\sqrt{\pi}} \int_0^{s_0} e^{-k^2 s^2}\, ds \ .$$

(The reduction of the double integral in (55) to (56) is achieved by setting $\varphi = q'/p'$.)

> L'analyse précédente s'applique également à la durée moyenne d'un grand nombre d'associations formées de trois individus ou de quatre individus, etc. [p.426]

Todhunter presents three alternative solutions to the problem. In the first of these he supposes that q'/p', a ratio of observed frequencies obtained from mortality tables, is the probability of a specified individual's being alive at age $a + x$. Then the probability of a specified pair being alive is $\pi \equiv \left(q'/p'\right)^2$, and thus the probability that of the n original marriages i are still unbroken is

$$\binom{n}{i} \pi^i (1-\pi)^{n-i} \ .$$

The replacement of the probability that a specified individual is alive at age $a + x$ by the observed ratio $q' : p'$ is seen by Todhunter as an assumption "analogous to what we have called an inverse use of James Bernoulli's theorem" [art.1036].

For his second alternative Todhunter relies on "the usual principles of inverse probability as given by Bayes and Laplace" [art.1036]. To this end he uses the formula given earlier by Laplace, viz. $P = \int_0^1 yz\, dx \Big/ \int_0^1 y\, dx$, with

$$y = \binom{p'}{q'} x^{q'} (1-x)^{p'-q'}$$

$$z = \binom{n}{i} \left(x^2\right)^i \left(1-x^2\right)^{n-i} \ .$$

Then the desired probability P is given by

$$P = \binom{n}{i} \int_0^1 x^{q'} (1-x)^{p'} \left(x^2\right)^i \left(1-x^2\right)^{n-i} dx \Big/ \int_0^1 x^{q'} (1-x)^{p'}\, dx \ . \qquad (57)$$

An exact evaluation of this ratio, in terms of Eulerian integrals of the first kind[122], yields

$$P = \sum_{k=0}^{n-i} \binom{n-i}{k} B\left(q' + 2i + k + 1, p + n - i + 1\right) \quad .$$

Tables of the gamma-function allow the finding of a numerical value of P for given values of the variables. Todhunter, however, suggests a different method of evaluating (57) above; he replaces $\binom{n}{i} \left(x^2\right)^i \left(1 - x^2\right)^{n-i}$ by

$$\left[2\pi n x^2 \left(1 - x^2\right)\right]^{-1/2} \exp(-r^2/2n\, x^2 \left(1 - x^2\right))$$

where r is not large, and shows eventually that the probability that the number of surviving marriages lies in the interval $\left[na^2 - \tau \sqrt{2na^2 \left(1 - a^2\right)}\,, na^2 + \tau \sqrt{2na^2 \left(1 - a^2\right)}\right]$, where $a = q'/p'$, is approximately given by

$$\frac{2}{\sqrt{\pi}} \int_0^\tau e^{-t^2}\, dt + \left[2\pi n a^2 \left(1 - a^2\right)\right]^{-1/2} \exp(-\tau^2) \quad .$$

Todhunter's final solution requires that one knows, from observation, that of m_1 marriages at age a, n_1 last to age $a + x$. The probability, then, that of n marriages i survive for the same period is, as in the preceding solution,

$$P = \binom{n}{i} \int_0^1 x^{n_1+i}(1 - x)^{m_1+n-n_1-i}\, dx \Big/ \int_0^1 x^{n_1}(1 - x)^{m_1-n_1}\, dx \quad ,$$

a ratio which may be evaluated as before.

In concluding our discussion of this section of the *Théorie analytique des probabilités*, we note that Laplace's solution requires the estimation of φ as q'/p' (as in Todhunter's first solution), and then the use of the formula used by Todhunter in his second solution.

The problem considered in Chapter IX, "Des bénéfices dépendants de la probabilité des événements futurs", is stated at the outset of §38 as follows:

> concevons que l'arrivée d'un événement procure le bénéfice ν, et que sa non-arrivée cause la perte μ. Une personne A attend l'arrivée d'un nombre s d'événements semblables, tous également probables, mais indépendants les uns des autres; on demande quel est son avantage. [p.428]

Laplace shows firstly that A's advantage is zero if $\nu q = \mu(1 - q)$, where q is the probability of the occurrence of each event, and then goes on to show that, if s is large, the probability that A's real benefit lies within the limits $s[q\nu - (1-q)\mu] \pm r\sqrt{s}(\mu + \nu)$ is

$$\sqrt{\frac{2}{\pi q(1-q)}} \int_0^\infty e^{-r^2/Q}\, dr + \frac{1}{\sqrt{2s\pi q(1-q)}} e^{-r^2/Q} \ ,$$

where $Q = 2q(1-q)$. This analysis is then extended to the case in which the (initial) probabilities of the s events, as well as the attendant gains and losses, are different[123].

Further modification is undertaken, in §39, in supposing that, at each trial ("événement"), A has any number whatever of chances to hope or fear (this is illustrated by an example about the drawing of balls from an urn); and Laplace then passes on to the case in which the probabilities of the events are unknown[124]. He supposes that of m similar expected events, n have occurred, and that A expects s similar events, each of which will procure him a gain ν if it occurs, and a loss μ if it does not occur. If we represent by $(n/m)s + z$ the number of the s events which will occur, the probability that z lies within $[-kt, +kt]$ is, by his §30,

$$\frac{2}{\sqrt{\pi}} \int_0^\infty e^{-t^2}\, dt \ ,$$

where $k^2 = 2ns(m-n)(m+s)/m^3$. (Todhunter's solution is again slightly different — see his Article 1040.) This latter integral is in fact also the probability that the real benefit to A lies within the limits

$$\left[\frac{n\nu}{m} - \frac{(m-n)\mu}{m}\right] s \pm kt(\nu + \mu) \ .$$

(The rest of this chapter is devoted to questions concerning life annuities, and as such need not concern us here.)

At the beginning of Chapter X, "De l'espérance morale", Laplace recalls the difference, already indicated in his §2, between mathematical and moral expectation ("l'espérance"). He reminds us that, in that article, he had cited a principle [viz. x being the physical fortune of an individual, the increase in his moral fortune is $k\,dx/x$], the principal useful results flowing from which he proposes to examine here. As a consequence of his preliminary investigations he finds the following:

(i) the game which is mathematically the fairest is always disadvantageous;

(ii) it is better to expose one's fortune in lots to independent risks than to expose it all to the same risk.

At the start of §42 Laplace states that the principle he uses to calculate moral expectation was proposed by Daniel Bernoulli, who had stated it in connexion with the St Petersburg paradox (a problem which Laplace now examines)[125]. Of this principle Laplace writes

> ainsi la supposition la plus naturelle que l'on puisse faire est celle d'un avantage moral réciproque au bien de la personne intéressée. [p.449]

Laplace's eleventh chapter[126] is entitled "De la probabilité des témoignages". Considering firstly the case of a single witness, who asserts that the number i was drawn from an urn containing n numbers, he notes that any one of four hypotheses may be entertained, viz.

> Ou le témoin ne trompe point et ne se trompe point; ou il ne trompe point et se trompe; ou il trompe et ne se trompe point; enfin, ou il trompe et se trompe à la fois. [p.455]

Let us denote by p the probability of the veracity of the witness, and by r the probability that he is not mistaken, and let the hypotheses given above be denoted respectively by H_1, H_2, H_3, H_4 with R the announcement of the number i. Under the assumptions of suitable uniformity and independence of $\{H_i\}$ and R, Laplace deduces that

$$\Pr[H_1 \vee H_2 \mid R] = \frac{pr/n + (1-p)(1-r)/n(n-1)}{pr/n + p(1-r)/n + (1-p)r/n + (1-p)(1-r)/n}.$$

It should be noted, though, that this is referred to as "la probabilité de la sortie du n^o i" [p.457]: this is yet another example of the difficulty one experiences in reading Laplace[127].

In similar vein Laplace considers the assertion that a white ball is drawn from an urn known to contain one white and $(n-1)$ black balls, while in the following section he considers the case of the drawing of a ball from one urn, its being placed in a second, and the subsequent drawing of a ball from this urn. Each stage of this procedure is attested to by different witnesses, and a probability similar to that given above is deduced.

In §47 Laplace considers the case of simultaneous testimony, deducing in general that, when prior probability of 1/2 is assigned to both the truth and the falsehood of the report, the posterior probability that the report is true is

$$p^r / [p^r + (1-p)^r] \quad ,$$

where r denotes the number of witnesses and p is the probability that each tells the truth.

Attention is also given to the case in which two witnesses assert that different numbers are drawn, while the case of r witnesses is addressed via a finite difference equation. Subsequent examples, in similar vein, and the three *Additions* to this chapter, contribute nothing else to our discussion. We pass on therefore to the Supplements.

The first Supplement, entitled "Sur l'application du calcul des probabilités a la philosophie naturelle", and dated 15 November 1816, is to a large extent made up of the contents of two earlier memoirs[128], as we have already mentioned (see §6.12). The material of the first of these memoirs, one which bears the same title as this Supplement, is in the main repeated, only some general comments and an application of some probability formulae to the length of a seconds' pendulum being omitted: the contents of the other memoir, "Sur le calcul des probabilités appliqué a la philosophie naturelle", are repeated in their entirety. These memoirs, and the appropriate sections of the first Supplement, deal with errors of observation, and as such do not concern us.

The second-last section, "De la probabilité des jugements", is, however, more directly pertinent[129]. The ideas have already been broached in the thirteenth chapter of the *Essai philosophique sur les probabilités*, but far more detail can be found here. After several general remarks which, although interesting in themselves, are not germane to our present discussion, Laplace considers (in §1, p.526) the following question: suppose that the probability of an offence is such that the citizens have more to fear from the infringements that might arise from its impunity than the errors of the tribunals — in which case the interest of society necessitates the sentence of the accused. Let a denote this degree of probability, and suppose that the judge who sentences an accused declares thereby that the probability of his offence is at least a. Let X $(\geq \frac{1}{2})$ be the probability of this opinion of the judge, varying by infinitely small degrees equal to x and equally probable *a priori.* Suppose too that the tribunal is composed of $p+q$ judges, of whom p convict and q acquit the accused.

For a given value x of X, the probability that the opinion of the tribunal is equitable[130] will be proportional to $x^p(1-x)^q$, while the probability that it is inequitable will be proportional to $(1-x)^p x^q$. Thus, by Article 1, the probability of the goodness of the judgement (an event which we shall denote by G) will be

$$\Pr[G \mid X = x;\ p, q] = x^p(1-x)^q \,/[x^p(1-x)^q + (1-x)^p x^q] \quad . \tag{58}$$

At this stage we find an argument in inverse probability appearing: the probability of X, given that p judges convict and q acquit the accused, is then[131]

$$\Pr[x < X < x + dx \mid p, q] = f_X(x)\,dx$$

$$= [x^p(1-x)^q + (1-x)^p x^q]\,dx \Big/ \int_{1/2}^{1} [x^p(1-x)^q + (1-x)^p x^q]\,dx$$

$$= [x^p(1-x)^q + (1-x)^p x^q]\,dx \Big/ \int_{0}^{1} x^p(1-x)^q dx \quad . \tag{59}$$

Thus the probability of the goodness of the judgement relative to x being the product of (58) and (59), the dx being introduced in Laplace's usual way, we find finally that the probability of the goodness of the judgement relative to all values of x is

$$\begin{aligned}\Pr[G] &= \int_{1/2}^{1} \Pr[G \mid X = x;\ p, q]\, f_X(x)\,dx \\ &= \int_{1/2}^{1} x^p(1-x)^q\,dx \Big/ \int_{0}^{1} x^p(1-x)^q\,dx \quad .\end{aligned}$$

The probability of the error to be avoided on the goodness of the judgement is then

$$\begin{aligned}1 - \Pr[G] &= \int_{0}^{1/2} x^p(1-x)^q\,dx \Big/ \int_{0}^{1} x^p(1-x)^q\,dx \\ &= I_{1/2}(p+1, q+1) \,/ B(p+1, q+1) \ ,\end{aligned}$$

the numerator being an incomplete beta-function[132]. This reduces to

$$\frac{1}{2^{p+q+1}}\left[1 + \frac{p+q+1}{1} + \frac{(p+q+1)(p+q)}{1.2} + \cdots + \frac{(p+q+1)(p+q)...(p+2)}{1.2...q}\right] ,$$

which becomes $2^{-(p+1)}$ in the case of unanimity (i.e. $q = 0$).

Then follows a section (§2) in which Laplace gives approximations for large p and q in two cases:

(i) when $p - q$ is large, in which case the probability of the error, as given by Article 28, is

$$\frac{(p+q)^{p+q+3/2}}{2^{p+q+3/2} p^{p+\frac{1}{2}} q^{q+\frac{1}{2}} (p-q)\sqrt{\pi}} \left\{1 - \frac{p+q}{(p-q)^2} - \frac{[(p+q)^2 - 13pq]}{12pq(p+q)}\right\} ;$$

(ii) when $p-q$ is small relative to p, in which case the probability of the error, by Article 19, is

$$\frac{1}{\sqrt{\pi}}\int_{t^2}^{\infty} e^{-x^2}\,dx \ ,$$

where $t^2 = (p-q)^2(p+q)/8pq$.

Each of these cases is illustrated by a numerical example: Pearson [1978, pp.692-693] presents some exact calculations for various small values of p and q.

There are a number of comments that one might make in connexion with this section[133]. Firstly, the value of x is assumed the same for each judge, this value being always taken to be at least $\frac{1}{2}$. Further, a factor $\binom{p+q}{p}$ is missing (there seems no reason not to suppose the judges interchangeable), but this in fact cancels out in the end, and so its omission does not affect the final result. Finally, Pearson (loc. cit.) has suggested that the factors $x^p(1-x)^q$ and $(1-x)^p x^q$ should be multiplied respectively by the probabilities of the guilt and innocence of the accused.

The final section of this Supplement is entitled "Sur une disposition du Code d'instruction criminelle": it contains nothing useful.

The second Supplement, dated February 1818, is entitled "Application du calcul des probabilités aux opérations géodésiques". It is described by Todhunter [1865, art.1050] as "very interesting, and considering the subject and the author it cannot be called difficult". Inspired by the desire to extend his application of the probability calculus to natural philosophy, Laplace proposes here to consider the question of triangulation.

> Cette application consiste à tirer des observations les résultats les plus probables et à determiner la probabilité des erreurs dont ils sont toujours susceptibles. [p.531]

After some general remarks on matters geodetic and the applicability of some of his earlier results, Laplace points out that the formulae to be discussed here are applicable to future observations; yet, suitably modified, they may also be applied to past data.

As a first application, Laplace considers a great arc $AA'A''\ldots$ on a sphere, around which a chain of triangles $ACC', CC'C'', C'C''C''', \ldots$ is formed, the sides $CC', C'C'', C''C''', \ldots$ intersecting this arc in $A, A', A'', \ldots$ Let $A, A^{(1)}, A^{(2)}, \ldots$ denote the angles $CAA', C'A'A'', C''A''A''', \ldots$, and let $C, C^{(1)}, C^{(2)}, \ldots$ denote the angles $ACC', CC'C'', C'C''C''', \ldots$ (Laplace states that "Je ne donne point de figure, parce qu'il est facile de la tracer d'après ces indications" [p.535]: nevertheless he supplies the figure in the

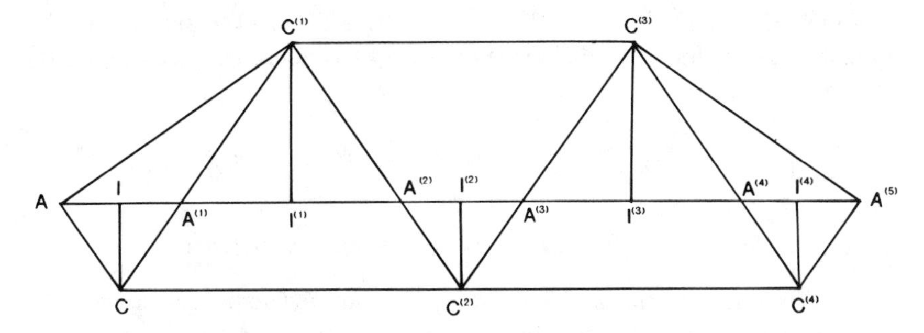

Figure 6.4: Laplace's sketch for triangulation.

third Supplement.) From these data one has

$$A + A^{(1)} + C - \alpha = \pi + t$$

where α is the error in the observed angle C and t is the excess of the angles of the spherical triangle ACA' over π. Setting up a series of such equations[134] Laplace shows that, angle A being completely known, the error in angle $A^{(n)}$ is

$$\alpha^{(n-1)} - \alpha^{(n-2)} + \cdots \pm \alpha$$

(+ for n odd, $-$ for n even).

Proposing to find the probability that this error lies within given limits, Laplace supposes[135] firstly that the probability of any error α is proportional to $\exp(-h\alpha^2)$. He then derives the probability that the error in $A^{(n)}$ lies within the limits $\pm r\sqrt{n}$ as

$$2\sqrt{\frac{3h}{2\pi}} \int \exp(-3hr^2/2)\, dr$$

or

$$2\sqrt{\frac{h}{\pi}} \int \exp(-hr^2)\, dr \quad ,$$

depending on whether or not the three angles of each triangle are corrected.

Laplace next turns his attention to the determination[136] of h. Considering the different values of h as *causes* of the observed event, he shows that the (posterior) probability of h will be, "par le principe de la probabilité des causes tirée des événements observés" [p.539],

$$h^{n/2} \exp(-h\theta^2/3)\, dh \Big/ \int_0^\infty h^{n/2} \exp(-h\theta^2/3)\, dh \quad ,$$

where $\theta^2 = T^2 + T^{(1)^2} + \cdots + T^{(n-1)^2}$, and the T's are the excesses in the triangles. The value of h which it is necessary to choose (as Laplace has it) is the mean, i.e.

$$\int_0^\infty h^{(n+2)/2} \exp(-h\theta^2/3)\, dh \bigg/ \int_0^\infty h^{n/2} \exp(-h\theta^2/3)\, dh$$

or $3(n+2)/2\theta^2$, which, for large n, is approximately $3n/2\theta^2$.

> Cette quantité est la valeur de h qui rend l'événement observé le plus probable, la probabilité de cet événement, *a priori*, étant proportionelle à $h^{n/2} \exp(-h\theta^2/3)$. [p.540]

Thus the probability that the error in angle $A^{(n)}$ lies within the limits $\pm r\sqrt{n}$ becomes

$$\frac{3\sqrt{n}}{\theta\sqrt{\pi}} \int \exp(-9nr^2/4\theta^2)\, dr \ ,$$

and the probability that it lies within the limits $\pm 2\theta r/3$ is

$$\frac{2}{\sqrt{\pi}} \int_0 \exp(-r^2)\, dr \ . \tag{60}$$

Similar expressions are derived elsewhere in this Supplement — see pp.543, 544, 546 and 548.

More generally, Laplace supposes next that the law of the probability of the error α is $\varphi(\alpha)$, rather than the $\exp(-h\alpha^2)$ considered before. He supposes also that the same errors, be they positive or negative, are equally probable, and that φ is defined over $[-\infty, +\infty]$[137]. The probability that a certain error falls within given limits is also given by (60), with different limits of integration to those found before.

The final section of this Supplement, entitled "Sur la probabilité des résultats déduits, par des procédés quelconques, d'un grand nombre d'observations", contains nothing pertinent.

The third Supplement, "Application des formules géodésiques de probabilité a la méridienne de France", is chiefly devoted to numerical applications of the formulae of the second Supplement. Neither the third nor the fourth Supplement (untitled) seems to contain any pertinent remarks. This last Supplement, written in 1825, is mainly the work of Laplace's son: it is devoted chiefly to generating functions[138].

6.15 Appendix 6.1

I propose here to consider the relationship between some results of Bayes and those given in Laplace's *Mémoire sur la probabilité des causes par les événements.*

Firstly, framing our thoughts in the "urn" situation, we see that Bayes had the idea of a single urn, whose composition was to be examined. Laplace, however, entertained the idea of a *population* of urns, and hence could ask which of them was the "cause" of the sample. A closely related point is this: Bayes (at least as far as communicated by Price), considered only the "estimation" of a probability, while Laplace's aim was definitely to predict future behaviour[139].

Let us now turn to the connexion between the Rules of Bayes's Essay and Laplace's expression (4). The first rule (p.399 of the Essay and §2.4 of the present work) states that if all that is known about an event is that it has happened p times and failed q times in $p+q$ or n trials, the chance that one is right in guessing that the probability of its happening in a single trial lies between any two degrees of probability X and x, is given by $(n+1)\binom{p+q}{p}$ multiplied by the difference between the series

$$\frac{X^{p+1}}{p+1} - q\,\frac{X^{p+2}}{p+2} + \frac{q(q-1)}{2}\,\frac{X^{p+3}}{p+3} - \ \&c.$$

and

$$\frac{x^{p+1}}{p+1} - q\,\frac{x^{p+2}}{p+2} + \frac{q(q-1)}{2}\,\frac{x^{p+3}}{p+3} - \ \&c.$$

Let us also recall that Price stated that Bayes, noting the impracticability of this formula for large values of p and q, had deduced his second Rule (see §2.4). We note here, from this Rule, the expression

$$\Sigma = \frac{n+1}{n} \times \frac{\sqrt{2pq}}{\sqrt{n}} \times Ea^p b^q \times \left[mz - \frac{m^3z^3}{3} + \frac{n-2}{2n}\,\frac{m^5z^5}{5} \right.$$
$$\left. - \frac{(n-2)(n-4)}{(2n)(3n)}\,\frac{m^7z^7}{7} + \frac{(n-2)(n-4)(n-6)}{(2n)(3n)(4n)}\,\frac{m^9z^9}{9}\ \&c.\right]$$

where $E = \binom{p+q}{p}$.

Price's investigations led him to conclude that

> In all cases when z is small, and also whenever the disparity between p and q is not great 2Σ is almost exactly the true chance required. And I have reason to think, that even in all other cases, 2Σ gives the true chance nearer than within the limits now determined. [Bayes 1764, art.28]

It is my aim now to show the correctness of this thought, by comparing 2Σ with the limit derived by Laplace. To effect this comparison, and correcting Bayes's definition of m^2, we note that

$$
\begin{aligned}
2\Sigma &= 2\frac{(p+q+1)}{(p+q)}\frac{\sqrt{2pq}}{\sqrt{p+q}}\frac{(p+q)!}{p!\,q!}\left(\frac{p}{p+q}\right)^p\left(\frac{q}{p+q}\right)^q\left[mz-\frac{m^3z^3}{3}\right.\\
&\qquad\left.+\frac{n-2}{2n}\frac{m^5z^5}{5}-\frac{n-2}{2n}\cdot\frac{n-4}{3n}\cdot\frac{m^7z^7}{7}+\cdots\right]\\
&= 2\frac{(p+q+1)!}{p!\,q!}\frac{p^pq^q}{(p+q)^{p+q}}\frac{1}{m}\left[mz-\frac{m^3z^3}{3}+\frac{n-2}{2n}\cdot\frac{m^5z^5}{5}\right.\\
&\qquad\left.-\frac{n-2}{2n}\cdot\frac{n-4}{3n}\cdot\frac{m^7z^7}{7}+\cdots\right].
\end{aligned}
$$

Now, for large values of p and q, $n-k\sim n$ for moderate values of k. Thus

$$
\begin{aligned}
2\Sigma &\approx 2\frac{(p+q+1)!}{p!\,q!}\frac{p^pq^q}{(p+q)^{p+q}}\left[z-\frac{m^2z^3}{3}+\frac{m^4z^5}{2.5}-\frac{m^6z^7}{2.3.7}+\cdots\right]\\
&\approx 2\frac{(p+q+1)!}{p!\,q!}\frac{p^pq^q}{(p+q)^{p+q}}\int_0^z\left(1-m^2x^2+\frac{m^4x^4}{2}-\frac{m^6x^6}{2.3}+\cdots\right)dx\\
&\approx \frac{(p+q+1)!}{p!\,q!}\frac{p^pq^q}{(p+q)^{p+q}}\int_0^z 2e^{-m^2x^2}\,dx\ .
\end{aligned}
$$

Using Stirling's formula and the approximation

$$(p+q+1)^{p+q+3/2}=e(p+q)^{p+q+3/2}\ ,$$

an equality which comes from Laplace [1774, p.32], who "deduces" it from

$$\left(1+\frac{1}{p+q}\right)^{p+q+3/2}=e\ ,$$

we find that

$$2\Sigma\approx\frac{(p+q)^{3/2}}{\sqrt{2\pi pq}}\int_0^z 2e^{-(q+p)^3x^2/2pq}\,dx\ ,$$

which we recall is exactly the approximation derived by Laplace (see (3) above).

Thus Price's conjecture as to the goodness of 2Σ as an approximation to the desired probability is seen to be well founded (note also the footnote on pp.316-317 of Bayes [1764]).

Let us return now to the probability $Q(p, q; m, n)$ defined in §6.3. The question that presents itself is, what is the relation between this probability and that derived by Bayes? The answer is by no means as simple as it would at first blush seem[140].

In his discussion of this matter, Pearson writes[141]

> ... Bayes' Theorem as we now usually state it is of the following nature. We suppose past experience to be represented by p successes and q failures in n trials, and we ask what is the chance of r successes and s failures in further m trials. We hold this problem answered by the expression
>
> $$C_{r,s} = \frac{(r+s)!}{r!\ s!} \frac{\int_0^1 x^{p+r}(1-x)^{q+s}\,dx}{\int_0^1 x^p(1-x)^q\,dx}$$
>
> [1978, p.366]

What I would stress, by giving this quotation, is that at this time (viz. the 1920's) the idea that Bayes had proved the so-called "discrete Bayes's Theorem" was by no means all-pervading. (Pearson later goes on to state that his quoted result is really "Condorcet's and Laplace's extension of Bayes.")

But let us return to our muttons. Recall that Bayes quite definitely considered only the probability of an event's happening in a *single* trial: it seems, therefore, that we should look more closely at

$$Q(p, q; 1, 0) = \int_0^1 x^{p+1}(1-x)^q\,dx \bigg/ \int_0^1 x^p(1-x)^q\,dx$$

(which is Pearson's $C_{1,0}$). This, however, is still not Bayes's result — that, we see, corresponds in fact (taking the limits of integration x_1 and x_2 in the numerator to be 0 and 1 respectively) to $m = 0 = n$, which looks rather odd, since $Q(p, q; 0, 0)$ has no (nontrivial) meaning when applied to a *future* occurrence. It seems, therefore, that Bayes's phrase "a single trial" does not — nay, *cannot* — refer to a future event: to what, then, does it refer?

That this question has been viewed as occasioning some difficulty in the past may be seen by referring to Pearson [1978, pp.368-369]. In these lectures on the history of statistics, Pearson even went so far as to say of Todhunter "like Price he does not show what Bayes means by the 'single throw' " (op. cit. p.369). That Todhunter was aware of a possible confusion (although he himself supported Price) is shown by his assertion that Lubbock and Drinkwater-Bethune believed that Bayes (or maybe Price) confounded the probabilities of Bayes and Laplace. The pertinent section from this little-known tract on probability (written c.1830) reads as follows:

> Bayes, or perhaps we should rather say Price, seems to have confounded the probability thus determined [i.e. in Bayes's Proposition 10], with the probability that an event which has been already observed m [sic] times in $p+q$ experiments, will happen again [i.e. $Q(p,q;1,0)$]. The difference between the two is obvious ... [p.48].

I must confess to agreeing with these authors (that the confusion was caused by Price and not Bayes) rather than the more distinguished historian.

Pearson [1978, p.369] suggests that the discussion given by Timerding [1908] might in fact be even more obscure than Todhunter's. Timerding suggested that Bayes's table be replaced by a box with a sliding drawer, with balls being dropped into the box, some falling inside and some outside the drawer. One would then be determining, by Bayes's Proposition 9, the probability that the drawer was pushed in a certain distance. However, it is probably true that Timerding in fact saw no problem at all and proposed his model as an illuminating alternative to Bayes's.

If Bayes's original experiment is recalled, it will be remembered that his first postulate concerns the throwing of a ball on a square level table ("at random", we might say); and it is in terms of this original toss that "successes" and "failures" are then defined. It is to this first toss of a ball that Pearson [1978, p.367] attributes the reference "a single trial", and what Bayes is finding is the probability of the chance of this event's lying between any two degrees of probability *after* $p+q$ further throws (or throws with another ball) have been made.

Of course, all these "problems" about the meaning of the phrase "single trial" vanish in the light of Bayes's own words. For in the statement of his second postulate (see §2.4) Bayes makes quite explicit the sense in which the awkward words will be used. Writing of the balls thrown upon the table $ABCD$, he says

> I suppose that the ball W shall be 1st thrown, and through the point where it rests a line os shall be drawn parallel to AD, and meeting CD and AB in s and o; and that afterwards the ball O shall be thrown $p+q$ or n times, and that its resting between AD and os after a single throw be called the happening of the event M in a single trial. [p.385]

Indeed, if we recall Bayes's statement of his problem (see §2.4), we see that his insistence on the words "single trial" (or some equivalent formulation), while perhaps often unnecessary, serves to make quite clear that which is his concern, as contrasted with the statement, at the outset of his problem, about "the number of times" that some event has happened or failed.

However, it seems to me that it might be possible to reconcile Bayes's and Laplace's investigations, in a manner that has perhaps not been sufficiently stressed before. As we have already seen (see §6.3), Laplace, in his discussion of this problem, wrote

> la probabilité que x est le vrai rapport du nombre des billets blancs au nombre total des billets est par le principe de l'article précédent égale à
>
> $$x^p(1-x)^q\,dx \Big/ \int x^p(1-x)^q\,dx$$
>
> [1774, p.30],

where the integral is taken from 0 to 1. If we integrate this expression from x_1 to x_2, we obtain the probability that x, the true ratio of white to total number of tickets, lies between x_1 and x_2, *given* that p white and q black tickets have been drawn — i.e. $\Pr[x_1 \le x \le x_2 \mid p \text{ white } \& \ q \text{ black}]$. But this is exactly Bayes's result! What Laplace went on to find (viz. our $Q(p, q; m, n)$) has, I think, been mistakenly confused with this result.

Summarizing, one might say

(i) Bayes found $\Pr[x_1 \le x \le x_2 \mid p \text{ successes } \& \ q \text{ failures}]$, while

(ii) Laplace found — or, more correctly, *almost* found — the same thing, and, in addition, found an expression for $\Pr[m \text{ future successes } \& \ n \text{ future failures} \mid p \text{ successes } \& \ q \text{ failures}]$.

I believe that neither Bayes nor Laplace confused these probabilities, yet this early clarity was soon lost. Pearson in fact says

> There is great obscurity about the whole matter, but if my view be correct Bayes had certainly not reached 'Bayes' Theorem'. [1978, p.368]

While one must agree with the first part of this statement, one might hesitate about accepting the second clause, until one remembers the expression that Pearson refers to as "Bayes's Theorem" (as we have already said, this is our $Q(p, q; m, n)$, which is due to Laplace and not even attempted by our reverend originator).

6.16 Appendix 6.2

Laplace begins the fifteenth article of his *Mémoire sur les probabilités* (see §6.5) with the following words:

> supposons qu'un événement donné de puisse être produit que par les n causes $A, A', \ldots, A^{(n-1)}$; soient x la probabilité qui en résulte pour l'existence de A; x' celle de l'existence de A'; x'' celle de l'existence de A'', etc. [pp.415-416]

It is clear from the ensuing discussion that $x, x', \ldots$ are intended to denote *conditional* probabilities given E: we shall denote $\Pr[A_i \mid E]$ by x_i. Further, let $a_i = \Pr[E \mid A_i]$.

Laplace next states that

> la probabilité d'un second événement semblable au premier sera égale au produit de a [our a_1] par la probabilité x [$= x_1$] de la cause A [$= A_1$], plus au produit de a' [$= a_2$] par la probabilité x' [$= x_2$] de la cause A' [$= A_2$], plus etc.; d'ou il suit que l'on aura
>
> $$ax + a'x' + a''x'' + \cdots$$
>
> pour cette probabilité ... [p.416].

To verify this, notice firstly that, under the assumption of a discrete uniform prior,

$$\begin{aligned}
\Pr[E_2 \mid E_1] &= \Pr[E_1E_2]/\Pr[E_1] \\
&= \sum_i \Pr[E_1E_2 \mid A_i]\Pr[A_i]/\sum_i \Pr[E_1 \mid A_i]\Pr[A_i] \\
&= \sum_i \Pr[E_1E_2 \mid A_i]/\sum_i \Pr[E_1 \mid A_i] \\
&= \sum_i \Pr[E_1E_2 \mid A_i]/\sum_i a_i \ .
\end{aligned}$$

Assuming the conditional independence of the E's and using the fact that the events are "semblable", we have

$$\begin{aligned}
\Pr[E_2 \mid E_1] &= \sum \Pr[E_1 \mid A_i]\Pr[E_2 \mid A_i]/\sum a_i \\
&= \sum a_i^2/\sum a_i \ .
\end{aligned}$$

Furthermore,

$$\begin{aligned}
\Pr[E_2 \mid E_1] &= \Pr[E_1 E_2]/\Pr[E_1] \\
&= \sum \Pr[E_1 E_2 \mid A_i]\Pr[A_i]/\Pr[E_1] \\
&= \sum \Pr[E_1 \mid A_i]\Pr[E_2 \mid A_i]\Pr[A_i]/\Pr[E_1] \\
&= \sum \Pr[E_2 \mid A_i]\Pr[E_1 A_i]/\Pr[E_1] \\
&= \sum \Pr[E_2 \mid A_i]\Pr[A_i \mid E_1] \\
&= \sum a_i x_i \ .
\end{aligned}$$

Thus, under the assumptions

(i) $\Pr[A_1] = \Pr[A_2] = \cdots = \Pr[A_n] = 1/n$, and

(ii) E_1 and E_2 are similar and conditionally independent with respect to each of the A_i,

we see that

$$\sum a_i x_i = \Pr[E_2 \mid E_1] = \sum a_i^2 / \sum a_i \ .$$

Similar expressions are derived for $\Pr[E_3 E_2 \mid E_1]$, etc., under a suitable extension of assumption (ii) above.

Finally, using the fact that $\sum_{i=1}^n x_i = 1$ and the n equations

$$\begin{aligned}
\sum a_i x_i &= \sum a_i^2 / \sum a_i \\
\sum a_i^2 x_i &= \sum a_i^3 / \sum a_i \\
&\vdots \\
\sum a_i^n x_i &= \sum a_i^{n+1} / \sum a_i \ ,
\end{aligned}$$

we find that $x_i = a_i / \sum a_i$, or $\Pr[A_i \mid E] = \Pr[E \mid A_i] / \sum \Pr[E \mid A_i]$, as required.

6.17 Appendix 6.3

The *Suite* to the *Mémoire sur les approximations des formules qui sont fonctions de très grands nombres* opens with the following words:

Ce Mémoire étant une suite de celui qui a paru sur le même objet dans le Volume précédent, je conserverai l'ordre des articles et des numéros. J'ai donne, dans le premier article, une méthode générale pour réduire en séries très convergentes les fonctions différentielles qui renferment des facteurs éléves à de grandes puissances. Dans le second article, j'ai ramené à ce genre d'intégrales toutes les fonctions données par des équations linéaires aux différences ordinaires ou partielles, finies et infiniment petites; et je suis ainsi parvenu, dans le troisième article, à déterminer les valeurs approchées de plusiers formules qui se rencontrent fréquemment dans l'Analyse, mais dont l'application devient très pénible lorsque les nombres dont elles sont fonctions sont considérables. Il me reste présentement à faire voir l'usage de cette analyse dans la théorie des hasards. [p.295]

Chapter 7

Poisson to Venn

I hope the gentle reader will excuse me for dwelling on these & the like particulars, which, however insignificant they may appear to grovelling vulgar minds, yet will certainly help a philosopher to enlarge his thoughts and imagination, and apply them to the benefit of public as well as private life.

Jonathan Swift, Travels into several Remote Nations of the World, by L. Gulliver.

7.1 Siméon-Denis Poisson (1781-1840)

Only two works by this author seem relevant to the present study: the first of these is a memoir published in 1830 (read 8th February 1829), and the second is the book *Recherches sur la probabilité des jugements en matière criminelle et en matière civile, précédées des règles générales du calcul des probabilités* of 1837.

Poisson begins his memoir, entitled "Sur la proportion des naissances des filles et des garçons", with some observations on the ratios of male to female births in various parts of France over some ten years, concluding this introduction with a description of Laplace's *Théorie analytique des probabilités* as an

> ouvrage aussi éminemment remarquable par la variété des questions qui y sont traitées, que par la généralité des méthodes que Laplace a imaginées pour les résoudre. [p.243]

The first (non-introductory) section of the work is entitled "Probabilité de la répétition d'un événement dont la chance est donnée." Here Poisson proves the *Bernoulli weak law of large numbers*[1], it being shown that if A and B are complementary events of constant probabilities p and q respectively, then the probability U that, in n trials, the number of occurrences of A lies between the limits

$$N \pm u\sqrt{2(n+1)pq}$$

where N is the greatest integer not exceeding $n - p$, is given by

$$U = 1 - \frac{2}{\sqrt{\pi}} \int_u^\infty e^{-t^2}\,dt + e^{-u^2} \Big/ \sqrt{2\pi npq} \quad . \tag{1}$$

This is followed by the statement:

> le rapport x'/n du nombre de fois que l'événement A arrivera au nombre total des épreuves, différera donc de moins en moins de la probabilité p de cet événement; et l'on pourra toujours prendre n assez grand pour qu'il y ait la probabilité U que la différence $(x'/n) - p$ sera aussi petite que l'on voudra; ce qui est, comme on sait, le théorème de Jacques Bernouilli sur la répétition, dans un très-grand nombre d'épreuves, d'un événement dont la chance est donnée *à priori*. [p.261]

There then ensues, in the tenth article of this section, a discussion of the case in which n is very large and p is very small, so that np is either a fraction or of moderate size. This leads to what is now termed the *Poisson distribution*[2], it being found that, under the above conditions and with $pn = \omega$, the probability that A occurs no more than x times in n trials is

$$\sum_{k=0}^{x} e^{-\omega}\omega^k \big/ k! \ .$$

Section II, "Probabilités des événements simples et des événements futurs d'après les événements observés" has its main thrust described in the first article (Article 12 of the paper) as follows:

> Jusqu'ici nous avons supposé connue *à priori*, la chance p de l'événement A, et nous en avons conclu la probabilité d'un événement futur, relatif à la répétition de A sur un très-grand

> nombre d'épreuves; mais dans les applications du calcul des hasards aux phénomènes naturels, et particulièrement dans la question indiquée par le titre de ce Mémoire, la valeur de p doit, au contraire, se déduire autant qu'il est possible, des événements observés en très-grands nombres, pour servir ensuite à calculer la probabilité des événements futurs. C'est ce problème qui va maintenant nous occuper. [p.265]

Supposing that p, the unknown probability of A, is susceptible only of values in the set $\{v_1, v_2, \ldots, v_m\}$ with

$$R_n = \Pr[p = v_n], \qquad n \in \{1, 2, \ldots, m\} ,$$

Poisson takes $V_1, V_2, \ldots, V_m$ as

> les probabilités correspondantes d'un événement composé C, en sorte que V_n désigne la probabilité de C en fonction de v_n qui aurait lieu s'il était certain qu'on eût $p = v_n$. [pp.265-266]

The event C having been observed, Poisson proposes to determine R_n, by, firstly, replacing each V_n by N_n/μ, where μ and the N_n are integral, and then identifying the question with that of the drawing of balls from urns, with the n-th urn containing μ balls of which N_n are white. What is required is the probability that a drawn ball, found to be white (event C), was taken from the n-th urn, a probability which Poisson finds to be given by

$$R_n = V_n / \sum_1^m V_n \quad . \tag{2}$$

This ratio of course is really $\Pr[p = v_n \mid C]$ rather than an absolute probability as Poisson suggests, and indeed many of the probabilities given below should be similarly conditioned.

If C' is another compound event depending on A with

$$V_n' = \Pr\left[C' \mid p = v_n\right] ,$$

then the probability of C' is given by

$$T = \sum_1^m V_n' R_n = \sum_1^m V_n' V_n \bigg/ \sum_1^m V_n \quad . \tag{3}$$

In the next article Poisson supposes p to be susceptible of any value in [0,1], and deduces that the formulae (2) and (3) then become

$$\Pr[v < p < v + dv] = V\,dv \bigg/ \int_0^1 V\,dv$$

and

$$\Pr\left[C'\right] = \int_0^1 V'V\,dv \bigg/ \int_0^1 V\,dv$$

respectively. Furthermore

$$Z \equiv \Pr\left[a < p < b\right] = \int_a^b V\,dv \bigg/ \int_0^1 V\,dv \quad ,$$

and if Q is "la probabilité que l'événement C' répondra à l'une des valeurs de p comprises entre ces limites" [p.268], then

$$Q = \int_b^a V'V\,dv \bigg/ \int_0^1 V\,dv \quad .$$

So much is standard: Poisson's unique contribution is now to note that, if M and M' denote respectively the maximum and minimum of V' over the complement of $\{p : a \leq p \leq b\}$, then

$$Q + M'(1 - Z) < T < Q + M(1 - Z) \ ,$$

an expression which yields, for $(1 - Z)$ negligible,

$$T \sim Q$$

"ce qui en simplifiera le calcul" [p.269].

As a first illustration of the use of these formulae Poisson considers the case in which the event C is the occurrence of A s times in m trials. Then

$$V = \binom{m}{s} v^s(1 - v)^{m-s} \quad ,$$

and hence

$$R = v^s(1 - v)^{m-s}\,dv \bigg/ \int_0^1 v^s(1 - v)^{m-s}\,dv \tag{4}$$

$$Z = \int_a^b v^s(1 - v)^{m-s}\,dv \bigg/ \int_0^1 v^s(1 - v)^{m-s}\,dv \quad . \tag{5}$$

The integrand achieves its maximum value $G = (s/m)^s(1 - s/m)^{m-s}$ on our taking $v = s/m$ ($\equiv g$, say). On setting

$$v^s(1 - v)^{m-s} = G\,e^{-t^2}$$

one finds, on taking logarithms, that $v = g + g' t + g'' t^2 + \cdots$, where

$$g' = \sqrt{2(m-s)s/m^3} \quad ; \quad g'' = 2(m-2s)/3m^2 \quad ,$$

and hence, on neglecting terms of order $1/m$, one has

$$\begin{aligned} \int_0^1 v^s (1-v)^{m-s}\, dv &= \int_{-\infty}^{\infty} G\, e^{-t^2} (dv/dt)\, dt \\ &= G\, g' \sqrt{\pi} \quad . \end{aligned} \tag{6}$$

Attention is next turned to the numerator of Z in (5) above. If one takes

$$a = g - g' z, \quad b = g + g' z, \quad v = g + g'\theta$$

one obtains, to the same order of magnitude as before,

$$t = \theta - \theta^2\, g'' \big/ g' \quad ,$$

and hence

$$e^{-t^2} = e^{-\theta^2} \left(1 + 2g''\, \theta^3 / g'\right) \quad .$$

Hence

$$\begin{aligned} \int_a^b v^s (1-v)^{m-s}\, dv &= \int_{-z}^{z} G\, g' e^{-\theta^2} \left(1 + 2g'' \theta^3 \big/ g'\right) d\theta \\ &= 2G\, g' \int_0^z e^{-\theta^2}\, d\theta \quad . \end{aligned}$$

It then follows from (5) and (6) that[3]

$$\begin{aligned} Z &= \Pr\left[(s/m) - g' z < p < (s/m) + g' z\right] \\ &= \frac{2}{\sqrt{\pi}} \int_0^z e^{-\theta^2} d\theta \quad , \end{aligned}$$

while from (4) one has

$$\begin{aligned} R &= \Pr\left[(s/m) + g'\theta < p < (s/m) + g'\theta + d\theta\right] \\ &= \frac{1}{\sqrt{\pi}} e^{-\theta^2} \left(1 + 2g'' \theta^3 \big/ g'\right) d\theta \quad . \end{aligned}$$

On taking $z = 3$ Poisson finds that $1 - Z = 0.00002209$, a quantity small enough to allow one to make the approximation $T \sim Q$ as mentioned above, where

$$Q = \frac{1}{\sqrt{\pi}} \int_{-z}^{z} \Pi e^{-\theta^2} \left(1 + 2g^{''}\theta^3 \Big/ g^{'}\right) d\theta$$

and where Π is the function previously denoted by $V^{'}$ with v replaced by $p = (s/m) + g^{'}\theta$.

In Article 15 Poisson turns his attention to the case in which $C^{'}$ is the event that, in n trials, A occurs x or fewer times, where x and $n - x$ are very large. The problem is solved under the assumption that the difference between $x/(n+1)$ and s/m is of order $1/\sqrt{m}$, or more precisely

$$x/(n+1) = (s/m) - \gamma g^{'} \qquad (7)$$

where γ (positive) is either a fraction or a small number. After some manipulation it is found that

$$\Pr\left[N(A) \leq x\right] = \frac{1}{\sqrt{\pi}} \int_{c}^{\infty} e^{-v^2}\, dv + \Gamma \ ,$$

where $N(A)$ is the number of times A occurs in n trials, x satisfies (7), or

$$x = (n+1)s/m - (c/m)\sqrt{2(n+1)\,(1+\alpha^2)\,(m-s)s} \ ,$$

and where

$$\alpha = \sqrt{(n+1)/m}$$

$$c = \gamma\alpha/\sqrt{(1+\alpha^2)}$$

$$\Gamma = \frac{(m+s)\sqrt{2}}{3\sqrt{ns(m-s)(1+\alpha^2)}} e^{-c^2} \ .$$

An expression is also given for the probability that $N(A)$ falls between two given values.

Comparison of this latter result with (1) is then undertaken under various approximations. For example, if $u > 0$ and quantities of order $1/n$ are neglected,

$$U = 1 - \frac{2}{\sqrt{\pi}} \int_{u}^{\infty} e^{-t^2}\, dt + me^{-u^2} \Big/ \sqrt{2n(1+\alpha^2)(m-s)s} \ ,$$

while if n is very small with respect to m (of a magnitude comparable to $\sqrt{m}$ in fact), the probability that $N(A)$ lies within the limits

$$N \pm (u/m)\sqrt{2(n+1)(m-s)s}$$

(where N is the greatest integer not exceeding ns/m) is

$$U = 1 - \frac{2}{\sqrt{\pi}} \int_u^\infty e^{-t^2}\, dt + me^{-u^2} \Big/ \sqrt{2\pi n(m-s)s} \;, \tag{8}$$

which coincides with (1) if $p = s/m$ and $q = (m-s)/m$.

> Lors donc que le nombre n des événements futurs est très-petit eu égard au nombre m des événements observés, les limites du nombre de fois que A arrivera et leur probabilité pourront se calculer en prenant pour la probabilité p de A, le rapport s/m du nombre de fois que cet événement est arrivé au nombre total des observations, comme si cette valeur de p était certaine et donnée *à priori*. Mais il n'en est pas ainsi quand les deux nombres n et m sont du même ordre de grandeur. [p.278]

Mention is also made of the possibility of neglecting the last term in (8).

In the seventeenth article Poisson stresses the importance of ensuring that the assumption that the probability of the event A remains the same in each trial (past or future) is satisfied. This is illustrated by the drawing of balls from an infinite urn, an example which is then extended to take account of m similar urns[4], and it is shown that the probability of drawing a white ball is $\sum_1^m p_i/m$. This mean value can then be substituted for p in the preceding formulae.

This essentially concludes the main theoretical portion of the memoir. Poisson now compares the question of births to that of different urns. The event A corresponds to the birth of a boy, the probability p being susceptible of all values from zero to one. Since p may well vary with time and place (and also from one family to another), an average value of p is required, and

> c'est en supposant que cette moyenne ne variera pas, que l'on calcule la probabilité des naissances masculines pendant un autre intervalle de temps. [p.283]

Application of the formulae developed earlier in the memoir is made to natal data for various periods, and it is found that

> la chance d'une naissance masculine dépend des localités, en sorte qu'elle varie, pour une même année, d'une département à un autre, et pour un même département, d'une année à une autre. [p.286]

In Article 21 Poisson supposes that two events A and A' happen s and s' times respectively in m and m' trials, where $s, m-s, s'$ and $m'-s'$ are all very large. If p and p' denote the probabilities of A and A', and if p' exceeds p at least by a given quantity ω, it is shown, under approximations of the same sort as those used earlier and when $s'/m' - s/m = \omega$, that

$$T \equiv \Pr\left[p' - p > \omega\right] = \frac{1}{2} + \frac{\lambda' - \mu\lambda}{\sqrt{\pi(1+\mu^2)}} + \frac{\mu^2\lambda' - \mu\lambda}{2\sqrt{\pi(1+\mu^2)^3}} \qquad (9)$$

where $\lambda = (m-2s)\sqrt{2}\Big/3\sqrt{ms(m-s)}$ (and λ' is defined analogously) and

$$\mu^2 = (m'/m)^3(s/s')(m-s)/(m'-s') \ .$$

He notes further that $T \to \frac{1}{2}$ as m and $m' \to \infty$.

If one supposes in addition that $s/m = s'/m'$ then (9) becomes

$$\Pr\left[p' > p\right] = \frac{1}{2} + (m - m')\lambda\Big/\sqrt{m'(m+m')} \ ,$$

and in the case in which

$$s'/m' - s/m - \omega = \alpha\sqrt{2s(m-s)/m^3} \equiv \alpha f$$

where $|\alpha|$ is small, one obtains

$$\Pr\left[p' - p > \alpha f + s'/m' - s/m\right] = \frac{1}{2} + \frac{1}{\sqrt{\pi}}\int_0^c e^{-t^2}\,dt$$

where $c = \mu\alpha/\sqrt{(1+\mu^2)}$. This formula, Poisson notes, coincides with a result given by Laplace in his *Théorie analytique des probabilités*[5].

Further application of the preceding formulae to the question of births follows, it being found that

> Nous pouvons donc conclure qu'à l'époque actuelle et pour la France entière, la probabilité d'une naissance masculine n'éprouve que de très-petites variations d'une année à une autre, et prendre pour sa valeur, la moyenne des dix années que nous avons considérées, c'est-à-dire, 0,5159. [p.307]

This completes our study of the memoir: we now turn to Poisson's *Recherches sur la probabilité des jugements en matière criminelle et en matière civile, précédées des règles générales du calcul des probabilités* of 1837[6]. Although the major part (if indeed not all) of this work is of no little interest, we shall firmly confine ourselves to pertinent passages.

After commenting on the use made by Condorcet and Laplace of Bayes's Theorem in their work on the probability of judgement and testimony (to which work animadversion has already been made in the present treatise), Poisson expresses the doubts to which he was still subject on this matter after reading these authors, and which resulted in his approaching the matter from a different point of view.

> Le caractère distinctif de cette nouvelle théorie de la probabilité des jugements criminels étant donc de déterminer d'abord, d'après les données de l'observation dans un très grand nombre d'affaires de même nature, la chance d'erreur du vote des juges, et celle de la culpabilité des accusés avant l'ouverture des débats, elle doit convenir à toutes les espèces nombreuses de jugements. [p.25]

Poisson emphasizes the rôle of prior knowledge as follows:

> les règles qui servent à remonter de la probabilité d'un événement observé à celle de sa cause, et qui sont la base de la théorie dont nous nous occupons, exigent que l'on ait égard à toute présomption antérieure à l'observation, lorsque l'on ne suppose pas, ou qu'on n'a pas démontré qu'il n'en existe aucune. [p.4]

The first chapter of this work is entitled "Règles générales des probabilités". Poisson starts off with a precise statement of the way in which he will use the word "probability":

> La *probabilité* d'un événement est la raison que nous avons de croire qu'il aura ou qu'il a eu lieu. Quoiqu'il s'agisse, dans un cas, d'un fait accompli, et dans l'autre, d'une chose éventuelle; pour nous, la probabilité est cependant la même, lorsque tout est d'ailleurs égal dans ces deux cas, en eux-mêmes si différents. [p.30]

He further stresses the dependence of probability upon individual experience with the words

> La probabilité dépendant des connaissances que nous avons sur un événement, elle peut être inégale pour un même événement et pour diverses personnes [p.30],

and he points out further that the term "probability" will also be used with this meaning, "chance" being reserved[7]

> aux événements en eux-mêmes et indépendamment de la connaissance que nous en avons. [p.31]

He further defines

> La mesure de la probabilité d'un événement, est le rapport du nombre de cas favorables à cet événement, au nombre total de cas favorables ou contraires, et tous également possibles, ou qui ont tous une même chance [p.31],

and he indicates later on [p.33] the possibility of (indeed, the necessity for) extending this definition to incommensurable quantities.

Poisson next points out (though of course not in these symbols) that $\Pr[E] + \Pr[\overline{E}] = 1$, and follows this with the important observation that when we have no reason to believe in the occurrence of E rather than its complement $\overline{E}$, each should be assigned probability $\frac{1}{2}$. The usual product rule for the probability of the joint occurrence of two independent events is stated, and this is extended to the observation that the probability of m successive happenings of the event E is p^m (where $\Pr[E] = p$). The extension to non-independent (or dependent) events is made, i.e. $\Pr[E \,\&\, E_1] = \Pr[E]\Pr[E_1 \mid E]$, where E denotes the event[8] "qui doit arriver le premier" [p.41], and expressions are given for probabilities resulting from the withdrawal, both with and without replacement, from an urn. In the tenth article we find a result which we would today write as

$$\Pr[A] = \sum_j \Pr[A \,\&\, H_j] \quad ,$$

and this is illustrated in the eleventh article by typical "urn and balls" examples[9].

Mathematical expectation is defined (acceptably) as follows:

> Le produit d'un gain et de la probabilité de l'obtenir est ce qu'on appelle *l'espérance mathématique* de chaque personne intéressée dans une spéculation quelconque [p.71],

and this is contrasted in the twenty-fourth article with *espérance morale*, the difference being illustrated by the St Petersburg Paradox[10].

The second chapter, occupying nigh on a hundred pages, is entitled "Suite des règles générales; probabilités des causes et des événements futurs, déduites de l'observation des événements passés". Poisson begins by giving a precise definition of the way in which the word "cause" is to be used in the calculus of probabilities:

> on y considère une *cause* C, relative à un événement quelconque E, comme étant la chose qui donne à l'arrivée de E, la chance déterminée qui lui est propre. [p.79]

Furthermore,

L'ensemble des causes qui concourent à la production d'un événement sans influer sur la grandeur de sa chance, c'est-à-dire, sur le rapport du nombre de cas favorables à son arrivée au nombre total des cas possibles, est ce qu'on doit entendre par le *hasard.* [p.80]

Poisson now passes, in the twenty-eighth article, to a discrete form of Bayes's Theorem. He supposes that the occurrence of an event E may be attributed to any one of a number m of mutually exclusive and exhaustive causes, all of which, prior to observation, are equally probable. The question is the determination of the *a posteriori* probabilities of these causes. If we denote the sequence of causes by $\{C_n\}$, we have

$$\omega_n \equiv \Pr[C_n \mid E] = \Pr[E \mid C_n] / \sum_j \Pr[E \mid C_j] \ .$$

In the next article Poisson points out that, in finding the probabilities of several successive events, one ought to consider not only the effect that the occurrence of one has on the chance of the following event, but also sometimes the probabilities of the divers causes of the first event. The results of this article are extended in the following one to the case of an event E' following E, the desired probability (under a suitable, though unstated, assumption of conditional independence) being given by

$$\omega' \equiv \Pr[E' \mid E] = \sum_n \Pr[E' \mid C_n] \Pr[E \mid C_n] \Big/ \sum_n \Pr[E \mid C_n] \ .$$

Telle est la formule qui sert à calculer la probabilité des événements futurs, d'après l'observation des événements passés. [p.87]

In Article 32 Poisson applies his results to some simple examples. In the first of these (later generalized by Catalan — see §7.8) he considers the drawing of a white ball from an urn B known to contain m white or black balls. The probability ω_n that the urn contains n white balls is shown to be $2n/m(m+1)$, under the assumption that the possible initial compositions of the urn are equally probable. If now another white ball is drawn from the urn (event E'), the probability ω' defined above is found to be (i) $(2m+1)/3$, if sampling occurs with replacement, and (ii) $2/3$ if the sampling is without replacement. The case in which $(m-1)$ draws from m white or black balls have resulted in $(m-1)$ white balls is also considered.

In his next article Poisson considers the case in which m is unknown: all that is known is that $m \leq 3$ (say). If E denotes the event that x white balls have been drawn in a series of n draws (with replacement), with $0 < x < n$, one may suppose that any one of the following three hypotheses about the composition of the urn holds:

C_1. one white and one black ball;
C_2. one black and two white balls;
C_3. one white and two black balls.

Then

$$\Pr[C_1] = (1/2)^x(1/2)^{n-x} = 1/2^n$$

$$\Pr[C_2] = (2/3)^x(1/3)^{n-x} = 2^x/3^n$$

$$\Pr[C_3] = (1/3)^x(2/3)^{n-x} = 2^{n-x}/3^n \quad .$$

The probabilities ω_1, ω_2 and ω_3 are then easily found. If the event E' is the withdrawal of a further white ball from the urn, then

$$\omega' = \left[(1/2)3^n + (2/3)2^{n+x} + (1/3)2^{2n-x}\right] / \left(3^n + 2^{n+x} + 2^{2n-x}\right) \quad .$$

Detailed examination of the cases (i) $n = 2x$, (ii) $x = 2i$ and $n = 3i$, (iii) $n = 3x$ follows, and Poisson notes that, as the number of withdrawals increases, ω' tends in these three instances to 1/2, 2/3 and 1/3 respectively.

In his thirty-fourth article Poisson considers the case in which the causes are not initially equally probable, expressions of the usual form for ω_n and ω' being obtained (consideration is also given to the case of the occurrence of yet another event E'' following on the occurrence of E' which in turn followed E). The theory is followed in Article 35 by an example, and in the following articles application is made to the question of testimony, an important observation being the following:

> la probabilité d'un événement qui nous est transmis par une chaîne traditionelle d'un très grand nombre de témoins, ne diffère pas sensiblement de la chance propre de cet événement, ou indépendante du témoignage; tandis que l'attestation d'un grand nombre de témoins directs d'un événement rend sa probabilité très approchante de la certitude, lorsqu'il y a pour chacun de ces témoins plus d'un contre un à parier qu'il ne nous trompe pas (n° 37). [p.112]

In his forty-third article Poisson turns his attention to the case in which the number of causes to which an event E may be attributed is infinite. Supposing firstly that the observed event E is the drawing of a white ball from an urn containing an infinite number of white and black balls, Poisson considers firstly the case in which the initial distribution of x, the ratio of white balls to the total number, is uniform (as we would phrase it today), obtaining for the probability ω of X the ratio

$$X\,dx \Big/ \int_0^1 X\,dx \quad ,$$

where X denotes the probability that x, if it were certain, would give to the occurrence of E. Similarly, if E' is a future event depending on the same causes as E, with corresponding probability X', we have

$$\omega' = \int_0^1 X\,X'\,dx \Big/ \int_0^1 X\,dx \quad .$$

If, on the other hand, the initial values of x are not equally probable but follow some distribution Y, then

$$\omega \quad = XY\,dx \Big/ \int_0^1 XY\,dx$$

$$\omega' \quad = \int_0^1 X\,X'Y\,dx \Big/ \int_0^1 XY\,dx \quad .$$

In his next article Poisson shows effectively that

$$\lambda \equiv \Pr\left[\alpha < x < \beta \mid E\right] = \int_\alpha^\beta f(x)\,\varphi(x)\,dx \Big/ \int_0^1 f(x)\,\varphi(x)\,dx \quad ,$$

where, as before, x denotes the prior probability of E. As an illustration, he considers the case of sampling with replacement from an urn containing a vast number of balls (as many white as black), the sampling resulting in the obtaining of n white balls in n draws (event E). In this case $f(x) = x^n$, and hence the probability that the urn contains more white than black balls will be

$$\int_{1/2}^1 x^n\,dx \Big/ \int_0^1 x^n\,dx = 1 - (1/2)^{n+1} \quad .$$

In Article 45 Poisson supposes that E and E' are both events which are composed of the same simple event G, the chance that the probability of this event is x being $Y dx$ before the occurrence of E and ω thereafter. The notation is Poisson's: one might perhaps rather write the first part of this supposition as

$$\Pr\left[x < p_G < x + dx\right] = f(x)\,dx \quad ,$$

where $\int_0^1 f(x)\,dx = 1$. Poisson next suggests that, according to the rule of mathematical expectation, one ought to take as the unknown value of the chance of G, before the occurrence of E, the value

$$\gamma = \int_0^1 x\, f(x)\, dx \ .$$

As an example he considers the case in which p_G is uniformly distributed over the interval (0,1), obtaining $\gamma = 1/2$ and $\omega' = 2/3$ when E and E' are both the event G. He also deduces that the probability that G will occur on the second trial if it has failed to occur on the first, is

$$\omega' = \int_0^1 (1-x)x\, dx \Big/ \int_0^1 (1-x)\, dx = 1/3 \ .$$

In the forty-sixth article Poisson considers the case in which E is the event that G has occurred m times and $\overline{G}$ (or H, in Poisson's notation) n times, and the future event E' is the occurrence of G and of $\overline{G}$, m' and n' times respectively. We then have

$$\omega' = \binom{m'+n'}{m'} \int_0^1 x^{m+m'}(1-x)^{n+n'} f(x)\, dx \Big/ \int_0^1 x^m(1-x)^n f(x)\, dx$$

(here the events in E — and in E' — may occur in any order whatsoever). In the case in which $f(\cdot)$ is a uniform density, Poisson shows that

$$\omega' = \binom{m+m'}{m}\binom{n+n'}{n} \Big/ \binom{m+m'+n+n'+1}{m+n+1} \ ,$$

which reduces, when $n = 0 = n'$, to

$$\omega' = (m+1) \big/ (m+m'+1) \ .$$

There follows, in Article 47, a consideration of the case $x = r + z$, where r should be taken equal to γ. On letting $m' = 1$ and $n' = 0$, and changing the variable of integration from x to z, Poisson shows that, neglecting terms of small order,

$$\omega' = r + [m/r - n/(1-r)]h$$

where $h = \int_{-r}^{1-r} f(z)\, z^2\, dz$. In the subsequent article he uses this result to compare the probability of two similar outcomes (i.e. both G or both $\overline{G}$) with that of dissimilar outcomes.

Poisson next passes on to the statement of his generalization of Bernoulli's law of large numbers[11] (its proof being postponed to his third chapter), and illustrates its use with some numerical data from Buffon's *Arithmétique morale*.

The third chapter of Poisson's work is entitled "Calcul des probabilités qui dépendent de très grands nombres". In Article 71 the rule of succession receives further attention: assuming that E and F are complementary events with constant but unknown probabilities p and q, Poisson supposes that in $\mu = m+n$ trials E and F have occurred m and n times respectively. The probability $U^{'}$ that in $\mu^{'}$ further trials E and F will occur $m^{'}$ and $n^{'}$ times respectively is given by[12]

$$U^{'} = \binom{m+m^{'}}{m}\binom{n+n^{'}}{n} \Big/ \binom{\mu+\mu^{'}+1}{\mu+1} .$$

Using the approximation

$$n! \sim n^n e^{-n}\sqrt{2\pi n} ,$$

Poisson shows that

$$U^{'} \approx HK \frac{(m+m^{'})^{m+m^{'}}(n+n^{'})^{n+n^{'}}(\mu+1)^{\mu}}{m^m n^n (\mu+\mu^{'}+1)^{\mu+\mu^{'}}} ,$$

where $H = \binom{m^{'}+n^{'}}{m^{'}}$ and

$$K = \frac{\mu+1}{\mu+\mu^{'}+1}\sqrt{\frac{(m+m^{'})(n+n^{'})(\mu+1)}{m\,n(\mu+\mu^{'}+1)}} .$$

Under the further assumption that $m^{'}$ and $n^{'}$ are very small in comparison with m and n, Poisson deduces that

$$U^{'} = \binom{m^{'}+n^{'}}{m^{'}}\left(\frac{m}{\mu}\right)^{m^{'}}\left(\frac{n}{\mu}\right)^{n^{'}} ,$$

which agrees with the usual binomial probability for the occurrence of $m^{'}$ E's and $n^{'}$ F's in $\left(m^{'}+n^{'}\right)$ trials, where E and F have the given *a priori* probabilities $p = m/\mu$ and $q = n/\mu$. Poisson notes further that this pleasing state of affairs ceases to obtain if $m^{'}$ and $n^{'}$ are of comparable magnitudes to m and n: indeed, in this case we obtain the (approximate) probability

$$\frac{1}{\sqrt{1+h}}\binom{m+n}{m}\left(\frac{m}{\mu}\right)^{m^{'}}\left(\frac{n}{\mu}\right)^{n^{'}}$$

where $m^{'} = mh, n^{'} = nh$ and $\mu^{'} = \mu h$.

Several articles of this third chapter are devoted to Bernoulli's Theorem[13]: since we shall need to refer to Poisson's statement of it in the near future, we shall state it here. Thus let p and q be the known and constant chances of the events E and F; let μ (a large number) trials be made[14], and let N be the greatest integer not exceeding μq. Further, let u be a quantity such that $u\sqrt{2(\mu+1)pq}$ is an integer which is very small in relation to N. If m and n are the numbers of occurrences of E and F in the μ trials, then[15]

$$\begin{aligned} R &\equiv \Pr\left[N - u\sqrt{2\mu pq} \leq n \leq N + u\sqrt{2\mu pq}\right] \\ &= 1 - \frac{2}{\sqrt{\pi}}\int_u^\infty e^{-t^2}\,dt + \frac{1}{\sqrt{2\pi\mu pq}}e^{-u^2} \quad . \end{aligned} \tag{10}$$

Various developments are also given: for example, n may be restricted to lie in a half-open or open interval, or one may neglect the difference between μq and N, in which case it is found that (10) yields $\Pr[|n/\mu - q| \leq u\sqrt{2pq/\mu}]$.

Poisson turns his attention in his eighty-third article to the case in which the probabilities p and q are not given, although the ratios m/μ and n/μ have been observed. In this case, he states,

> les formules que nous avons trouvées feront connaître les valeurs très probables et très approchées des inconnues p et q. [p.209]

It thus follows that (10) yields the probability

$$\Pr[|p - m/\mu| \leq u\sqrt{2pq/\mu}]$$

(where it should be stressed that p and q are unknown). If R differs only slightly from 1, the terms p and q under the root sign may be replaced by m/μ and n/μ respectively, yielding

$$\Pr[|p - m/\mu| \leq (u/\mu)\sqrt{2mn/\mu}] = 1 - \frac{2}{\sqrt{\pi}}\int_u^\infty e^{-t^2}\,dt + e^{-u^2}\sqrt{\mu/2\pi mn}\,. \tag{11}$$

Poisson points out further that the same result[16] may be used in connexion with future events. Thus, if m, n and μ are large, if μ' further trials result in m' occurrences of E and n' occurrences of F, and if μ', although small in comparison to μ, is still very large, we have

$$\begin{aligned} &\Pr\left[|n'/\mu' - n\mu| \leq (u/\mu)\sqrt{2mn/\mu'}\right] \\ &\quad = 1 - \frac{2}{\sqrt{\pi}}\int_u^\infty e^{-t^2}\,dt + e^{-u^2}\mu \Big/ \sqrt{2\pi\mu' mn}\,. \end{aligned}$$

Finally, Poisson concludes this article by noting that the results obtained so far by an inversion of Bernoulli's Theorem, are inapplicable if μ and $\mu^{'}$ are of comparable magnitude: he proposes therefore to consider another line of approach.

Once again it is supposed that E and F, of constant probabilities p and q, have occurred m and n times in a large number $\mu = m + n$ of trials. Then m/μ and n/μ may be taken as the approximate values of p and q. Now in his proof of Bernoulli's Theorem Poisson had shown that

$$Q \equiv \Pr\left[n < \mu q - r\sqrt{2\mu pq}\right] = \frac{1}{\sqrt{\pi}}\int\limits_{r+\delta}^{\infty} e^{-t^2}\,dt + e^{-r^2}(q-p)/3\sqrt{2\pi\mu pq} \quad (12)$$

where $\delta = (p-q)r^2/3\sqrt{2(\mu+1)pq}$. The left-hand side of (12) is, of course, to be regarded as a function of *known* p and q: in the present discussion, however, Poisson glibly passes from this to

$$\Pr\left[q > n/\mu + r\sqrt{2pq/\mu}\right]$$

where q is "la chance inconnue ... de l'événement F" [p.211]. Replacing p and q on the right-hand side of the last inequality by their limiting values m/μ and n/μ, he obtains

$$\Pr\left[q > n/\mu + (r/\mu)\sqrt{2mn/\mu}\right] = Q \ .$$

It then follows that

$$\frac{-dQ}{dr}\,dr = \Pr\left[n/\mu + (r/\mu)\sqrt{2mn/\mu} < q < n/\mu + ((r+dr)/\mu)\sqrt{2mn/\mu}\right],$$

or, as Poisson has it,

> $-\frac{dQ}{dr}$ exprimera la probabilité infiniment petite que l'on a précisément
>
> $$q = n/\mu + (r/\mu)\sqrt{2mn/\mu}$$
>
> pour toutes les valeurs de r positives et très petites par rapport à $\sqrt{\mu}$, comme le suppose l'expression de Q. [p.211]

Similarly, from

$$\begin{aligned} Q^{'} &\equiv \Pr\left[q > (n/\mu) - (r^{'}/\mu)\sqrt{2mn/\mu}\right] \\ &= 1 - \frac{1}{\sqrt{\pi}}\int_{r'-\delta'}^{\infty} e^{-t^2}\,dt + e^{-r^{'2}}(q-p)/3\sqrt{2\pi\mu pq} \ , \end{aligned}$$

it follows that $(dQ'/dr')\,dr'$ equals the probability that

$$q = n/\mu - (r'/\mu)\sqrt{2mn/\mu} \ .$$

Here $\delta' = (p-q)r'^2/3\sqrt{2\mu pq}$.

Working from the integral forms of Q and Q', and replacing p and q by m/μ and n/μ in both δ and δ', Poisson finds that

$$-\frac{dQ}{dr} = \frac{1}{\sqrt{\pi}}e^{-r^2} - 2(m-n)r^3\, e^{-r^2} \Big/ 3\sqrt{2\pi\mu mn}$$

$$\frac{dQ'}{dr'} = \frac{1}{\sqrt{\pi}}e^{-r'^2} + 2(m-n)r'^3\, e^{-r'^2} \Big/ 3\sqrt{2\pi\mu mn}$$

on neglecting terms having μ as divisor. It follows that the expression[17]

$$V = \frac{1}{\sqrt{\pi}}e^{-\nu^2} - 2(m-n)\nu^3\, e^{-\nu^2} \Big/ 3\sqrt{2\pi\mu mn}$$

yields $V\,d\nu$ for the probability that $q = n/\mu + (\nu/\mu)\sqrt{2mn/\mu}$. Since $p = 1-q$ and $m = \mu - n$, it is immediate that this expression is also the probability that $p = m/\mu - (\nu/\mu)\sqrt{2mn/\mu}$.

Todhunter [1865, art.997] has pointed out that

$$\int_{-\tau}^{\tau} V\,d\nu = \frac{2}{\sqrt{\pi}}\int_0^{\tau} e^{-t^2}\,dt \ ,$$

which is different from the result given in (11) above. He finds it "curious" that Poisson failed to comment on the difference between the results obtained by these two methods: however, since the results require the substitution of m/μ for p and n/μ for q at different stages in the proof, one need not be surprised that different answers are obtained — especially since both are in fact approximations.

To conclude this article Poisson considers the finding of the probability of a future event E', an event which consists in the occurrence of certain numbers of E's and F's. Denoting by Π the probability of E' for given values of $\Pr[E]$ and $\Pr[F]$ (so that Π is a function of p and q), he concludes that the desired probability is given (approximately) by

$$\Pi' = \int \Pi V\,d\nu \ ,$$

and he notes further that

> Ce résultat s'accorde avec celui qui a été obtenu plus directement, dans le second paragraphe de mon mémoire sur *la proportion des naissances des deux sexes.* [p.214]

Having discussed the relevant theory, Poisson now turns his attention to some examples. In his eighty-fifth article he considers the probability Π' that μ' further trials will yield m' occurrences of E and n' occurrences of F. Under the assumption that the ratio $m' : n'$ is approximately the same as that of $m : n$, he deduces, in a manner similar to that used earlier, that

$$\Pi' = \frac{1}{\sqrt{\pi}} U' \int \exp\left(-\nu^2 - \mu'^3\, \nu^2 / 2m'n'\right) d\nu \ ,$$

where $U' = \sqrt{\mu'/2\pi m'n'}$. On replacing m' and n' by mh and nh, and neglecting sufficiently small terms, one finds that

$$\Pi' = \frac{1}{\sqrt{1+h}} U' \ ,$$

a result which should be compared with that given in Article 71.

In the next article Poisson considers, in the notation of the preceding example, the probability Π' that $|n'/\mu' - n/\mu|$ does not exceed $\alpha/\sqrt{\mu'}$, where α is defined by $m' = mh - \alpha\sqrt{\mu'}$ (the details will not be given here). The matter is pursued further in the following article.

In Article 88 Poisson turns his attention to a further development of the theory[18]. Suppose that the complementary events E and F, of unknown chances p and q, have occurred m and n times in a large number $\mu = m+n$ of trials. Suppose further that the complementary events E_1 and F_1, of unknown chances p_1 and q_1, have occurred m_1 and n_1 times in a large number $\mu_1 = m_1 + n_1$ of trials. The aim is to determine the probability of an inequality between p and p_1, q and q_1, corresponding to differences between the ratios m/μ and m_1/μ_1, n/μ and n_1/μ_1. Setting

$$m_1/\mu_1 - m/\mu = \delta \ ,$$

Poisson shows, by approximations similar to those carried out before, that for ϵ positive and small,

$$\lambda \equiv \Pr\left[p_1 \geq p + \epsilon\right] = \int\!\!\int V V_1 \, d\nu \, d\nu_1 \ ,$$

where V is the expression introduced earlier, and V_1 is the corresponding formula *mutatis mutandis.* Neglect of the second terms in V and V_1 occasions

$$\lambda = \frac{1}{\pi} \int\!\!\int \exp\left(-\nu^2 - \nu_1^2\right) d\nu \, d\nu_1 \ .$$

After various substitutions Poisson obtains

$$\lambda = \begin{cases} \dfrac{1}{\sqrt{\pi}} \displaystyle\int_u^\infty e^{-t^2}\, dt\ , & \text{if } \epsilon - \delta > 0 \\ 1 - \dfrac{1}{\sqrt{\pi}} \displaystyle\int_u^\infty e^{-t^2}\, dt\ , & \text{if } \epsilon - \delta < 0 \end{cases}$$

where

$$u = \left| \frac{(\epsilon - \delta)\mu\, \mu_1 \sqrt{\mu\, \mu_1}}{\sqrt{2(\mu^3 m_1 n_1 + \mu_1^3 mn)}} \right| \ .$$

Indeed, the neglect of the second terms in V and V_1 in fact results in $\lambda = \Pr[p_1 > p + \epsilon]$.

As an illustration Poisson considers in Article 89 the results of a coin-tossing experiment carried out by Buffon[19]: once again we shall omit the details. The remaining articles of this chapter are devoted to the question of repeated draws, without replacement, from an urn containing white and black balls, the draws resulting in long sequences of lengths $\mu, \mu', \ldots$. The preceding theory is applied to find the probability that the number of trials in which the number of white balls drawn exceeds the number of black balls drawn, lies within given bounds, an expression of the form (10) being obtained. An application to elections follows.

In his fifth chapter, entitled "Application des règles générales des probabilités aux décisions des jurys et aux jugements des tribunaux", Poisson makes use of the Bayes-type formulae developed in earlier chapters.

Starting off in Article 114 with the simplest of situations, he examines the case of a single juror. He supposes that the probability k that an accused, on arraignment or indictment, is guilty, is based on preliminary information and the subsequent accusation. On denoting by u the probability that the juror is not mistaken in his decision, we find that the probability γ that the accused will be convicted is given by

$$\gamma = ku + (1 - k)(1 - u) \ .$$

Then the probability that he is guilty, given that he has been convicted, is, by Article 34,

$$p = ku/[ku + (1 - k)(1 - u)] \ .$$

Similarly the probability q that he is innocent, given that he is acquitted[20], is

$$q = (1 - k)u/[(1 - k)u + k(1 - u)] \ .$$

Further developments of this single-juror case follow: they need not be our concern.

In the next article Poisson broadens the scope of his investigation to encompass the addition of a second juror. The concern here is to determine $\Pr[C_1C_2]$, the probability that the accused is convicted by each juror, $\Pr\left[C_1\overline{C}_2 \vee \overline{C}_1C_2\right]$, the probability that he is convicted by one and acquitted by the other, and $\Pr\left[\overline{C_1}\overline{C_2}\right]$, the probability that he is acquitted by both. Now $\Pr[C_1C_2] = \Pr[C_1]\Pr[C_2 \mid C_1]$, and, as in the preceding article,

$$\Pr[C_1] \equiv \gamma_1 = k\,u_1 + (1-k)(1-u_1) \ ,$$

the subscript referring to the first juror. For the second juror, however, k is replaced by p_1 ($\equiv p$ of the preceding article), and hence

$$\Pr[C_2 \mid C_1] \equiv \gamma_2 = p_1u_2 + (1-p_1)\,(1-u_2) \ .$$

It thus follows that

$$\Pr[C_1C_2] = k\,u_1u_2 + (1-k)\,(1-u_1)\,(1-u_2) \ .$$

A similar argument showing that

$$\Pr\left[\overline{C_1}\overline{C_2}\right] = k\,(1-u_1)\,(1-u_2) + (1-k)u_1u_2 \ ,$$

it hence follows that the probability that both jurors arrive at the same conclusion is

$$\Pr\left[C_1C_2 \vee \overline{C}_1\overline{C}_2\right] = u_1u_2 + (1-u_1)\,(1-u_2)$$

(independent of k). Further argument in the same vein yields

$$\Pr\left[C_1\overline{C}_2 \vee \overline{C}_1C_2\right] = (1-u_1)\,u_2 + (1-u_2)\,u_1 \ .$$

The probability $\Pr[G \mid C_1C_2]$ that the accused is guilty, given that he has been convicted by both jurors, will be, by an argument similar to one used in the previous article,

$$\Pr[G \mid C_1C_2] = p\,u_2\,/[p\,u_2 + (1-p)\,(1-u_2)] \ ,$$

and similarly

$$\begin{aligned}
\Pr\left[\overline{G} \mid \overline{C}_1\overline{C}_2\right] &= q\,u_2\,/[q\,u_2 + (1-q)\,(1-u_2)] \\
\Pr\left[G \mid \overline{C}_1C_2\right] &= (1-q)u_2\,/[(1-q)u_2 + q\,(1-u_2)] \\
\Pr\left[\overline{G} \mid C_1\overline{C}_2\right] &= (1-p)u_2\,/[(1-p)u_2 + p\,(1-u_2)] \ .
\end{aligned}$$

On writing these last two expressions in terms of k, u_1 and u_2 one finds that $\Pr\left[G \mid \overline{C}_1 C_2\right] = k$ and $\Pr\left[\overline{G} \mid C_1 \overline{C}_2\right] = 1 - k$ when $u_1 = u_2$, as might well be expected.

In the following articles similar results are obtained for more than two jurors: we shall not present intermediate results here, but shall pass immediately to the case in which the accused is convicted by at least $(n-i)$ votes and acquitted by at most i, when $(n-i)$ and i are very large[21]. Denoting this probability by c_i, and letting d_i be the probability that the accused is acquitted by at least $(n-i)$ votes and convicted by at most i, we have, as in the preceding cases,

$$c_i = k\, U_i + (1-k) U_i \quad ; \quad d_i = k\, V_i + (1-k) V_i$$

where

$$U_i = \sum_{j=0}^{i} \binom{n}{j} u^{n-j} (1-u)^j \text{ and } V_i = \sum_{j=0}^{i} \binom{n}{j} (1-u)^{n-j} u^j \ .$$

The methods of approximation introduced in Article 77 yield

$$U_i \quad = \frac{1}{\sqrt{\pi}} \int_\theta^\infty e^{-x^2}\, dx + e^{-\theta^2} (n+i)\sqrt{2} \Big/ 3\sqrt{\pi n i (n-i)}$$

$$V_i \quad = 1 - \frac{1}{\sqrt{\pi}} \int_\theta^\infty e^{-x^2}\, dx + e^{-\theta^2} (n+i)\sqrt{2} \Big/ 3\sqrt{\pi n i (n-i)} \ ,$$

where $\theta > 0$ satisfies

$$\theta^2 = i\, \ln\left[i/\nu(n+1)\right] + (n+1-i) \ln\left[(n+1-i)/u(n+1)\right]$$

with $\nu = 1 - u$.

Various developments of these formulae follow: once again they need not concern us.

Poisson begins his Article 124 with the following observation:

> Les formules précédentes donneraient les solutions complètes de toutes les questions relatives à l'objet de ce chapitre, si avant le jugement, la probabilité k de la culpabilité était connue, et que l'on connût aussi, pour chaque juré et dans chaque affaire, la probabilité qu'il ne se trompera pas; ou bien, si cette chance de ne pas se tromper a plusiers valeurs possibles, il faudrait que toutes ces valeurs fussent données, ainsi que leurs probabilités respectives; ou bien encore, quand ces valeurs sont en nombre infini et ont chacune une probabilité infiniment petite, il serait nécessaire que nous connussions la fonction qui exprime la loi de leurs probabilités. [p.345]

In an attempt to eliminate these unknown elements Poisson supposes in Article 125 that the jurors have the same chance of being mistaken, a chance U which has probability density function φ. It follows that

$$\lambda_i \equiv \Pr\left[\ell < U < \ell'\right]$$

$$= \frac{k\int_\ell^{\ell'} u^{n-i}(1-u)^i\varphi(u)\,du + (1-k)\int_\ell^{\ell'} u^i(1-u)^{n-i}\varphi(u)\,du}{k\int_0^1 u^{n-i}(1-u)^i\varphi(u)\,du + (1-k)\int_0^1 u^i(1-u)^{n-i}\varphi(u)\,du} \quad . \quad (13)$$

Various special cases of this formula are then considered[22]: in the case in which $n = 2i$, λ_i is seen to be independent of k, as in the case in which $\varphi(1-u) = \varphi(u)$ and $\ell' = 1 - \ell$ with $\ell < \frac{1}{2}$.

The *a posteriori* probability of guilt is found in the following article to be

$$\zeta_i = \frac{k\int_0^1 u^{n-i}(1-u)^i\varphi(u)\,du}{k\int_0^1 u^{n-i}(1-u)^i\varphi(u)\,du + (1-k)\int_0^1 u^i(1-u)^{n-i}\varphi(u)\,du} \quad , \quad (14)$$

and this reduces to k when $n = 2i$ or $\varphi(1-u) = \varphi(u)$. Analogous formulae are given for the case in which one knows merely that the accused has been convicted by a majority of at least m, or $(n-2i)$, votes.

In Articles 128 and 129 Poisson passes to the case in which $(n-i)$ and i are very large numbers. After various approximations he arrives at

$$\lambda_i = \Pr\left[(n-i)/n - \delta\sqrt{2i(n-i)/n^3} < U < (n-i)/n + \delta\sqrt{2i(n-i)/n^3}\right]$$

$$= \frac{1}{\sqrt{\pi}}\,\zeta_i \int_{-\delta}^{\delta} e^{-x^2}\,dx \quad ,$$

where $\delta > 0$ is very small in relation to $\sqrt{n}$ and

$$\zeta_i = k\varphi\left((n-i)/n\right)/[k\varphi\left((n-i)/n\right) + (1-k)\varphi\left(i/n\right)] \quad .$$

In the next two articles, with much labour, Poisson derives the following approximate results:

(i) the posterior probability of guilt, after conviction by at least $(n-i)$ votes, is

$$Z_i = k\int_\alpha^1 \varphi(u)\,du \Big/ \left[k\int_\alpha^1 \varphi(u)\,du + (1-k)\int_0^{1-\alpha} \varphi(u)\,du\right]$$

where $\alpha = (n-i)/n$;

(ii) the probability, in conviction, that the chance U lies between α and $1-\alpha$ (i.e. between $(n-i)/n$ and i/n), is

$$Y_i = K\left(c\int_c^\infty e^{-x^2}\,dx + \tfrac{1}{2} - e^{-c^2}\right)\sqrt{2i(n-i)/\pi n^3}\ ,$$

where

$$K = \frac{k\,\varphi(\alpha) + (1-k)\,\varphi(1-\alpha)}{k\int_\alpha^1 \varphi(u)\,du + (1-k)\int_0^{1-\alpha}\varphi(u)\,du}$$

and $c > 0$ satisfies $(n-i)^i i^{n-i} = i^i(n-i)^{n-i}e^{-c^2}$.

Noting in the next article the need for some specific choice of φ, Poisson writes

> L'hypothèse que Laplace a faite pour cet objet, consiste à supposer que la fonction φu soit zéro pour toutes les valeurs de u moindres que $\frac{1}{2}$, et qu'elle ait une même valeur pour toutes celles de u qui surpassent $\frac{1}{2}$. [p.363]

Under this assumption the formula (14) becomes

$$\zeta_i = \frac{k\int_{1/2}^1 u^{n-i}(1-u)^i\,du}{k\int_{1/2}^1 u^{n-i}(1-u)^i\,du + (1-k)\int_0^{1/2} u^{n-i}(1-u)^i\,du}\ ,$$

a result which coincides with that given by Laplace[23] when $k = \frac{1}{2}$, and which yields

$$1 - \zeta_i = (1/2)^{n+1}\sum_{j=0}^{i}\binom{n+1}{j}\ .$$

Similarly, if $0 < \delta < \frac{1}{2}$, $k = \frac{1}{2}$, $\ell = \frac{1}{2}$ and $\ell' = \frac{1}{2} + \delta$, the formula (13) becomes

$$\lambda_i = \int_{(1/2)-\delta}^{(1/2)+\delta} u^{n-i}(1-u)^i\,du \Big/ \int_0^1 u^{n-i}(1-u)^i\,du\ .$$

Some general remarks and numerical examples conclude the work.

7.2 John William Lubbock (1803-1865) & John Elliot Drinkwater-Bethune (1801-1851)

An undated and anonymous tract, *On Probability*, was published in the early part of the nineteenth century "under the Superintendence of the Society for the Diffusion of Useful Knowledge". There seems little doubt now, however, that this slim volume was the work of Lubbock and Drinkwater-Bethune[24], and from Example 9 (concerned with the odds on certain horses winning the Gold Cup or the St Leger) we can place its date[25] as post May 1828.

In their introductory remarks the authors state that

> It is usual to apply the word belief to the past, and the word expectation to the future; but the theory of probability is in all respects the same, whether it be applied to past or to future events. [art.3]

If Shafer [1982] is correct, then $\Pr[E_1 \mid E_2]$ and $\Pr[E_2 \mid E_1]$ (where E_1 and E_2 are two events with E_2 subsequent to E_1) require rather different consideration, and this perhaps casts some suspicion on part of the above quotation.

The authors define "probability" in the usual way, though a certain measure of subjectivism appears. The definition runs as follows:

> the probability of any event is the ratio of the favourable cases to all the possible cases which, in our judgement, are similarly circumstanced with regard to their happening or failing. [art.4]

A basic constituent of modern subjective probability is *coherence*[26]. We can perhaps see a precursor of this concept in the following sentence:

> Since the sum of the probabilities of any number of conflicting events is equal to unity, we have an equation of condition between the odds; and whenever they do not satisfy this equation, it is possible to bet with the certainty of gain. [art.13]

We now come to that part of the tract devoted to Bayes's Theorem. At the outset Lubbock and Drinkwater-Bethune give a precise definition of the probability of a hypothesis as "the number of cases which favour this hypothesis divided by the whole number of cases possible" [art.45]. They next derive a discrete form of Bayes's Theorem, stating it as follows:

> The probability of any hypothesis is the probability of the observed event upon this hypothesis multiplied by the probability of the hypothesis antecedently to the observation divided by the sum of the products which are formed in the same manner from all the hypotheses. [art.45]

This is followed by a "bag and balls" example in which the ratio of the number of white balls to the total number of balls (white or black) may be any of the quantities $x, 2x, \ldots, ix$ with equal probability $1/i$. The authors deduce the posterior probabilities of these hypotheses after one white ball has been drawn, and also find the probability that a future draw will yield a white ball (sampling occurring with replacement). By a rather laborious argument, involving the expansion of $\exp(kx)$ for various values of k, this latter probability is shown to be $(2i+1)x/3$, which, for large i, is approximately $2ix/3$. This in turn "if ... the ratio of the white balls may be any ratio between 0 and unity" [art.47], becomes 2/3, a result which is of course more readily obtainable by considering

$$\int_0^1 x^2\,dx \Big/ \int_0^1 x\,dx \quad .$$

Lubbock and Drinkwater-Bethune next consider the case in which the ratio of white balls to the total number may be any one of $\Delta x, 2\Delta x, \ldots, i\Delta x$. It is shown that, if m white and n black balls have been drawn (in any order), the probability of drawing a further m' white and n' black balls becomes, in the limit as $\Delta x \to 0$ and $i\Delta x \to 1$ (and under an assumption of equi-possibility),

$$\binom{m'+n'}{m'} \int_0^1 x^{m+m'}(1-x)^{n+n'}\,dx \Big/ \int_0^1 x^m(1-x)^n\,dx \quad .$$

These integrals are then evaluated, and the result obtained is illustrated by that hoary example of the probability of the sun's rising once more, if it has risen 2,000,000 times.

> This probability, which is already very great, must be very considerably increased, if the discoveries of physical astronomy are taken into account. [art.48]

The authors now pass on to note that if $p+q$ draws have resulted in p white and q black balls, the probability of a further white is $(p+1)/(p+q+2)$, that of one more black being $(q+1)/(p+q+2)$. Furthermore, these fractions approximate more and more closely to $p/(p+q)$ and $q/(p+q)$ as p and q increase, an observation which is stated to be the "converse of Bernoulli's theorem" [art.49].

As an illustration of the foregoing theory, the case is considered of an individual who has made $(m+n)$ assertions, of which m were true and n false. The probability of his telling the truth in a further case is then $v = (m+1)/(m+n+2)$. If p denotes the *a priori* probability of the event whose happening is asserted, then the probability that the event did occur given that the witness asserts it occurred, is

$$pv/\left[pv + (1-p)(1-v)\right] \quad , \tag{15}$$

a fraction which is greater than p when $v > \frac{1}{2}$. This remark is extended to the case of $(n+1)$ individuals reporting independently, it being shown that

> the assertion of the $(n+1)$-th individual increases the probability of the event arising from the testimony of the other n individuals, only when his veracity is greater than $\frac{1}{2}$. [art.49]

Finally, in this matter, it is pointed out that when there are no data by which the veracity v of the individual may be determined, the expression (15) should be replaced by

$$\int_0^1 pv/[pv + (1-p)(1-v)]\, dv \ .$$

This is illustrated by a juratorial example.

An extension is now made to sampling with replacement from a bag containing balls of i different colours. If $m_1, m_2, \ldots, m_i$ balls have already been drawn, the probability of drawing a further $n_1, n_2, \ldots, n_i$ balls (again under an "equally probable" *a priori* assumption) will be

$$\int \cdots \int x_1^{m_1+n_1} x_2^{m_2+n_2} \ldots x_i^{m_i+n_i}\, dx_i \ldots dx_1 \Big/ \int \cdots \int x_1^{m_1} \ldots x_i^{m_i}\, dx_i \ldots dx_1 \ ,$$

the i-fold integrals being taken over the set

$$\{(x_1, \ldots, x_i) : \ 0 \le x_j \le 1 - x_1 - \cdots - x_{j-1} \ , \ j \in \{1, 2, \ldots, i\}\} .$$

Hence, as a special case, one finds that the probability of one further trial's yielding a ball of the r-th colour is[27]

$$(m_r + 1) / (m_1 + \cdots + m_i + i)$$

where $r \in \{1, 2, \ldots, i\}$.

In section 33 of his *Mémoire sur les probabilités* of 1778 Laplace gave a multinomial generalization of the Bayes prior: consider the outcomes $X_1, X_2, \ldots, X_N$ of a k-category multinomial with prior $\mathbf{p} = (p_1, p_2, \ldots, p_k)$. Then the probability of the frequency count $\mathbf{n} = (n_1, n_2, \ldots, n_k)$ is

$$\Pr[\{n_1, \ldots, n_k\}] = \frac{N!}{\Pi n_i!} \int \Pi p_i^{n_i}\, dF(\mathbf{p}) \ ,$$

the integral being taken over $\left\{\mathbf{p} : \sum_1^k p_i = 1\right\}$, with $dF(\mathbf{p}) = dp_1 \ldots dp_{k-1}$. As in the case of the binomial distribution, it follows, as Lubbock and Drinkwater-Bethune show, that

$$\Pr[X_{n+1} \in \ i\text{th category} \mid \mathbf{n}] = (n_i + 1) / (N + k) \ .$$

This same result was justified by W.E. Johnson [1932] in a manner similar to that advanced by Bayes in his Scholium, and it in fact follows (see Zabell [1982]) from Johnson's Sufficientness Postulate, viz.

$$\Pr\left[X_{n+1} \in i\text{th category} \mid \mathbf{n}\right] = f\left(n_i, \mathbf{n}\right) \ ,$$

that there exists $\kappa > 0$ such that

$$f\left(n_i, \mathbf{n}\right) = \left(n_i + \kappa\right) / \left(N + k\kappa\right) \ ,$$

a formula whose connexion with the continuum of inductive methods discussed by Carnap [1952] is evident.

In Article 80 Lubbock and Drinkwater-Bethune mention Bayes's Essay and correctly state the main result; viz. the probability that the happening of an event, which has already occurred p times in $(p+q)$ experiments, has a probability between A and a $(A < a)$ is

$$\int_A^a x^p(1-x)^q\,dx \Big/ \int_0^1 x^p(1-x)^q\,dx.$$

They go on to make the astute observation that

> Bayes, or perhaps we should rather say Price, seems to have confounded the probability thus determined, with the probability that an event which has been already observed m [sic] times in $p+q$ experiments, will happen again. The difference between the two is obvious. [art.80]

This remark, as we have mentioned in an earlier chapter, is not accepted as correct by Todhunter [1865, art.551].

In 1830 two papers by Lubbock on annuities were printed in the *Transactions of the Cambridge Philosophical Society*, the first of these (and the only one to concern us here) also containing some thoughts on probability. In the second article of this paper it is supposed that, from a bag containing a number (possibly – or necessarily – infinite) of balls of p different colours, $m_1 + m_2 + \cdots + m_p$ are drawn, where m_1 balls are of the first colour, m_2 are of the second, etc. If x_i denotes the probability that a ball of the i-th colour is drawn in one trial, then the probability of the given event is

$$x_1^{m_1} \times x_2^{m_2} \times \cdots \times x_p^{m_p}$$

multiplied by the coefficient of this term in the expansion of

$$\left(x_1 + x_2 + \cdots + x_p\right)^{m_1+m_2+\cdots+m_p} \ .$$

If this event is observed, "the probability of this system of probabilities" [p.144] will be

$$\binom{m_1+\cdots+m_p}{m_1,\ldots,m_p}\prod_1^p x_i^{m_i}\Big/\int\cdots\int\binom{m_1+\cdots+m_p}{m_1,\ldots,m_p}\prod_1^p x_i^{m_i}\,dx_1\ldots dx_p\ ,$$

the integration here, and elsewhere, being over the set

$$\{(x_1,\ldots,x_p):\ \ 0\le x_i\le 1-x_1-\cdots-x_{i-1}\ ,\ \ i\in\{1,2,\ldots,p\}\}.$$

It follows in the usual way that the probability that $n_1+n_2+\cdots+n_p$ subsequent trials yield n_1 balls of the first colour, n_2 of the second, etc., will be

$$\binom{n_1+\cdots+n_p}{n_1,\ldots,n_p}\int\cdots\int\prod_1^p x_i^{m_i+n_i}\,dx_1\ldots dx_p\Big/\int\cdots\int\prod_1^p x_i^{m_i}\,dx_1\ldots dx_p.$$

Lubbock's evaluation of this last probability is easily seen to be expressible as

$$\prod_1^p\binom{m_i+n_i}{m_i}\Big/\binom{M+N+p-1}{N}\ , \tag{16}$$

where $M=\sum_1^p m_i$ and $N=\sum_1^p n_i$.

Several useful observations follow: firstly, "this probability is the same as if the simple probability of drawing a ball of the pth colour were m_p+1, with the difference of notation" [p.146]; secondly, if $n_p=1$ and all other n_i are zero, the chance that a ball of the p-th colour is drawn is

$$(m_p+1)/\,(m_1+\cdots+m_p+p)\ \ ;$$

and thirdly, the probability that the index of the colour drawn lies between $(n-1)$ and $(n+q+1)$ is

$$(m_n+m_{n+1}+\cdots+m_{n+q}+q)\,/(m_1+m_2+\cdots+m_p+p)\ \ .$$

While the second and third of these observations are correct, the first requires some attention. A similar remark in Lubbock and Drinkwater-Bethune's 1830 tract (for $p=2$) shows that the expression m_p+1 should in fact be $(m_p+1)/\,(M+p)$. For then the probability that N draws yield $n_1,n_2,\ldots,n_p$ balls is given by the multinomial probability

$$\binom{n_1+\cdots+n_p}{n_1,\ldots,n_p}(m_1+1)^{n_1}\ldots(m_p+1)^{n_p}\Big/(m_1+\cdots+m_p+p)^{n_1+\cdots+n_p}$$

which corresponds, with the substitution of square brackets for the parentheses, to Lubbock's formulation of (16) as

$$\binom{n_1+\cdots+n_p}{n_1,\ldots,n_p}[m_1+1]^{n_1}\ldots[m_p+1]^{n_p}\Big/[m_1+\cdots+m_p+p]^{n_1+\cdots+n_p}$$

where $[x+1]^n=(x+1)(x+2)\ldots(x+n)$.

Several applications to annuities follow: we shall not pursue these here. Attention is also paid to the following problems:

1. Find the probability of getting n_1 balls of the first colour in n_1+N (further) trials, the colour of the other N balls being anything other than the first colour.

2. Let $M=m_2+m_3+\cdots+m_p+p-2$, and suppose that $n_1:N::m_1:M$. Find the chance that the number of balls of the first colour in n_1+N trials lies between the limits n_1 and $n_1\pm z$. (To solve this Lubbock uses a result from Laplace's *Théorie analytique des probabilités.*)

3. Suppose that the "law of possibility" of x_p is $\varphi(x_p)$ (i.e. no longer uniform). How are the earlier results modified? (Not very much can in fact be said, since no specific form of φ is assumed.)

4. Find the probability of any future event when the results of the preceding trials are uncertain. (This example is followed by an application to the veracity of witnesses.)

7.3 Bernard Bolzano (1781-1848)

In 1837 Bolzano's *Wissenschaftslehre* appeared. Here the definition of logical probability proposed by the author is seen as being in complete agreement with that given by Laplace and Lacroix, and it is, moreover, a definition in which probability is clearly seen as a relation between propositions[28]. But despite the importance of this book as a contribution to inductive probability, and of the discussion of confidence, belief, and subjective probability to be found there, there is little that is directly relevant to our present theme. Indeed, the only pertinent point seems to be a brief use of the rule of succession in §379. Here Bolzano states that if the proposition A has occurred α times in n cases, the probability that A is present in a further case is $(\alpha+1)/(n+2)$.

7.4 Augustus de Morgan (1806-1871)

Augustus de Morgan[29] was born in Madura, India, the year of his birth being the solution of a conundrum he himself proposed[30], viz. "I was x years of age in the year x^2." Despite a physical defect (according to MacFarlane [1916], "one of his eyes[31] was rudimentary and useless" [p.19]) de Morgan was an indefatigable author: Peter Heath, in his 1966 edition of some of de Morgan's logical works, says that his output was "probably the largest of any mathematician of his time" [p.ix]. Yet among this vast number[32] only some half-a-dozen are devoted to probability in itself[33], and even in these there is little that seems directly relevant.

The origin of the name "inverse probability" has been traced by A. Fisher (1877-1944) to de Morgan's *Essay on Probabilities* of 1838. However, in an anonymous review of Laplace's *Théorie analytique des probabilités* in 1837, a review[34] generally attributed to de Morgan, we find a slightly earlier reference to the notion in the words that in the science of probability

> the problems which most naturally present themselves in practice are of an inverse character, as compared with those which an elementary and deductive course first enables the student to solve. [p.239]

A yet earlier occurrence of the term is to be found in an outline of some lectures given at the *École Polytechnique* in Paris in the late eighteenth or early nineteenth century. It is not certain whether these lectures were given by Fourier or Garnier, though they were certainly drawn up by the former. The outline is today to be found in Fourier's papers in the *Bibliothèque Nationale* — see Crepel [1989c], where we find the following summary:

> **Méthode inverse des probabilités. Règles**
> De la probabilité des causes prise des événemens, mesure de cette probabilité.
> De la probabilité des événemens futurs dont les causes sont ignorées.
> De la probabilité des événemens prise des événemens observés.
> Remarques analytiques sur le calcul des fonctions de très grands nombres.
> Des cas où les événemens observés indiquent les causes avec beaucoup de vraisemblance.
> [Crepel 1989c, p.37]

The next work of de Morgan's that warrants attention is his *An Essay on Probabilities and their Application to Life Contingencies and Insurance*

Offices of 1838, a volume described by Fisher [1926, p.16] as "the first [English] work of importance" after the publication of de Moivre's *The Doctrine of Chances*, and by Heath [1966, p.ix] as a first-rate elementary text-book. The book is most interesting to read, and can be well recommended.

In the sixteen-page preface we find a discussion of the difficulties which beset early investigators in probability, among which de Morgan mentions

> the not having considered, or, at least, not having discovered, the method of reasoning from the happening of an event to the probability of one or other cause [p.vi],

and on the same page he specifically refers to "the want of an inverse method", further elaboration being given as follows:

> De Moivre, nevertheless, did not discover the inverse method. This was first used by the Rev. T. Bayes, in *Phil. Trans.* liii.370.; and the author, though now almost forgotten, deserves the most honourable remembrance from all who treat the history of this science. [p.vii]

In Chapter I, entitled "On the notion of probability and its measurement; on the province of mathematics with regard to it, and reply to objections", de Morgan speaks of the principle of *the want of sufficient reason* [p.10] and its occurrence in some simple situations. Also in this introductory chapter we find the sentiment

> causes are likely or unlikely, just in the same proportion that it is likely or unlikely that observed events should follow from them. The most probable cause is that from which the observed event could most easily have arisen [p.27],

an opinion which is discussed in a later chapter[35]. This form, in which the *a priori* probabilities cancel, is described by Keynes [1921, chap. XVI, §14] as the "uninstructed view".

Asserting that probability questions may be of two different types, viz.

> 1. Where we know the previous circumstances and require the probability of an event.
> 2. Where we know the event which has happened, and require the probability which results therefrom to any particular set of circumstances under which it might have happened.
> The first I call direct, and the second inverse, questions [pp.31-32],

de Morgan devotes his second chapter to discussion of questions of the first type, and the third to those of the second: it is to this third chapter that we now turn our attention.

At the outset de Morgan, having outlined the typical "argument from event to cause", provides a precise definition of a *cause* as "simply a state of things antecedent to the happening of an event" [p.53], and moreover limits himself to cases involving merely a finite number of antecedent possible states. As a first illustration he considers the case of four urns, A containing three black balls, B containing one white and two black, C containing two white and one black, and D three white. Under the assumption that each urn has probability 1/4 of being chosen, he deduces that, after a white ball has been drawn, the *a posteriori* probabilities of A, B, C and D are 0, 1/6, 2/6 and 3/6. As a second illustration he examines the case in which two urns contain different numbers of balls, and answers a question like that posed above.

Next we find the following basic postulate:

> When an event has happened, and the state of things under which it happened must have been one out of the set A, B, C, D, &c., take the different states for granted, one after the other, and ascertain the probability that, such state existing, the event which did happen would have happened. Divide the probability thus deduced from A by the sum of the probabilities deduced from all, and the result is the probability that A was the state which produced the event: and similarly for the rest. [p.55-56]

Note the tacit assumption that the intial circumstances are equally probable. This principle is followed by some more simple examples involving lotteries and testimony, and de Morgan then turns his attention to a problem in which the urns have unequal probabilities of being drawn: suppose that two urns contain 3 white and 4 black, and 2 white and 7 black balls respectively, and that the first urn is three times as likely to be drawn as the second. The method of solution proposed is curious: de Morgan introduces two further urns, each of the same composition as the first, which then results in a situation capable of being handled by the earlier principle. He follows this by giving (in words) the rule

$$\Pr[H_i \mid E] = \Pr[E \mid H_i]\Pr[H_i] / \sum_j \Pr[E \mid H_j]\Pr[H_j] \quad ,$$

and illustrates it with a further simple example.

A discussion of what is essentially the rule of succession provides a heuristic for the following principle:

> Having given an observed event A, to find the probability which it affords to the supposition that a coming event shall be B, find the probability which A gives to every possible preceding state; multiply each probability thus obtained by the chance which B would have from that state, and add the results together. [pp.60-61]

This may be written symbolically as

$$\Pr[B \mid A] = \sum_i \Pr[H_i \mid A] \Pr[B \mid H_i] \quad ,$$

an expression which is true provided that A and B are conditionally independent given each H_i. This is followed by some further lottery examples, following which the general result is stated that if A and B have happened m and n times respectively, the probability that the next event will be an A is $(m+1)/(m+n+2)$, this being based on the consideration that the antecedent probabilities of the events may be anything whatever[36]. Similarly, the probability that a further $(p+q)$ events will result in p occurrences of A and q of B is given as

$$\binom{m+p}{p}\binom{n+q}{q} \Big/ \binom{m+n+p+q+1}{p+q} \quad ,$$

and this is illustrated once again by a lottery.

In Chapter IV, "Use of the tables at the end of this work", we find the suggestion that, when the chances of A and B are known, we may well suppose that in a large number of trials A and B will occur in proportion to their respective probabilities. Several problems of a direct nature (i.e. when the *a priori* probabilities are supposed known) are solved using the tables, which are seen to be based on cumulative frequencies for the probability density function defined by

$$y = \frac{2}{\sqrt{\pi}} 10^{-x^2/\alpha}, \quad x > 0 \qquad (17)$$

where $\alpha = \ln 10$.

Attention is next turned to inverse problems, the first considered being the following:

> In $a+b$ trials A has happened a times and B b times: from which, if a and b be considerable numbers, it is safe to infer that it is a to b nearly for A against B. What is the presumption that the odds for A against B really lie between $a-k$ to $b+k$ and $a+k$ to $b-k$? [p.83]

Denoting by $P(A)$ the (unknown) probability of A, we may write the quaesitum as

$$\Pr\left[|P(A) - a/(a+b)| < k/(a+b)\right] \ ,$$

a probability which de Morgan states, though not in so many words, may be approximated by integration of the function y given in (17) above from 0 to $k \Big/ \sqrt{2ab/(a+b)}$. This in fact, making allowances for the different densities, coincides with Laplace's inversion of Bernoulli's Theorem (cf. §6.14.2 and Keynes [1921, chap. 30]), according to which

$$\Pr\left[|P(A) - a/(a+b)| < \gamma\sqrt{2ab/(a+b)^3}\right] \approx \frac{2}{\sqrt{\pi}} \int_0^{\gamma} e^{-t^2}\, dt \ .$$

Similar problems, with different limits, are also considered. In each case de Morgan's commendable intent to provide results readily accessible to the enquiring layman leads to the avoidance of mathematical notation as far as possible, resulting in the provision of "Rules" with practically no justification.

In another work also published in 1838 (but read February 26, 1837) and entitled "On a question in the theory of probabilities", de Morgan corrects "an oversight" made by both Laplace and Poisson in their discussion of Bernoulli's Theorem. Both these savants deduced that, if A_n denotes the number of times an event A occurs in n trials,

$$\Pr\left[|A_n - np| < \ell \mid p\right] = \int_0^{\gamma} e^{-t^2}\, dt + \frac{1}{\sqrt{2\pi v w}} e^{-\gamma^2} \qquad (18)$$

where $\gamma = \ell\sqrt{n/2vw}$, with $n = v + w$ where v and w "are proportional to the chances of arrival or non-arrival in a single trial" (de Morgan [1838b, p.423]). De Morgan points out that both Laplace and Poisson inferred that (18) therefore represented the same probability when p was unknown.

After detailed attention to the approximation of the ratio

$$\int_a^b x^p(1-x)^q\, dx \Big/ \int_0^1 x^p(1-x)^q\, dx \ ,$$

the approximation following Laplace's method, de Morgan finds that

$$\Pr\left[\omega - \mu\alpha < P(A) < \omega\right] = \frac{1}{\sqrt{\pi}} \int_{-\mu}^{0} e^{-t^2}\, dt + \beta\left(e^{-\mu^2} - 1\right)$$

$$\Pr\left[\omega - \mu\alpha < P(A) < \omega + \mu\alpha\right] = \frac{1}{\sqrt{\pi}} \int_{-\mu}^{\mu} e^{-t^2}\, dt$$

$$\Pr\left[\omega < P(A) < \omega + \mu\alpha\right] = \frac{1}{\sqrt{\pi}} \int_{0}^{\mu} e^{-t^2}\, dt + \beta\left(1 - e^{-\mu^2}\right) \ ,$$

where $\alpha = \sqrt{2\omega(1-\omega)/n}$ and $\beta = \sqrt{2}(1-2\omega)/3\sqrt{n\pi\omega(1-\omega)}$.

In an *Addition* to this paper de Morgan notes that (18) is unnecessarily complex: if ℓ is replaced by $\left(\ell + \frac{1}{2}\right)$ one finds, neglecting terms of the same order as those disregarded by Laplace, that

$$\Pr\left[v - \ell < A_n < v + \ell\right] = \frac{2}{\sqrt{\pi}} \int_0^{\gamma} e^{-t^2}\, dt \ ,$$

where $\gamma = \left(\ell + \frac{1}{2}\right)\sqrt{n/2vw}$.

In his extensive article "Theory of probabilities" in the *Encyclopaedia Metropolitana*[37] of 1843 de Morgan once again discusses "*inverse* principles, in which we reason from known events to probable causes" [§18]. Two such principles are given: the first runs as follows:

> Principle IV. Knowing the probability of a compound event, and that of one of its components, we find the probability of the other by dividing the first by the second. This is a mathematical result of the last too obvious to require further proof. [§18]

If what is meant is $\Pr[A] = \Pr[A \ \&\ B]/\Pr[B]$, then some independence assumption is lacking. (The same assertion is given as Principle III in the *Essay on Probabilities*, again with no mention of independence.) The second principle mentioned is the following:

> Principle V. When an event has happened, & may have happened in 2 or 3 different ways, that way which is most likely to bring about the event is most likely to have been the cause. [§19]

We also find here (essentially) the result

$$\Pr[H_i \mid E] = \Pr[E \mid H_i] / \sum_j \Pr[E \mid H_j] \ , \tag{19}$$

illustrated by a "balls and urn" example of the usual sort.

Some references to inverse methods may also be found in de Morgan's *Formal Logic* of 1847. The first runs as follows:

> This inversion of circumstances, this conclusion that the circumstances under which the event did happen, are most probably those which would have been most likely to bring about the event, is of the utmost evidence to our minds [pp.188-189],

while on page 190 we find a verbal expression of (19) above.

7.5 Irenée Jules Bienaymé (1796-1878)

In a memoir[38] published in 1838 and devoted to a direct proof of a result by Laplace on the probability of the mean of observations, Bienaymé opens with some historical remarks. Having mentioned Bernoulli's and de Moivre's results, he refers to the inverse problem in the following words:

> La solution ne fut donnée que soixante ans plus tard, par Bayes, savant anglais peu connu, sans doute parce qu'une mort trop prompte interrompit ses travaux, mais qui paraît avoir possédé à un très-haut degré les qualités du géomètre. [p.514]

He then cites a numerical example from Bernoulli, and states (wrongly) that Bayes had discussed an inverse to this particular numerical result.

The fundamental result used in this memoir is the following:

> si un événement a été observé p fois sur un grand nombre $p+p_1$ d'épreuves, la probabilité que la possibilité de cet événement est comprise dans les limites
>
> $$(1) \qquad \frac{p}{p+p_1} \pm c\sqrt{\frac{2pp_1}{(p+p_1)^2}}$$
>
> est égale à l'intégrale définie
>
> $$(2) \qquad \frac{2}{\sqrt{\pi}}\int_0^c e^{-t^2}\,dt$$
>
> [pp.517-518].

Bienaymé next supposes that γ, γ_1 are two arbitrary functions of the observed events, relative respectively to the event A, which has happened p times, and to the event B, which has happened p_1 times. If x, x_1 denote "les possibilités inconnues de ces deux événements" [p.518], we wish to determine the value of the quantity

$$v = \gamma x + \gamma_1 x_1 \quad .$$

Taking for this quantity the mean of the products of γ, γ_1 multiplied respectively by p, p_1, he suggests that one try to find the probability that

$$v' = (\gamma p + \gamma_1 p_1)/(p+p_1)$$

does not differ from the true value v by any given amount; and he notes further that this question reduces to that given in the preceding quotation when $\gamma = 1, \gamma_1 = 0$.

He then considers, preserving the previous notation, the finding of the probability that $v = \gamma x + \gamma_1 x_1$ lies between two values a' and a. Noting that (for given x) the probability that A and B occur p and p_1 times respectively is

$$\binom{p+p_1}{p} x^p (1-x)^{p_1} \quad ,$$

Bienaymé suggests that one should find x and $1-x$ from the expression for v, whence

$$x = (v-\gamma_1)/(\gamma-\gamma_1) \quad , \qquad (1-x) = (\gamma-v)/(\gamma-\gamma_1) \quad .$$

Then, "dans l'hypothèse d'une valeur assignée à v" [p.521], the compound event has probability

$$\binom{p+p_1}{p} \left(\frac{v-\gamma_1}{\gamma-\gamma_1}\right)^p \left(\frac{\gamma-v}{\gamma-\gamma_1}\right)^{p_1} \quad .$$

Thus the probability "de l'hypothèse d'une valeur de v" [p.521] will be

$$(v-\gamma_1)^p(\gamma-v)^{p_1} dv \Big/ \int_\gamma^{\gamma_1} (v-\gamma_1)^p(\gamma-v)^{p_1}\, dv \quad ,$$

where $\gamma_1 > \gamma$. Finally, the probability that the true value of v lies between a' and a is found by integrating this last expression between the given limits.

Attention is next paid to the case of three events, a procedure analogous to that detailed above being followed. At the conclusion of the exercise the curious statement is made that

> l'on est dès lors complétement certain que la solution est indépendante de la loi de probabilité des divers événements simples. [p.529]

An extension is also made to n events.

7.6 Mikhail Vasil'evich Ostrogradskiĭ (1801-1861)[39]

In his "Extrait d'un mémoire sur la probabilité des erreurs des tribunaux" [1838], a memoir read on the 12th June 1834, Ostrogradskiĭ[40] talks of the "formule de Laplace" which arises in connexion with a specific example, as

$$\int_0^{1/2} x^m(1-x)^n\, dx \Big/ \int_0^1 x^m(1-x)^n\, dx \quad .$$

In a memoir read on the 23rd October 1846, entitled "Sur une question des probabilités", Ostrogradskiĭ considers the following problem[41]: suppose an urn contains an unknown number of white and black balls. A sample of size ℓ is drawn from the urn and is found to contain n white and m black balls. What is the probability that the total number of white balls in the urn does not exceed some number q?

Assuming that all possible constitutions of the sample are equally probable, the chance of getting n white and $(\ell - n)$ black balls will be $1/(\ell + 1)$. Suppose that the urn contains s balls, x of which are white. Denoting by $U(\alpha, \beta)$ [respectively $S(\alpha, \beta)$] the event that the urn [the sample] contains α white and β black balls, we have[42]

$$\Pr\left[S(n, \ell - n) \mid U(x, s - x)\right] = \binom{x}{n}\binom{s-x}{\ell-n} \Big/ \binom{s}{\ell} \quad .$$

Thus, again under an assumption of equiprobability,

$$\begin{aligned}
&\Pr\left[U(x, s-x) \mid S(n, \ell - n)\right] \\
&\quad = \Pr\left[S(n, \ell - n) \mid U(x, s - x)\right] \Pr\left[U(x, s-x)\right] / \Pr\left[S(n, \ell - n)\right] \\
&\quad = \frac{\binom{x}{n}\binom{s-x}{\ell-n}}{\binom{s}{\ell}} \frac{1}{s+1} \Big/ \frac{1}{\ell+1} \\
&\quad = \binom{x}{n}\binom{s-x}{\ell-n} \Big/ \binom{s+1}{\ell+1} \quad .
\end{aligned}$$

It thus follows that the probability that the number of white balls (say W) in the urn does not exceed some specified q, when there are n white and $m = \ell - n$ black balls in the sample, is given by

$$\Pr\left[W \leq q \mid S(n, \ell - n)\right] = \sum_{j=0}^{q} \binom{j}{n}\binom{s-j}{\ell-n} \Big/ \binom{s+1}{\ell+1} \quad .$$

Gnedenko is quoted by Maistrov [1974] as evaluating Ostrogradskiĭ's contributions to probability as follows:

> In spite of the fact that in his definition of probability Ostrogradskiĭ committed methodological errors, slipping towards a philosophy of subjectivism, the general direction of his creative work in probability theory should be evaluated as instinctively materialistic. [p.187]

7.7 Thomas Galloway (1796-1851)

Writing in the seventh edition of the *Encyclopædia Britannica* of 1839, Galloway comments[43] on the noteworthiness of Bayes's two papers in the *Philosophical Transactions* for 1763 and 1764. In Section V, entitled "Of the probability of future events deduced from experience", he gives the expressions

$$\omega_i = \lambda_i P_i \,/\textstyle\sum \lambda_i P_i$$

$$\omega = x^m(1-x)^n\,dx \Big/ \int_0^1 x^m(1-x)^n\,dx \quad ,$$

and from the latter deduces the rule of succession. There is no mention of Bayes or Price here.

7.8 Eugène Charles Catalan (1814-1894)

In four papers, spread over some forty years, Catalan considered essentially the same problem, one which we may loosely phrase as follows: how does the probability of drawing a white ball from an urn change under various modifications of the contents?

The first paper, published in 1841 in the *Journal de Liouville*, is entitled "Deux problèmes de probabilités"; and the first of these problems runs as follows:

> Une urne A contient b boules blanches et n boules noires. On en extrait, par hasard, m boules que l'on place, sans les connaître, dans une seconde urne B, laquelle renferme alors m boules, blanches et noires, en proportion inconnue. On tire de cette urne, successivement, p boules; et il arrive que toutes sont blanches. Quelle est la probabilité que, faisant un tirage de plus, on obtiendra encore une boule blanche? [p.75]

Three different cases present themselves, depending on whether m is less than, between or greater than b and n. Being persuaded that all cases lead to the same result[44], Catalan proposes to consider in detail only the first; his argument runs as follows:

Before the extraction of the p white balls, urn B's composition may be given by any one of the $m-p+1$ hypotheses

$$H_{m-p-i+1} : p+(i-1) \text{ white and } (m-p-i+1) \text{ black} ,$$

$i \in \{1, 2, \ldots, m-p+1\}$. Now the probability that B has $(m-i)$ white balls is proportional to the probability of withdrawing from A, $(m-i)$ white balls in m draws. Moreover, asserts Catalan,

> Elle est proportionnelle aussi à la probabilité d'extraire p boules blanches d'une urne qui en contiendrait $m-i$ blanches et i noires. [p.76]

Letting $b+n=s$, we find that

$$\Pr[H_i] = \binom{m}{i}(b)_{m-i}(n)_i \,/(s)_m$$

where $(x)_k = x(x-1)\ldots(x-k+1)$, while

$$\Pr[p \text{ white balls drawn} \mid H_i] = (m-i)_p \,/(m)_p \quad .$$

Hence

$$\Pr[H_i \mid p \text{ white balls drawn}] \propto \binom{m}{i}(b)_{m-i}(n)_i(m-i)_p \,/(s)_m(m)_p =$$

$$\propto [(m-p)!(b)_p \,/(s)_m]\,[(b-p)_{m-p-i}(n)_i/\,(m-p-i)!\,i!\,] \quad .$$

Denoting the last square-bracketed term by A_i, one finds that

$$\omega_i \equiv \Pr[H_i \mid p \text{ white balls drawn}] = A_i \Big/ \sum_{j=0}^{m-p} A_j \quad .$$

Recognizing that the denominator in this last expression is the coefficient of $u^{b-m+n}v^{m-p}$ in the expansion of $(u+v)^{b+n-p}$, Catalan deduces that

$$\omega_i = (b-p)_{m-p-i}(n)_i(m-p)!\,/(m-p-i)!\,i!\,(s-p)_{m-p} \quad .$$

We now pass on to consider the further drawing of a white ball from B. The probability of this event, if H_i obtains, being

$$\omega_i(m-p-i)\,/(m-p) \quad ,$$

it follows that

$$P \equiv \Pr[1 \text{ white} \mid p \text{ white drawn}] = \sum_{i=0}^{m-p} \omega_i(m-i-p)/(m-p) \quad ,$$

an expression which simplifies to

$$P = (b-p)/(s-p) \quad .$$

The independence of this result of the number of balls in B should be noted, and Catalan notes further that the introduction of the urn B is an unnecessary affectation: one might as well suppose an appropriate partitioning of the balls in the initial urn A.

The second problem is a natural development of the first:

> Une urne contient b boules blanches et n boules noires; une autre urne renferme b' boules blanches et n' boules noires. On tire au hasard p boules de la première urne, et p' boules de la seconde; et l'on réunit ces $p+p'$ boules dans une troisième urne. Quelle est la probabilité d'extraire de celle-ci une boule blanche? [p.79]

The desired probability P is given, by the previous result, as

$$P = bp/s(p+p') + b'p'/s'(p+p') \ ,$$

and the extension to m such urns is also given.

In 1877 Catalan published a paper entitled "Un nouveau principe de probabilités", in which he announced the following result[45]:

> La probabilité d'un événement futur ne change pas lorsque les causes dont il dépend subissent des modifications inconnues. [p.463]

Although there is a glimmering of this result in some of Poisson's work (not cited here), the principle merits the qualification "nouveau" in as much as this appears to be the first proof.

The proof presented by Catalan is somewhat curious: suppose that an urn A contains b white balls and $n-b$ balls of other shades. If p (which may be known or unknown) balls are drawn from A and, unobserved, placed in an urn B,

> les probabilités d'extraire une boule blanche, soit de cette urne B, soit de l'urne A, dont la composition a été modifiée, sont égales à b. [p.465]

On the other hand, one may leave the urn A in its original state, and consider the white ball drawn as coming from an isolated group of p balls within A. This group replaces urn B, while the remaining $(n-p)$ balls correspond to the modified urn A. "Le théorème est donc démontré" [p.465].

As an application of his theorem Catalan considers in the third section of this paper the following problem:

> Une urne A contient 4 boules blanches et 3 boules noires. On en tire, sans les compter ni les regarder, un certain nombre de boules. Quelle est la probabilité d'extraire une boule blanche, de l'urne A modifiée? [p.467]

The eighteen possible withdrawals are enumerated (the case in which no balls are drawn is omitted, and although Catalan states that he will not consider the case in which all balls are taken, this case appears in his table, while that of one black and four white is omitted: since both of these cases contribute nought to the final calculation, the error is of no consequence in the final analysis). The posterior probabilities of the various hypotheses (as to the nature of the withdrawals) are evaluated by Bayes's Theorem (in the discrete form); and the desired probability is found to be 4/7, as expected.

The third memoir of Catalan's to warrant our attention was published in 1886, under the title "Problèmes et théorèmes de probabilités". Here Catalan considers a generalization of the following problem considered by Poisson in Article 32 of his *Recherches sur la probabilité des jugements*:

> On sait qu'une urne renfermait m boules, blanches ou noires; on en a tiré une blanche: et l'on demande quelle est la probabilité de l'extraction d'une nouvelle boule blanche, la première n'ayant pas été remise dans l'urne. [p.3]

This problem Catalan proposes to solve, unlike Poisson, by a method "qui *supprime* les longs calculs nécessités par le *théorème de Bayes*" [p.3]. To this end the first section of the memoir is devoted to some combinatorial formulae, the second section beginning with the following problem:

> Une urne A contenait, primitivement, s boules. On en a tiré, au hasard, m boules blanches, m' boules non blanches. Quelle est la probabilité d'extraire, de l'urne modifiée, une nouvelle boule blanche? [p.7]

According to Bayes's Theorem, the probability ω_k that the urn contains $(m+k)$ white balls, supposing always that sampling is without replacement, is

$$\begin{aligned}\omega_k &= \binom{m+k}{m}\binom{m'+p-k}{m'} \Big/ \sum_{k=0}^{p} \binom{m+k}{m}\binom{m'+p-k}{m'} \\ &= \binom{m+k}{m}\binom{m'+p-k}{m'} \Big/ \binom{m+m'+p+1}{p} ,\end{aligned}$$

where $p = s - m - m'$. Now if k of the p balls remaining in the urn are white, then the probability of drawing a further white ball will be k/p, and hence the required probability P will be given by

$$P = \sum_{k=0}^{p} (k/p)\,\omega_k \ ,$$

an expression which some combinatorial prestidigitation reduces to

$$P = (m+1)/(m+m'+2) ,$$

independent of s. This result Catalan summarizes in the following theorem[46]:

> Si, d'une urne A, contenant s boules, il est sorti m boules blanches, m' boules non blanches; la probabilité de l'extraction d'une nouvelle boule blanche est égale à la probabilité d'extraire une boule blanche d'une urne B, contenant $m+1$ boules blanches et $m'+1$ boules noires. [p.9]

Some simple corollaries follow.

Recalling his aphorism "si un long calcul amène un résultat simple, il est inutile" [p.9], Catalan notes that, in the case of the drawing of a further white ball from an urn which has already yielded m white and m' non-white balls,

> La probabilité P, de cet événement, ne sera pas alterée, si les causes dont il dépend subissent des modifications inconnues. [p.9]

P will thus remain unaltered if 1,2,... or even $(s-m-m'-1)$ balls are set aside. One may therefore consider the replacement of urn A by a fictitious urn B initially containing $(m+m'+1)$ balls. After the drawing of the $(m+m')$ balls, two hypotheses may be entertained about the composition of B, viz.

$$H_1 : \quad m \text{ white and } (m'+1) \text{ non-white balls ; or}$$

$$H_2 : \quad (m+1) \text{ white and } m' \text{ non-white balls .}$$

The probabilities of these hypotheses being respectively proportional to $(m+1)$ and $(m'+1)$, one finds that

$$\omega_1 = (m+1)/(m+m'+2) , \quad \omega_2 = (m'+1)/(m+m'+2) ,$$

and since H_1 is incompatible with the drawing of a further white ball, H_2 necessarily holds. Thus ω_2 is in fact the desired probability P.

An extension of this result is obtained in the next problem:

> Une urne A contenait, primitivement, s boules. On en a tiré, au hasard, b blanches, n non blanches. Quelle est la probabilité P d'extraire b' blanches, n' non blanches, de l'urne modifiée? [p.10]

Proceeding as before Catalan obtains the value

$$P = \binom{b+b'}{b'}\binom{n+n'}{n'} \Big/ \binom{b+b'+n+n'+1}{b'+n'} ,$$

the same as the result given by "la méthode *classique*". Several particular cases follow.

In the next problem urns containing balls of any one of three colours are considered, and this is extended in the following problem to f possibilities.

In an *Addition* to his paper Catalan points out that a thing may be modified "soit en l'unissant à une chose de même nature, soit en supprimant quelqu'une de ses parties" [p.15]. His new principle, he observes, is not applicable in the case of modifications of the first type, and as an example he considers the question of the drawing of balls of various colours from an urn whose initial composition is known and to which a further n balls, of unknown shades, are added. Indeed, if the urn initially contained a white, b black, and c red balls, the probability, after the addition of the n balls, of drawing from the urn of size $s = n + a + b + c$, a, b and c balls coloured white, black and red respectively, is

$$P = \binom{n+s+2}{n} \Big/ \binom{n+2}{2}\binom{s}{n} ,$$

independent of the actual values a, b, c.

In 1888 Catalan's paper "Sur une application du théorème de Bayes, faite par Laplace" appeared — a paper, as we shall see, in which many of his earlier results are rehearsed. Here he states the "Principe" given in Laplace's memoir of 1774, and notes that Laplace stated this result "sans nommer Bayes" — a fact which is perhaps hardly surprising, since the proposition is not in fact found in Bayes's Essay. Laplace then, as Catalan notes, applied this result (in his "Problème 1") to the problem of finding the probability P of drawing a white ball from an urn containing an infinite number of white and black balls, if $(p+q)$ draws have already resulted in p white and q black, the solution being given by

$$P = (p+1)/(p+q+2) . \tag{20}$$

In musing on this result Catalan was apparently struck by a multitude of questions, among which he mentions the following:

(i) Why was Laplace not struck by the simplicity of this result?

(ii) Why did he not perceive that his calculation, so simple in the case of an *infinite* number of balls, would become prolix and tedious if one supposed the number of balls to be ten thousand, for example?

(iii) Why did he not ask if his formula (20) would not hold in the case of *any number whatsoever*, greater than $(p+q)$, of balls?

Here Catalan proposes to consider the following general problem:

> Une urne A contenait, primitivement, s boules. On en a tiré, au hasard, m boules blanches, m' boules non blanches. Quelle est la probabilité d'extraire, de l'urne modifiée, une nouvelle boule blanche? [p.256]

The event expected ("l'événement attendu") is then defined as the drawing of a white ball from the urn of $(s-m-m')$ balls of various colours in unknown proportions. Basic to the solution presented is the following observation (from his paper of 1877):

> La probabilité P, de cet événement, ne sera pas altérée, si les causes dont il dépend subissent des modifications inconnues. [p.256]

It thus follows that p is unchanged if $1,2,\ldots,(s-m-m'-1)$ balls from the original urn are placed, unseen, to one side.

This, however, as we have noted before, is tantamount to replacing the original urn A by a fictitious urn B containing $(m+m'+1)$ balls, of which m are white and m' non-white. The urn B may then have either of the following compositions:

$$H_1: \quad m \text{ white and } (m'+1) \text{ non-white balls ; or}$$

$$H_2: \quad (m+1) \text{ white and } m' \text{ non-white balls ,}$$

with $\omega_1 \equiv \Pr[H_1] \propto (m'+1)$ and $\omega_2 \equiv \Pr[H_2] \propto (m+1)$. Thus

$$\omega_1 = (m'+1)/(m+m'+2) \quad \text{and} \quad \omega_2 = (m+1)/(m+m'+2) \ .$$

Since H_1 is incompatible with the observed event, the second must in fact obtain. Thus the desired probability is

$$P = \omega_2 = (m+1)/(m+m'+2) \ ,$$

which agrees with that given in (20) above.

As a final relevancy from this paper we may cite the extension made to sampling from an urn containing balls of k colours. If m_i balls of colour i have been obtained, the probability that the next draw will yield a ball of j-th colour is

$$(m_j+1)/(m_1+\cdots+m_k+k), \quad j \in \{1,2,\ldots,k\} \ .$$

As a postscript Catalan points out that if the balls (b white, n black) from an urn A are distributed, unseen, among urns $B_1, B_2, \ldots, B_k$, the probability of drawing a white ball from any of these auxiliary urns will be $b/(b+n)$, unless $k > b+n$.

Several comments on this paper come to mind. The first is to note that a similar discussion of the finite urn was given by Terrot (see §7.16), with later and more detailed discussion by Keynes [1921, chap. XXX, §11] and Burnside [1928], though the latter two authors concentrate mainly on the case of sampling with replacement, while Catalan's concern is with sampling without replacement.

Secondly, as Burnside (op. cit.) has pointed out, the assumption that all of n results are equally likely is not the same as requiring that each two of the n results are equally likely. The latter has been shown by this author to be the appropriate assumption to be made in questions of the type discussed by Catalan, and it appears that this should be taken into account in the latter's work.

Thirdly, the extension to balls of k colours was, as we have already seen, given by Lubbock and Drinkwater-Bethune [c.1830]. Ignorance of this extension led Kneale [1949, pp.203-204] to a vain attempt at confutation of the rule of succession.

7.9 Jacob Friedrich Friess (1773-1843)

In the second chapter, "Berechnung der Wahrscheinlichkeit, wenn die Theilung der Sphäre in ihre gleichmöglichen Fälle selbst erst errathen werden muß, ober Bestimmung der Wahrscheinlichkeit *a posteriori*" of the first section "Reine Theorie der Wahrscheinlichkeitsrechnung" of his book *Versuch einer Kritik der Principien der Wahrscheinlichkeitsrechnung* of 1842, Friess[47] gives the expression

$$\binom{m+n}{n} \int_0^1 x^m (1-x)^n \, dx \ ,$$

and points out (though not in so many words) that this holds for a uniform prior. He also deduces the rule of succession. No mention of Bayes or Laplace is to be found here.

7.10 Antoine Augustin Cournot (1801-1877)

In his *Exposition de la Théorie des Chances et des Probabilités* of 1843 Cournot gives the following form of "la règle attribuée à Bayes":

> les probabilités des causes ou des hypothèses sont proportionelles aux probabilités que ces causes donnent pour les événements observés. La probabilité de l'une de ces causes ou hypothèses est une fraction qui a pour numérateur la probabilité de l'événement par suite de cette cause, et pour dénominateur la somme des probabilités semblables relatives à toutes les causes ou hypothèses. [p.158]

Thus understood, he goes on to point out,

> la règle de Bayes est un théorème qui ne donne lieu à aucune équivoque, et dont on ne peut contester la justesse [p.158],

although a scant three pages before, in talking of this rule on which Condorcet, Laplace and their successors had wished to build the theory of *a posteriori* probabilities, he had drawn attention to the ambiguities and the grave errors resulting from the misuse of this rule — the rectification of which misuse called for a distinction between objective and subjective probabilities[48].

As an illustration of the use of Bayes's Theorem in the subjective theory Cournot considers three players whose probabilities of winning a game are in the ratio 3:2:1. These probabilities will vary from one individual to another, depending on knowledge. In this subjective setting Bayes's rule

> n'a donc d'autre utilité que celle de conduire à une fixation de paris, dans une certain hypothèse sur les choses que connaît et sur celles qu'ignore l'arbitre. [p.160]

Some slight misunderstanding of Bernoulli's Theorem seems evident in Cournot's work here. For in writing of the need for the determination "par l'expérience, ou *à posteriori*" [p.154] of chances according to data, he says that one is led to such a determination by Jacques Bernoulli's principle,

> car si, en désignant par x la chance inconnue de la production d'un événement, par n le nombre de fois que cet événement est arrivé en m épreuves, on peut toujours obtenir une probabilité P que l'écart fortuit $x - n/m$ tombe entre les limites $\pm\ell$ (le nombre ℓ et la différence $1 - P$ tombant au-dessous de toute grandeur assignable, pourvu que les nombres m, n soient suffisament grands), il est clair que, si rien ne limite le nombre des épreuves, la probabilité x peut être déterminée avec une précision indéfinie; qu'on peut arriver, par exemple, à être sûr qu'il n'y a pas, entre le rapport n/m donné par l'expérience et le nombre inconnu x, une différence d'un cent-millième. [pp.154-155]

In view of the assumption here that x is unknown, the description seems more applicable to Bayes's Theorem than Bernoulli's, though it is not clear whether Cournot viewed the former as anything more than an extension of the latter.

7.11 John Stuart Mill (1806-1873)

The only pertinent result by this author, a writer justly famous for his economic, philosophic and logical works, is to be found in his *A System of Logic, Ratiocinative and Inductive: Being a Connected View of the Principles of Evidence and the Methods of Scientific Investigation.* First published in 1843, this work went through eight editions in all in Mill's lifetime, each edition being carefully revised[49]. So substantial were the alterations adopted, after the publication of the first edition, in Book III, Chapter XVIII, "Of the Calculation of Chances", that Mill had the revision, together with that of Chapter XXV, published as a separatum. Our comments here will be restricted to the first edition with reference, where relevant, to the eighth edition of 1872.

Mill begins Book III, Chapter XVIII, by recalling Laplace's definition of probability (given in the *Essai philosophique sur les probabilités*) as having reference partly to our ignorance and partly to our knowledge. Laplace's requirement that events should be mutually exclusive and exhaustive and equally possible is found by Mill to be unsatisfactory:

> To be able to pronounce two events equally probable, it is not enough that we should know that one or the other must happen, and should have no ground for conjecturing which. Experience must have shown that the two events are of equally frequent occurrence. [1843: §2]

This view was afterwards withdrawn[50], Mill concluding that

> the theory of chances, as conceived by Laplace and by mathematicians generally, has not the fundamental fallacy which I had ascribed to it [1872: §1]

and this in turn is based on his belief that

> the probability of an event is not a quality of the event itself, but a mere name for the degree of ground which we, or some one else, have for expecting it. [1872: §1]

However Laplace's definition is perhaps not altogether suitable when it comes to the application of the doctrine of chances to a scientific purpose, and Mill points out that the knowledge required in such a case "is that of the comparative frequency with which the different events in fact occur" [1872: §3]; and he professes further the (perhaps somewhat unorthodox) opinion that[51]

> The probability of events as calculated from their mere frequency in past experience, affords a less secure basis for practical guidance, than their probability as deduced from an equally accurate knowledge of the frequency of occurrence of their causes. [1872: §4]

In §3 (§5 of the eighth edition) Mill quotes and proves the sixth principle in Laplace's *Essai philosophique sur les probabilités.* The discussion is limited to two possible causes of an event, the posterior probabilities being arrived at via frequentist considerations.

In the first edition Mill finds it "necessary to point out another serious oversight in Laplace's theory" [§3]. He finds the preceding proposition untenable when its application is extended to cover *hypotheses* rather than *causes*, on the ground that the substitution of

> mere suppositions affording no ground for concluding that the effect would be produced, in the room of causes capable of producing it [1843: §3],

would invalidate the theorem. This argument appears to rest upon the assumption that

$$\Pr[A \mid M] = \Pr[A]\Pr[M \mid A] \quad ,$$

and it is then hardly surprising that he concludes that "the proposition, as thus stated, is an absurdity." This passage was dropped from later editions.

7.12 Mathurin-Claude-Charles Gouraud (1823 - ?)

In 1848 Gouraud[52] published his *Histoire du Calcul des Probabilités depuis ses origines jusqu'à nos jours*, a work which Todhunter [1865] describes as

> a popular narrative entirely free from mathematical symbols, containing however some important specific references. Exact truth occasionally suffers for the sake of a rhetorical style unsuitable alike to history and to science; nevertheless the general reader will be gratified by a lively and vigorous exhibition of the whole course of the subject. [p.x]

Gouraud correctly attributes [p.47] the contributions made by Bernoulli and de Moivre to the result generally known by the former's name; and he later [pp.61-62] contrasts this result with that proved by Bayes. He then points out [pp.62-63] the use and development of Bayes's Theorem made by Laplace, "le sublime géomètre" [p.64]. He perhaps errs, however, in referring to Condorcet's *Essai* of 1785 as containing the

> principe récemment entrevu par Bayes et démontré par Laplace. [pp.95-96]

The same error is repeated towards the end of this *Histoire*, where we find the words

> Le Principe entrevu par Bayes et analytiquement démontré par Laplace, qui consiste à conclure la probabilité des causes et de leur action future de la simple observation des événements passés. [p.146]

We have in fact already discussed the distinct contributions made by Bayes and Laplace in this respect.

Reference is made to Price's actuarial work and Condorcet's memoir of 1781-1784, stressing the latter's work on (a) the determination of the probability of future events from the observation of past events, and (b) testimony. Several pages are devoted to a discussion of Laplace's *Théorie analytique des probabilités*, including the *Essai philosophique sur les probabilités*, and despite fulsome praise of this work, Gouraud considers Laplace's historical comments to be too short in respect of certain passages and to have some regrettable omissions.

7.13 Robert Leslie Ellis (1817-1859)

An early exponent of the frequency interpretation of probability, Ellis[53] published in 1849 in the *Transactions of the Cambridge Philosophical Society* [1844] a paper entitled "On the foundations of the theory of probabilities". Although aimed at showing "the inconsistency of the theory of probabilities with any other than a *sensational* philosophy" [p.1], the paper contains some comments on inverse probability.

Thus, writing of the application of probability to inductive results, Ellis notes that, if a certain event has been observed to occur on m occasions, "there is a presumption that it will recur on the next occasion" [p.4], a presumption estimated by $(m+1)/(m+2)$. This, however, prompts two questions, viz.

> What shall constitute a "next occasion?" What degree of similarity in the new event to those which have preceded it, entitles it to be considered a recurrence of the same event? [p.4]

questions which Ellis considers with special reference to a simple example appearing in de Morgan's *Essay on Probabilities* [1838, p.64].

Finding the

> assertion ... that 3/4 is the probability that any observed event had on an *à priori* probability greater than $\frac{1}{2}$, or that three out of four observed events had such an *à priori* probability [p.5]

to be completely lacking in precision, Ellis proposes the following frequency explanation[54]. Suppose that a large number h of trials are performed, in each of which the probability of a certain event is $1/m$. Then let a second sequence of h trials be carried out, the probability now being $2/m$, &c. After all these sequences, approximately $h\sum_{i=1}^{m} i/m$ of the sought events will have occurred, of which $h\sum_{i=1}^{m/2}(\frac{1}{2}+i/m)$ had an *a priori* probability greater than 1/2. The ratio of the second to the first of these series gives $(3m+2)/(4m+4)$, which has the limit 3/4. Similarly, if p events are taken in succession from each trial, rather than the single events considered before, then we are led to consideration of the ratio

$$h\sum_{i=1}^{m/2}((1/2)+i/m)^p \Big/ h\sum_{i=1}^{m}(i/m)^p \quad ,$$

a ratio which tends to

$$\int_{1/2}^{1} x^p\,dx \Big/ \int_0^1 x\,dp = 1-(1/2)^{p+1} \quad .$$

Ellis concludes his essay with the following observations:

> The principle on which the whole depends, is the necessity of recognizing the tendency of a series of trials towards regularity, as the basis of the theory of probabilities. I have also attempted to show that the estimates furnished by what is called the theory *à posteriori* of the force of inductive results are illusory. [p.6]

7.14 William Fishburn Donkin (1814-1869)

In 1851 Donkin[55] published in three parts in the *Philosophical Magazine*, an article entitled "On certain questions relating to the theory of probabilities." He begins by taking it as "generally admitted ... that the subject-matter of calculation in the mathematical theory of probabilities is *quantity of belief*' [p.353], an observation which puts him squarely in the non-frequentist camp[56].

The law on which the whole theory is based is stated to be the following:

> When several hypotheses are presented to our mind, which we believe to be mutually exclusive and exhaustive, but about which we know nothing further, we distribute our belief equally amongst them. [p.354]

This being granted, the rest of the theory "follows as a deduction of the way in which we must distribute it in complex cases, *if we would be consistent*" (loc. cit.). Further evidence of Donkin's subjective views, perhaps more in the style of Harold Jeffreys than Bruno de Finetti, is furnished by the observation that probability is

> always *relative* to a particular state of knowledge or ignorance; but it must be observed that it is *absolute* in the sense of not being relative to any individual mind; since, the same information being presupposed, all minds *ought* to distribute their belief in the same way. [p.355]

Perhaps the most important result in the paper — certainly the most fundamental — is the following[57]:

> Theorem — If there be any number of mutually exclusive hypotheses, $h_1, h_2, h_3, \ldots$, of which the probabilities relative to a particular state of information are $p_1, p_2, p_3, \ldots$, and if new information be gained which changes the probabilities of some of them, suppose of h_{m+1} and all that follow, *without having otherwise any reference to the rest*, then the probabilities of these latter have the *same ratios* to one another, *after* the new information, that they had *before*; that is
>
> $$p_1' : p_2' : p_3' : \cdots : p_m' = p_1 : p_2 : \cdots : p_m,$$
>
> where the accented letters denote the values after the new information has been acquired. [p.356]

Whether this[58] might not preferably be termed an *axiom*[59] is arguable: indeed, Donkin himself seems to suggest this, since he finds it "certainly as evident before as after any proof which can be given of it" [p.356]. Boole, in his *An Investigation of the Laws of Thought* [1854], adds this result as an eighth principle to his list of similar fundamentals taken mainly from Laplace, and it can also be related to Burnside's [1928, p.4] modification of the usual "equally likely" definition of probability, in terms of which "*each two* of the n results are assumed to be equally likely" (emphasis added). Specializing this result, Donkin considers as a theorem the case in which the new information obtained is to the effect that some of the hypotheses must be rejected, or others admitted, or both.

From these two theorems the following results ensue[60]

(a) $\Pr[H \,\&\, h] = \Pr[h \mid H]\ \Pr[H]$;

(b) $\Pr[H_i \mid h] = \Pr[h \mid H_i] \Pr[H_i] \Big/ \sum\limits_i \Pr[h \mid H_i] \Pr[H_i]$;

(c) If $\{H_i\}_1^n$ are mutually exclusive and exhaustive and S_1 and S_2 are two (independent) states of information, then

$$\Pr[H_i \mid S_1, S_2] = \Pr[H_i \mid S_1] \Pr[H_i \mid S_2] \Big/ \sum_i \Pr[H_i \mid S_1] \Pr[H_i \mid S_2] ,$$

where the H_i are *a priori* equally likely and S_1 and S_2 are conditionally independent given each H_i;

(d) extension of (c) to several independent sources of information.

The last two of these results are an early contribution to the problem of the assessment of probabilities on different (and on combined) data.

In the course of discussion of some miscellaneous examples illustrating the use of these theorems, Donkin distinguishes between *a priori*, *provisional* and *a posteriori* probabilities. The first of these terms refers to "probabilities derived from information which we possess antecedently to the observation of the phænomenon considered" [p.360], while the last is defined in the usual way. *Provisional* probability is illustrated as follows: suppose an approximate value p_0 of p is assigned, with belief as to the precision of the approximation expressed by $\varphi(p)$, where

(i) $\varphi(p)$ is maximized by $p = p_0$;

(ii) $\int_0^1 \varphi(p)\, dp = 1$;

(iii) $\varphi(p)\, dp$ "is my belief that the true value would turn out to lie between p and $p + dp$" [p.362].

Then the provisional probability of p is just its expectation.

As an illustration of the distinction necessary to be observed between provisional and *a posteriori* probability, Donkin considers the following problem:

> An event E has been observed, which can only have resulted from some one or other of the causes $C, C', \ldots$ of which any one would necessarily produce it, and no two could coexist. It is required to assign the probability that it has resulted from C. [p.362]

If one now defines $\Pr[C \mid H] = a$, $\Pr[C \mid \overline{H}] = b$, $\Pr[\cup' C \mid H] = \alpha$, $\Pr[\cup' C \mid \overline{H}] = \beta$, where H(respectively $\overline{H}$) denotes that a hypothesis H is true (respectively false), and $\cup' C$ denotes the existence of some cause other than C, one can show, using the definition of E, that

$$\Pr[C \mid E, p] \equiv \Pr[C \mid E] = \frac{pa + (1-p)b}{p(a+\alpha) + (1-p)(b+\beta)} \quad . \tag{21}$$

If our "provisional" value of p is

$$\omega = \int_0^1 p\,\varphi(p)\,dp \ ,$$

then we may give the solution, from (21), as

$$(k\omega + \ell)/(\omega + m) \ , \tag{22}$$

where k, ℓ and m are all known. However, one may also argue that

$$\varphi(p)[(kp+\ell)/(p+m)]\,dp$$

expresses the quantity of our belief that the value of p lies between p and $p + dp$, and that C caused E. Then the solution of our problem is

$$\int_0^1 \varphi(p)(kp+\ell)/(p+m)\,dp \ . \tag{23}$$

The distinction between these two solutions is noted by Donkin thus:

> [(23)] expresses a real *provisional solution*; that is, it expresses our belief in the existence of C, *influenced by the consideration that we do not possess a definitive knowledge of p.* Whereas [(22)] expresses a solution obtained by *treating a provisional value of p as if it were definitive*; or it is what would be the definitive solution of the problem to a person whose state of information (antecedently to the event E) was such that ω was to him the *definitive à priori* probability of H. [pp.363-364]

He asserts further that while (23) is right in *principle*, (22) is right in *result*, a claim which is justified on the following wise: denoting by P the random variable taking on the value p, we have (in a notation different to Donkin's)

$$\Pr[p < P < p + dp \ \&\ E] = [p(a+\alpha) + (1-p)(b+\beta)]\varphi(p)\,dp$$

$$\Pr[p < P < p + dp \mid E] = \frac{p(a+\alpha)+(1-p)(b+\beta)}{\omega(a+\alpha)+(1-\omega)(b+\beta)}\varphi(p)\,dp$$

$$\Pr[p < P < p + dp \ \&\ C \mid E] = \Pr[p < P < p + dp \mid E]\Pr[C \mid E, p] \quad .$$

Therefore

$$\begin{aligned}\Pr[C \mid E] &= \int_0^1 \Pr[p < P < p + dp \ \&\ C \mid E]\,dp \\ &= (k\omega + \ell)/(\omega + m) \ ,\end{aligned}$$

which is the same as (23).

This part of the paper concludes with an examination of the probabilities of whether some arrangement of chessmen on a board was produced by accident or design.

In the second part of his paper Donkin stresses the importance of keeping clear the distinction between *a priori* and *a posteriori* probabilities, illustrating his remarks with the following application of the discrete Bayes's Theorem: let p be the *a priori* probability of an event that a witness asserts has happened, and let v and w be the *a priori* probabilities that he chooses to assert it supposing it to be true or false respectively. Then the probability, after his assertion, that the event really happened is $pv/[pv+(1-p)w]$.

This second part concludes with a discussion of the probability of the existence of binary stars: this is considered in §4.3 of the present tractate.

7.15 George Boole (1815-1864)

Although chiefly, and justifiably, remembered for his work in mathematical logic, Boole devoted considerable time to other branches of mathematics[61]. His work on probability is contained in the main in some dozen papers and in his book *An Investigation of the Laws of Thought* of 1854.

In the first of these papers, an essay among the Boole manuscripts in the Royal Society Library and entitled "Sketch of a theory and method of probabilities founded upon the calculus of logic", we find the statement of a general problem which was to be the chief object of Boole's attention in his writings on probability, viz.

> Given the probabilities of any events, simple or compound, to ascertain the probability of any other event. [Boole 1952, p.158]

The solution presented here lacks the clarity evident in later writings, and we shall accordingly postpone any discussion of it for the time being.

In 1851 Boole published a paper on Michell's problem of the distribution of fixed stars. This paper is considered in the context of that problem in §4.3 of the present work: it is sufficient to note here that we find again the statement of the general problem (in terms of probabilities of propositions rather than events) [Boole 1952, p.251]. This problem, and its solution, are further stated as follows:

> Given the probability (p) of the truth of the proposition, If the condition A is satisfied, the event B will not happen. Required the probability P of the proposition, If the event B does happen, the condition A has not been satisfied. The result which I obtain is
>
> $$P = c(1-a)/[c(1-a) + a(1-p)] ,$$
>
> where c and a are arbitrary constants, whose interpretation is as follows: viz. a is the probability of the fulfilment of the condition A, c is the probability that the event B would happen if the condition A were not satisfied. [Boole 1952, pp.255-256]

In general, Boole's solutions contain arbitrary constants, specification of which usually yields bounds within which the desired probability must lie.

In another paper of 1851, "Further observations on the theory of probabilities", the general problem and its solution of the previous paper are further discussed. Boole takes exception to Herschel's doctrine that $P = p$ (cf. Herschel [1857, p.421]), an opinion which seems implicitly sanctioned by Laplace in his *Essai*[62], and also queries de Morgan's choice of the constants a and c as $\frac{1}{2}$ and 1 respectively.

In *The Cambridge and Dublin Mathematical Journal*, Vol. VI, November 1851, Boole proposed the following question:

> If an event E can only happen as a consequence of some one or more of certain causes $A_1, A_2, \ldots, A_n$, and if generally c_i represent the probability of the cause A_i, and p_i the probability that if the cause A_i exist the event E will exist, then the series of values $c_1, c_2, \ldots, c_n$, $p_1, p_2, \ldots, p_n$, being given, required the probability of the event E. [Boole 1952, p.268]

It should be noted here that the A_i are not assumed mutually exclusive. This result, which has become known as Boole's "challenge problem", occasioned some discussion in mathematical circles[63], Boole's own solution appearing in *The Philosophical Magazine* in 1854 and in the same year in *An Investigation of The Laws of Thought.* After what Keynes [1921, chap. XVII, §2] describes as "calculations of considerable length and great difficulty", Boole arrives at the solution (for two causes) as that value of u solving the quadratic

$$\frac{[1-c_1(1-p_1)-u][1-c_2(1-p_2)-u]}{1-u} = \frac{(u-c_1p_1)(u-c_2p_2)}{c_1p_1+c_2p_2-u}$$

and which also satisfies

$$\max\{c_1p_1, c_2p_2\} \leq u \leq \min\{1-c_1(1-p_1), 1-c_2(1-p_2), c_1p_1+c_2p_2\} \ .$$

Various special cases follow. Keynes (loc. cit.) finds this solution wrong[64], the correct answer[65] in fact being

$$u = (c_1p_1+c_2p_2)/\,(1+z) \ ,$$

where z is the probability after the event that both causes were present. Keynes obtains the limits given by Boole for a solution independent of z. The extension[66] to n causes is given by Boole in *An Investigation of The Laws of Thought* as Problem VI of Chapter XX. Further discussions of this point were given by Wilbraham [1854][67] and by Boole in his two replies to Wilbraham in 1854.

Many, if not most, of Boole's writings on probability are concerned with his general problem and various developments thereof. One paper, of 1857, for which he was awarded the Keith Prize, is concerned with the probabilities of testimonies: this paper Keynes [1921, chap. XVI, §6] considers to be Boole's "most considered contribution to probability".

We now pass on to *An Investigation of the Laws of Thought on which are founded the mathematical theories of logic and probabilities*, published in 1854. Eschewing, albeit with difficulty, any discussion of the complete work, apart from drawing attention to the consideration of general probabilistic principles[68], we pass on immediately to the twentieth chapter[69], entitled "Problems relating to the connexion of causes and effects." Several of the problems discussed here are developments of those already discussed in this monograph, and no more need be said on this point, apart from recording that we find, on p.357, a discrete Bayes's formula.

More relevant is Problem X, viz.

> The probability of the occurrence of a certain natural phænomenon under given circumstances is p. Observation has also recorded a probability a of the existence of a permanent cause of that phænomenon, i.e. of a cause which would always produce the

> event under the circumstances supposed. What is the probability that if the phænomenon is observed to occur n times in succession under the given circumstances, it will occur the $n+1^{\text{th}}$ time? What also is the probability, after such observation, of the existence of the permanent cause referred to? [p.358]

Boole provides two methods of solution to the first question. The first of these is complicated: the second, attributed to Donkin[70], runs as follows: let $\Pr[E] = p$, $\Pr[C] = a$ and $\Pr[E \mid \overline{C}] = x$. Then $p = a + (1-a)x$, and hence $x = (p-a)/(1-a)$. The *a priori* probability of the occurrence of the event n times being 1 (if C exists) or x^n (if $\overline{C}$ obtains) we have

$$\Pr[C \mid x_1, \ldots, x_n] = a / [a + (1-a)x^n]$$

$$\Pr\left[\overline{C} \mid x_1, \ldots, x_n\right] = (1-a)x^n / [a + (1-a)x^n] \quad .$$

Hence the probability of another occurrence is

$$\{a/[a + (1-a)x^n]\}\, 1 + \{(1-a)x^n / [a + (1-a)x^n]\}\, x \quad .$$

On replacing x by its value $(p-a)/(1-a)$ we obtain the result

$$\left[a + (p-a)^{n+1} \big/ (1-a)^n\right] \Big/ \left[a + (p-a)^n \big/ (1-a)^{n-1}\right] \quad ,$$

the solution to the second question being a divided by the above denominator.

Boole now proceeds to consider the usual mode of approach to such problems, whereby the "*necessary* arbitrariness of the solution" [p.368] is evaded[71]. This is exemplified by the case of the sun's rising[72]: let p be an unknown probability and c (infinitesimal and constant) be the probability that the probability of the sun's rising lies between p and $p + dp$. Then the probability that the sun will rise m times in succession is

$$c \int_0^1 p^m \, dp \quad ,$$

and hence the probability of one further rise, given m rises in succession, is

$$c \int_0^1 p^{m+1} \, dp \Big/ c \int_0^1 p^m \, dp = (m+1)/(m+2) \quad .$$

Boole however rejects the principle "of the equal distribution of our knowledge, or rather of our ignorance" [p.370], on account of its arbitrary nature[73]. He notes that different hypotheses may lead to the same result[74], while other hypotheses, as strictly involving this principle, may conduct to other conflicting conclusions. As an illustration of the latter possibility Boole considers the drawing of balls from a bag containing an infinite number of black or white balls, under the assumption that "*all possible constitutions of the system of balls are equally probable*" [p.370]. We seek the probability of getting a white ball on the $(m+1)$th drawing given that the m previous draws all yielded white balls.

This problem Boole solves in two ways: the first (and shorter) of these relies on his logical approach to probability[75], while the second proceeds in the more usual style as follows: suppose initially that the urn contains μ balls and that sampling proceeds with replacement, all constitutions of the system being *a priori* equally likely. Then the probability of obtaining r white and $p-r$ black balls in p drawings, irrespective of order and under the assumption that the urn contains n white balls, is

$$\binom{p}{r}\left(\frac{n}{\mu}\right)^r\left(1-\frac{n}{\mu}\right)^{p-r} .$$

Since the probability that exactly n balls are white is $\binom{\mu}{n}/2^\mu$, (the number of possible constitutions of the system being 2^μ), it follows that the (unconditional) probability of obtaining r white balls is

$$\sum_{n=0}^{\mu}\binom{p}{r}\binom{\mu}{n}\left(\frac{n}{\mu}\right)^r\left(1-\frac{n}{\mu}\right)^{p-r} \Big/ 2^\mu .$$

Using the Heaviside D operator and the Newton Series Boole shows that, for large values of μ, this probability reduces to $\binom{p}{r}/2^p$. On our setting $p=r=m$, the probability that the $(m+1)$th drawing will yield a white ball after the first m draws have yielded white, is found to be $\frac{1}{2}$.

In Chapter XXI the general method discussed earlier in the work is applied to the question of the probability of judgements[76]. Perhaps all that need be said here is to repeat Boole's statement that "It is apparent that the whole inquiry is of a very speculative character" [p.379].

7.16 Charles Hughes Terrot (1790-1872)

In 1853 Terrot[77] published a paper under the title "Summation of a compound series, and its application to a problem in probabilities." It is the application which is of particular interest here, concerning as it does the rule of succession.

The series referred to in the title of the paper may be written

$$\sum_{j=0}^{m-q-p} (m-q-j)_p(q+j)_q = p!\,q! \sum_{j=0}^{m-q-p} \binom{m-q-j}{p}\binom{q+j}{j}$$

$$= p!\,q! \binom{m+1}{p+q+1}$$

using an identity from Feller [1957] *. Having established this result, Terrot turns his attention in the second section of his paper to the following problem:

> Suppose an experiment concerning whose inherent probability of success we know nothing, has been made $\overline{p+q}$ times, and has succeeded p times, and failed q times, what is the probability of success on the $\overline{p+q+1}^{\text{th}}$ trial. [p.542]

To realize this problem Terrot considers the case of a bag containing m balls, all either black or white, but in unknown proportions[78]. From this bag p white and q black balls have been drawn. Then the following four cases present themselves [Terrot 1853, p.543]:

1. m may be given, and the balls drawn may have been replaced in the bag;
2. m may be given, and the balls drawn not replaced;
3. m may be infinite or indefinite, and the balls replaced;
4. m may be infinite or indefinite, and the balls not replaced.

In this paper Terrot solves the second case (in which the fourth is subsumed) and makes an attempt at the first case (the third has the well-known solution $(p+1)/(p+q+2)$).

Denoting by E the observed event, and by H_i the hypothesis that the bag contains initially $(m-q-i)$ white and $(q+i)$ black balls ($i \in \{0, 1, \ldots, m-q-p-1\}$), we have

$$\Pr[E \mid H_i] = p!\,q! \binom{m-q-i}{p}\binom{q+i}{i} \Big/ (m)_{q+p} \,,$$

where order is taken into account[79]. Under the assumption that all (possible) initial compositions of the bag are equally probable, we have, by an

*Recall that $(x)_n = x(x-1)\ldots(x-n+1)$.

application of a discrete form of Bayes's Theorem,

$$\Pr[H_i \mid E] = \binom{m-q-i}{p}\binom{q+i}{i} \Big/ \sum_{j=0}^{m-q-p} \binom{m-q-j}{p}\binom{q+j}{j} .$$

Since

$$\Pr[\text{white ball drawn} \mid E \ \&\ H_i] = (m-p-q-i)/(m-p-q) ,$$

it follows that

$$\Pr[\text{white ball drawn} \ \&\ H_i \mid E] = \Pr[\text{white ball drawn} \mid E \ \&\ H_i]\Pr[H_i \mid E]$$

$$= \frac{(m-p-q-i)}{(m-p-q)}\binom{m-q-i}{p}\binom{q+i}{i} \Big/ \sum_{j=0}^{m-q-p} \binom{m-q-j}{p}\binom{q+j}{j}$$

$$= \frac{(m-p-q-i)}{(m-p-q)}\binom{m-q-i}{p}\binom{q+i}{i} \Big/ \binom{m+1}{p+q+1} .$$

Thus

$$\Pr[\text{white ball drawn} \mid E]$$

$$= \sum_{i=0}^{m-q-p-1} \frac{(m-p-q-i)}{(m-p-q)}\binom{m-q-i}{p}\binom{q+i}{i} \Big/ \binom{m+1}{p+q+1}$$

$$= \frac{p+1}{m-p-q} \sum_{i=0}^{m-q-p-1} \binom{m-q-i}{p+1}\binom{q+i}{i} \Big/ \binom{m+1}{p+q+1}$$

$$= \frac{p+1}{m-p-q} \binom{m+1}{p+q+2} \Big/ \binom{m+1}{p+q+1}$$

$$= (p+1)/(p+q+2) .$$

This, the solution to Terrot's second case, being independent of m, is clearly also the answer to the fourth case.

Terrot now turns his attention to the first case, noting firstly that the main object here is the summation of the series

$$(m-1)^p \times 1^q + (m-2)^p \times 2^q + \cdots + 1^p \times (m-1)^q .$$

He discusses in detail the specific case $p = 2, q = 3$: we shall give a more general discussion.

Suppose, then, that $(p+q)$ draws (with replacement) from m balls have resulted in p white and q black balls (event E). If there are r white balls in the bag, the probability of E is

$$\binom{p+q}{p}\left(\frac{r}{m}\right)^p\left(1-\frac{r}{m}\right)^q ,$$

while the probability that one further draw results in a white ball is r/m. Thus[80]

$$\Pr[\text{white ball drawn} \mid E] = \sum_{r=0}^{m}(r/m)^{p+1}(1-r/m)^q \Big/ \sum_{r=0}^{m}(r/m)^p(1-r/m)^q$$

$$= (1/m)\sum_{r=0}^{m} r^{p+1}(m-r)^q \Big/ \sum_{r=0}^{m} r^p(m-r)^q .$$

Having obtained this result, Terrot finally points out that in the limit as m tends to infinity this result approaches $(p+1)/(p+q+2)$, as is of course expected. This observation concludes the paper.

7.17 Anton Meyer (1802-1857)

In 1856 Meyer[81] published a paper entitled "Note sur le théorème inverse de Bernoulli", in which he noted, in addition to the theorem mentioned in the title, the results of Bayes and Laplace. His own note was devoted to the direct proof, given by Laplace, of this inverse Bernoulli theorem, and the main result runs as follows: let x_1 and x_2 be the unknown probabilities of two complementary events A_1 and A_2. If, in a large number $\mu = m_1 + m_2$ of trials, A_1 and A_2 occur m_1 and m_2 times respectively, then the probability that x_1 lies within the limits

$$\frac{m_1}{\mu} \pm \gamma \frac{\sqrt{2(\mu - m_1)m_1}}{\mu^2}$$

will be

$$P = \frac{2}{\sqrt{\pi}}\int_0^\gamma e^{-t^2}\,dt$$

to terms of order $1/\mu$. We shall not pause to discuss this result here, but shall pass on immediately to a longer work.

Meyer's *Essai sur une Exposition nouvelle de la Théorie analytique des Probabilités a posteriori* appeared in 1857. His avowed aim in writing this monograph is expressed in the foreword as follows:

> en écrivant cet essai, j'ai eu primitivement en vue la nécessité de rendre plus rigoureux les calculs, et de concentrer les méthodes et les principes dans l'exposition de la théorie des probabilités à posteriori.

Whether he was altogether successful in attaining this goal will become clear as we discuss that part of this work which is pertinent to our purpose.

The second part of this *Essai* is entitled "Théorèmes de Bayes et de Laplace sur la probabilité des causes." Here Meyer discusses, in addition to the two results mentioned in the title, theorems due to Bernoulli and Poisson and an inverse Bernoulli theorem. We shall discuss these results *seriatim.*

Denoting by $y = f(x)$ the probability of an event depending upon the unknown x (where x is called the "cause" of that event), Meyer states as a theorem due to Bayes the following result:

> $x = \left\{ \begin{smallmatrix} b \\ a \end{smallmatrix} \right.$ désignant les limites de toutes les valeurs possibles de x, si $y = fx$ est la probabilité de l'une quelconque des valeurs de x, regardée comme certaine, je dis que l'on aura une probabilité
>
> $$P = \int_\alpha^\beta y\,dx \bigg/ \int_a^b y\,dx \quad ,$$
>
> que l'inconnue x est comprise dans les limites α et β. [p.19]

Now it seems rather curious to attribute this result, in which no mention is made of the number of occurrences or failures of the event, to Bayes. In fact, the expression given seems to be only $\Pr[\alpha < X < \beta \mid a < X < b]$ where X is a random variable with probability density function f.

Two corollaries to this result are given. The first states

> La probabilité p d'une valeur unique de x est par conséquent exprimée par
>
> $$P = y\,dx \bigg/ \int_a^b y\,dx \quad .$$
>
> [p.20]

This is in fact just $\Pr[x < X < x + dx \mid a < X < b]$. The second corollary runs thus:

> Soit $z = \varphi x$ la probabilité d'un évènement futur, due à la cause x, et $y = fx$ la probabilité d'un évènement observé, soit P_i la

probabilité de l'évènement futur en vertu de la cause dont la probabilité est la valeur p ci-dessus, nous aurons évidemment

$$P_i = pz = zy\,dx \Big/ \int_a^b y\,dx$$

Donc si π exprime la probabilité que l'évènement futur arrivera en vertu de l'une des causes $x = \begin{cases} \beta \\ \alpha \end{cases}$, nous aurons

$$\pi = \int_\alpha^\beta zy\,dx \Big/ \int_a^b y\,dx \quad .$$

[pp.20-21]

This corollary is recognizable as an extension of Meyer's first theorem in the same way that Price's result extended Bayes's.

The second theorem, attributed to Laplace, that Meyer proves is the following:

x étant la cause inconnue d'un évènement composé, dont la probabilité est

$$y = (fx)^s \quad ,$$

si m désigne la valeur de x qui rend y un maximum, je dis qu'en supposant s très-grand, on aura, aux quantités prés de l'ordre $1/s$, une probabilité

$$P = \frac{2}{\sqrt{\pi}} \int_0^\gamma e^{-\tau^2}\,d\tau$$

que l'inconnue, ou la cause x, est comprise entre les limites

$$m \pm \frac{\gamma}{\sqrt{-s\left(\dfrac{f''x}{2fx}\right)_m}} = m \pm \frac{\gamma}{\sqrt{-\left(\dfrac{d^2 \log y}{2dx^2}\right)_m}} \quad .$$

[p.21]

(Here "log" denotes the natural logarithm.) The proof given (which makes use of Meyer's version of Bayes's Theorem) is long and involved, and will not be presented here. The result, however, seems correct[82]. The proof is succeeded by the following three remarks:

(i) if P remains constant, the limits contract as s increases;
(ii) the limits remaining constant, which requires that γ increases as s increases, the probability P tends to 1 as $s \to \infty$;
(iii) by increasing s one may therefore contract the limits and simultaneously increase P: for $s = \infty$, we have $x = m$ and $P = 1$.

Meyer is not reluctant to blow his own trumpet: before stating his second theorem he writes

> quoique mes déductions procédent au fond des idées de Laplace, elles sont à la fois plus claires et plus rigoureuses que celles de cet auteur. [p.21]

The third result cited is the inverse Bernoulli theorem[83], viz.

> x et $1-x$ désignant les probabilités simples et inconnues de deux évènements contraires A et B, en supposant que A arrive p fois, et B q fois en un très-grand nombre $\mu = p+q$ d'épreuves, je dis qu'on aura la probabilité
>
> $$P = \frac{2}{\sqrt{\pi}} \int_0^\gamma e^{-\tau^2}\, d\tau$$
>
> que x est compris entre
>
> $$\frac{p}{\mu} + \frac{\gamma}{\mu}\sqrt{\frac{2pq}{\mu}}$$
>
> [sic] [p.28].

Notice here that the probability x is supposed unknown, in contrast to its appearance in a "known" capacity in the (direct) Bernoulli Theorem. Once again Meyer makes use of his first theorem in the proof, and indeed this result appears essentially as a special case of the second theorem.

The fourth theorem is attributed to Bernoulli, and is stated as follows:

> x, et $1-x$ étant les probabilités simples, supposées constantes et connues des évènements contraires A et B, le rapport m/s du nombre de fois m que A arrivera le plus probablement en un très-grand nombre s d'épreuves, à ce nombre s, est, aux quantités prés de l'ordre $1/s$, compris entre les limites
>
> $$x \pm \frac{\gamma}{\sqrt{s}}\sqrt{2x(1-x)}$$

avec une probabilité

$$P = \frac{2}{\sqrt{\pi}} \int_0^{\gamma} e^{-t^2}\, dt + \frac{e^{-\gamma^2}}{\sqrt{2\pi s x(1-x)}} \ .$$

[p.30]

The proof of this result is unexceptionable: the theorem however is in fact *not* that given by Bernoulli — indeed Meyer's statement owes far more to de Moivre than to Bernoulli[84].

Finally Meyer discusses Poisson's theorem, which differs from Bernoulli's result in as much as the probabilities of the individual events are no longer required to be the same.

One might perhaps summarize this section of the monograph by saying that, while Meyer provides useful and accurate proofs of the theorems stated, he is somewhat less than careful in his eponymy.

7.18 Albert Wild

In 1862, in a work entitled "Die Grundsätze der Wahrscheinlichkeits-Rechnung und ihre Anwendung", Wild quotes Bayes on the probability of causes: the reference appears in connexion with a simple discrete form of Bayes's Theorem, but Wild does not attribute this result to Bayes. He passes on, in the section on "Die Wahrscheinlichkeit die Naturereignisse" to the formula

$$h = x^m(1-x)^n\, dx \Big/ \int_0^1 x^m(1-x)^n\, dx \ ,$$

and then gives the rule of succession. The extended form to r and s future occurrences of events of two (only possible) types is discussed. Finally we find Bayes's result

$$\int_a^b x^m(1-x)^n\, dx \Big/ \int_0^1 x^m(1-x)^n\, dx$$

and the limiting form

$$\frac{2}{\sqrt{\pi}} \int_0^t e^{-t^2}\, dt \ .$$

7.19 John Venn (1834-1923)

From one who was primarily a philosopher rather than a mathematician[85] one might be surprised to find statistical work emanating[86]. Yet in his book *The Logic of Chance*[87], first published in 1866, Venn strongly advocated the frequency concept of probability on which so much of "classical" statistics depends — a concept based on a series which "combines individual irregularity with aggregate regularity" [Venn 1962, p.4][88].

In the fourth chapter of his book, in which he considers modes of establishing certain properties of these series, Venn discusses (i) the meaning to be attached to the phrase "equally likely" and (ii) the Principle of Sufficient Reason, a rule in which he finds

> very great doubts whether a contradiction is not involved when we attempt to extract results from it. [p.82]

In Chapter VI, entitled "The subjective side of probability. Measurement of belief", Venn expresses the views of de Morgan and Donkin (according to which views probability is defined with reference to belief), exposes various difficulties that arise in trying to assimilate these views, and reiterates his opinion that

> all which Probability discusses is the statistical frequency of events, or, if we prefer so to put it, the quantity of belief with which any one of these events should be individually regarded, but leaves all the subsequent conduct dependent upon that frequency, or that belief, to the choice of the agents. [p.137]

Furthermore

> The subjective side of Probability therefore, though very interesting and well deserving of examination, seems a mere appendage of the objective, and affords in itself no safe ground for a science of inference. [p.138]

In Chapter VII Venn turns his attention to inverse probability, a concept which he had defined in an earlier chapter as "the determination of the nature of a cause from the nature of the observed effect" [p.109]. Arguing that the distinction between direct and inverse probability should be abandoned, Venn illustrates his point with the usual sort of "balls and bag" examples, and concludes that any such distinction either vanishes or[89]

> merely resolves itself into one of *time*, which, ... is entirely foreign to our subject. [p.185]

A ground for rejecting the inverse argument is the use of the entirely arbitrary "equally likely" assumption.

Venn now turns his attention to the rule of succession[90] (a term introduced by him himself), his eighth chapter[91] containing what Jaynes has described as[92]

> an attack on Laplace's rule of succession, so viciously unfair that even Fisher (1956) was impelled to come to Laplace's defense on this issue. [1976, p.242]

This rule, says Venn, is generally stated as follows:

> "To find the chance of the recurrence of an event already observed, divide the number of times the event has been observed, increased by one, by the same number increased by two." [p.196]

He states, without proof, the customary result $(m+1)/(m+2)$ for a "balls and bag" example, and goes on to say that

> Then comes in the physical assumption that the universe may be likened to such a bag as the above, in the sense that the above rule may be applied to solve this question:– an event has been observed to happen m times in a certain way, find the chance that it will happen in that way next time [p.197],

illustrating this with examples from Laplace and de Morgan. Venn concludes[93] that "It is hard to take such a rule as this seriously" [p.197].

Venn returns to the subject of inverse probability in his tenth chapter, pointing out the needments for deciding whether an event has been produced by chance or by design, i.e.

> (1) The relative frequency of the two classes of agencies, viz. that which is to act in a chance way and that which is to act designedly. (2) The probability that each of these agencies, if it were the really operative one, would produce the event in question. [p.249]

While the probability instanced in the second case is generally readily obtainable, the frequencies needed in (1) present a severe problem to an adherent to the frequency theory of probability, but Venn concludes that such problems "are at least intelligible even if they are not always resolvable" [p.258].

Like so many writers Venn devotes some thought (see his Chapters XVI and XVII) to the application of probability to testimony: his conclusion is that such problems ought not to be considered as questions in probability, a decision which is perhaps understandable in the light of a frequentist flame.

Venn's work on probability did not pass without comment. Thus Edgeworth [1884b], while agreeing in the main with Venn's objective approach, suggested that the latter's

> logical scepticism has often carried him too far from the position held by the majority of previous writers upon Chance. [p.224]

Pearson [1920a] draws attention to Venn's criticism of inverse probabilities, a criticism apparently based on an "objection to the principle of equal distribution of ignorance" [p.2], and one which Pearson finds curious in the light of Venn's approach to the problem of the effect of Lister's method. This argument receives further attention in the first appendix to Pearson's paper of 1928, while more recently Jaynes has pointed out a curiosity in Venn's thinking, viz.

> How is it possible for one human mind to reject Laplace's rule of succession; and then advocate a frequency definition of probability? Anybody who assigns a probability to an event equal to its observed frequency in many trials, is doing just what Laplace's rule tells him to do. [1976, p.242]

Support for Venn's approach was given by Fisher [1922], who, regarding inverse probability as a "fundamental paradox", paid tribute to the criticisms of Boole, Venn and Chrystal, as having "done something towards banishing the method, at least from the elementary text-books of Algebra" [p.311]. He also comments on the "decisive criticism" of these three authors of "the baseless character of the assumptions made under the titles of inverse probability and Bayes' Theorem" [p.326]. Fisher's remarks, in turn, have been critically examined by Zabell [1989].

Chapter 8

Laurent to Pearson

I have one concluding favour to request of my reader; that he will not expect to be equally diverted & informed by every line or every page of this discourse; but give some allowance to the author's spleen, & short fits or intervals of dullness, as well as his own.

Jonathan Swift, Gulliver's Travels & Other Writings.

8.1 Mathieu Paul Hermann Laurent (1841-1908)

In 1873 Laurent[1] published his *Traité du Calcul des Probabilités*, a work which was to be considered as "une véritable Introduction au Traité de Laplace" [pp.ix-x]. The work begins with definitions and general comments, and leaving these aside, we find the following statement of a "théorème fondamental dû au géomètre anglais Bayes":

> Soient $p_1, p_2, \ldots, p_i, \ldots$ les probabilités que des causes $C_1, C_2, \ldots, C_i, \ldots$, s'excluant mutuellement, donnent respectivement à l'événement E. Soient $q_1, q_2, \ldots, q_i, \ldots$ les probabilités de ces causes. Supposons maintenant que l'événement E ait été observé dans une épreuve, la probabilité ω_i que l'arrivée de l'événement observé est due à la cause C_i est donnée par la formule $\omega_i = p_i q_i / (p_1 q_1 + p_2 q_2 + \cdots + p_i q_i + \cdots)$. [p.57]

Laurent had earlier, by-the-by, given a precise definition of "cause", viz.

> Nous appellerons *cause* d'un événement, dont l'arrivée n'est pas certaine, ce qui lui donne sa probabilité. [p.47]

The above expression for ω_i is also given for equal q_i.

In the section entitled "Théorème inverse de celui de Bernoulli" Laurent supposes that an event E, of constant though unknown probability, has occurred α times in s trials. Then "en vertu du théorème de Bayes" [p.107]

$$P \equiv \Pr\left[|p - \alpha/s| < l\right] = \int_{\alpha/s-l}^{\alpha/s+l} x^{\alpha}(1-x)^{s-\alpha}\,dx \bigg/ \int_0^1 x^{\alpha}(1-x)^{s-\alpha}\,dx \quad .$$

He obtains further the limit

$$P = \frac{2}{\sqrt{\pi}} \int_0^{\gamma} e^{-\gamma^2}\,d\gamma$$

for sl/α and $sl/(s-\alpha)$ very small and of order $1/\sqrt{s}$.

This work is also to be noted for its extensive bibliography of the principal works on probability published to that date.

8.2 Cecil James Monro (1833-1882)

In 1874 Monro, in his paper "Note on the inversion of Bernoulli's theorem in probabilities", suggested that under the name "Bernoulli's Theorem" two results, the deductive and the inductive, should be comprehended. In the former the probability p of a given result on a single trial should be regarded as constant (i.e. known?), this not being so in the latter. If we denote by P the probability that, in the "deductive" case, the desired result is produced from $mp - l$ to $mp + l$ times (or from $x - l$ to $x + l$, where x is the largest integer not exceeding $(m+1)p$), and by P' the probability, in the "inductive" setting, that the facility of a given result which has been produced mp times in m trials (with a constant facility of production) lies between $p - l/m$ and $p + l/m$, then

$$P = 2\sqrt{h/\pi} \int_0^{l+1/2} e^{-h\lambda^2}\,d\lambda \quad ,$$

where $h = [2p(1-p)m]^{-1}$, as given by Laplace. Here it is assumed that l is of order $\sqrt{m}$ at most, and also that terms of order $1/m$ may be neglected.

Monro points out Laplace's two methods for the inversion of this result.

In the first[2] of these P' is set equal to P "by an implicit inference from the deductive theorem" [p.74], while in the second P' is (correctly) given by

$$P' = 2\sqrt{h/\pi}\int_0^l e^{-h\lambda^2}\,d\lambda$$

under the assumption of a uniform prior. Assuming that equal ranges contain equally probable values, Monro shows that

> the inversion is so far legitimate, that either theorem may be inferred from the other with little calculation, ... and accordingly that the two solutions are identical in principle. [p.75]

To this end he notes firstly that $l + 1/2$ may be substituted for l in the statement of the deductive theorem, since our concern is with integral values of λ. Secondly, as regards the inductive theorem,

> P is the probability that the facility lies between the limits $p \pm (l+1/2)/m$, and the second solution is correct for the limits $p \pm l/m$; provided always that a valid correspondence exists between the two theorems. [p.76]

To establish the desired correspondence, Monro denotes by u_n the probability of $n = m\omega$ results in m trials, each of facility x/m, and by U_n the probability given by x results in m trials that their (constant) facility is within $+d\omega$ of ω: "This supposition expresses the hypothesis of equally probable values of the facility within equal ranges" [p.76]: the required proviso is then established by comparing u_n in the deductive case with $\int U_n\,d\omega$, between $(n-1/2)/m$ and $(n+1/2)/m$, in the inductive. Now

$$\begin{aligned} u_n &= \binom{m}{n}\left(\frac{x}{m}\right)^n\left(\frac{m-x}{m}\right)^{m-n} \\ U_n &= \omega^x(1-\omega)^{m-x} \Big/ \int_0^1 \omega^x(1-\omega)^{m-x}\,dx \\ &= \frac{(m+1)!}{x!\,(m-x)!}\left(\frac{n}{m}\right)^x\left(\frac{m-n}{m}\right)^{m-x}. \end{aligned}$$

(Note the substitution of n/m for ω in the numerator of U_n.) Neglect of terms of order $1/m$ results in $U_n = (m+1)u_n$, and the required integration yields, to the desired degree of approximation, the stated result[3].

8.3 William Stanley Jevons (1835-1882)

Although well known for his work in economics and logic, Jevons[4] is less remembered for his statistical work. Of his writings the only one that seems relevant here is his book *The Principles of Science: a treatise on logic and scientific method*, published in two volumes in 1874 and in one volume in 1877. Of this work Keynes is somewhat scathing, saying

> There are few books, so superficial in argument yet suggesting so much truth, as Jevons's *Principles of Science.* [1921, chap. XXIII, §10]

Further, while stressing the important advance made by Jevons when he "emphasised the close relation between induction and probability", Keynes goes on to say

> Combining insight and error, he spoilt brilliant suggestions by erratic and atrocious arguments. His application of inverse probability to the inductive problem is crude and fallacious, but the idea which underlies it is substantially good. [loc. cit.]

Be that as it may: let us turn forthwith to Jevons's book itself.

The tenth chapter, entitled "The theory of probability", is devoted to a fairly general discussion of chance and probability, the latter being understood as having reference to our mental condition[5]. Because he finds difficulties with "belief", Jevons prefers to say that "the theory of probability deals with *quantity of knowledge*" [1877, p.199].

The method to be used in the theory has as basis the calculation of "the number of all the cases or events concerning which our knowledge is equal" [p.201]. Rules for the calculation of probabilities are given, and the importance of distinguishing between absolute and comparative probabilities is stressed. Boole's method is found to be "fundamentally erroneous" [p.206], Jevons siding with Wilbraham in this matter.

In this chapter are to be found some remarks on antecedent (or prior) probabilities, including the famous example[6] that the only odds which may be ascribed to "a Platythliptic Coefficient is positive" are evens [p.212]. Jevons also comments on Terrot's suggestion that the symbol $\frac{0}{0}$ should be used, rather than $\frac{1}{2}$, to express complete doubt, and goes on to say

> if we grant that the probability may have any value between 0 and 1, and that every separate value is equally likely, then n and $1 - n$ are equally likely, and the average is always 1/2. Or we may take $p.dp$ to express the probability that our estimate concerning any proposition should lie between p and $p + dp$.

> The complete probability of the proposition is then the integral taken between the limits 1 and 0, or again 1/2. [pp.212-213]

From the first sentence it seems to follow that 2/8 and 7/8 (say) are also equally likely, and their average is no longer 1/2: so some care is needed here. Keynes criticizes Jevons's views on this matter as follows:

> It is difficult to see how such a belief, if even its most immediate implications had been properly apprehended, could have remained plausible to a mind of so sound a practical judgement as his. [1921, chap.XX, §7]

In the twelfth chapter, "The inductive or inverse application of the theory of probability", we find a statement of Laplace's proposition for inverse application of the rules of probability[7], viz.

$$\Pr[H_i \mid E] \propto \Pr[E \mid H_i]$$

under the assumption of *a priori* equally probable causes H_i. (Here the symbol "|" is interpreted as "inferred from" on the left-hand side and "derived from" on the right-hand.) We also find here a discrete Bayes's Rule, formulated in symbols and also in words as follows:

> If it is certain that one or other of the supposed causes exists, the probability that any one does exist is the probability that if it exists the event happens, divided by the sum of all the similar probabilities. [p.243]

The next section of this chapter is devoted to some simple applications of the inverse method, chiefly of an astronomical nature. Again we find Keynes taking exception, albeit slight, to Jevons's use of the principle of the inverse method in scientific induction.

The *general inverse problem* is stated as follows:

> An event having happened a certain number of times, and failed a certain number of times, required the probability that it will happen any given number of times in the future under the same circumstances. [p.251]

As an illustration Jevons considers a "balls and ballot-box" example which he attributes to Condorcet: an urn contains four black or white balls, in unknown ratio; if four drawings (with replacement) have yielded three white balls, what is the probability that the next draw will also yield a white ball? Jevons first deduces the posterior probabilities of the hypotheses specifying the composition of the ballot-box, and then, in the usual manner, finds the required probability.

He next passes to the general solution of the inverse problem, presenting, though without proof, the customary expressions arising in the rule of succession. He then considers the extension to more than two possibilities[8]: thus if there are n events $A_1, A_2, \ldots, A_n$, and A_i has occurred r_i times, then the probability that the next event will be A_i is

$$(r_i + 1) \Big/ \sum_1^n (r_j + 1) \quad .$$

Furthermore,

> if new events may happen in addition to those which have been observed, we must assign unity for the probability of such new event. [p.258]

Thus, if there is one such new event, the probability that the next event will be A_i is

$$(r_i + 1) \Big/ \left[1 + \sum_1^n (r_j + 1)\right] \quad .$$

Jevons stresses the need for the incorporation of all additional information in the application of the method of inverse probabilities[9]. We also find here a comment to the effect that, if a coin is to be tossed for the first time, we should assign probability 1/2 to each of the two possible outcomes. However, the obtaining of a head on the first throw provides "very slight experimental evidence in favour of a tendency to show head" [p.260]. This is, of course, in accordance with the rule of succession, though it does seem to suggest that, after one toss, a coin will always be considered — even if only temporarily — as biased.

Jevons mentions the thoughts of James Bernoulli and de Moivre on the estimation of the probability of future events from past experience, although Bayes and Price were "undoubtedly the first who put forward any distinct rules on the subject" [p.261]. Mention is also made of the contributions of Condorcet "and several other eminent mathematicians" and of Laplace, who carried "the solution of the problem almost to perfection" [p.261].

Writing of subjects in which deduction is only probable, Jevons proposes the following scheme:

> (1) We frame an hypothesis.
> (2) We deduce the probability of various series of possible consequences.
> (3) We compare the consequences with the particular facts, and observe the probability that such facts would happen under the hypothesis. [p.267]

This reasonable scheme is however followed by a statement which immediately provides grounds for Keynes's criticism: Jevons writes

> The above processes must be performed for every conceivable hypothesis, and then the absolute probability of each will be yielded by the principle of the inverse method. [p.267]

This rule Jevons describes as "that which common sense leads us to adopt almost instinctively" [1877, p.243]: Keynes views it as a "fallacious principle" [1921, chap.XVI, §14].

8.4 Rudolf Hermann Lotze (1817-1881)

In 1874 Lotze published his *Logik*, in which he expressed the view that probability is subjective; indeed he says of probability that *

> sie bezeichnet, zunächst wenigstens, durchaus nur subjectiv das Maß des vernünftigen Zutrauens, welches wir im voraus zu dem Eintreten eines bestimmten Falles dan hegen dürfen, wenn uns nur die Anzahl aller unter den jedesmal gegebenen Bedingungen möglichen Fälle, aber kein sachlicher Grund gegeben ist, der für Nothwendigkeit des einen von ihnen mit Ausschluß der anderen entscheide. [chap. IX, art.282.1]

Lotze further seems to suggest that a cause C whose likelihood is greater than that of any other (conditional on the occurrence of some event E) should be regarded as the cause of that event, when he writes

> Wenn gegebene Thatsachen aus mehreren verscheidenen Ursachen ableitbar sind, so ist diejenige Ursache die wahrscheinlichste, unter deren Voraussetzung die aus ihr berechnete Wahrscheinlichkeit der gegebenen Thatsachen die größte wird. [chap.IX, art.282.4]

Though whether this is supposed to imply that $\Pr[C \mid E]$ is necessarily the greatest is not clear.

His discussion of the rule of succession bears note: after stating that $(m+1)/(m+2)$ is the probability that an event E will occur one further time if it has been observed m times without exception, Lotze provides the following proof of his result: in this fraction, viz. $(m+1)/(m+2)$,

> der Nenner enthält die Summe der denkbaren Falle, den nach m wirklichen Fällen kommen immer 2 denkbare, Wiederholung und Nichtwiederholung des E, hinzu. [chap. IX, art.282.5]

*Quotations are from the third edition of 1912.

And this deduction, he further asserts,

> mir scheint sie nicht viel weniger überzeugend, als die undurchsichtigere analytische Behandlung, durch die man sie gewöhnlich gewinnt. [chap. IX, art.282.5]

8.5 Bing's paradox

In 1879 F. Bing published a paper entitled "Om aposteriorisk Sandsynlighed" in which the concept of *a posteriori* probability received close examination[10]. The paper opens with a statement of the "equally possible" definition of probability, which is followed by a discussion of some "balls and bags" type examples. Then follows a statement of the discrete Bayes's Theorem, in illustration of which Bing considers the following problem[11]:

> A blindfolded person withdraws marbles from a bag; some of these marbles are found to be white; others black. Knowing that the marbles in the bag are either white or black, a question arises as to the probability of the bag's having a particular content, e.g. equal numbers of black and white marbles. [p.5]

Bing points out that one may assume an equally probable prior distribution on the contents of the bag; yet while the answer then obtained is certainly valid, it is *not* the answer to the question asked. To illustrate this latter point, Bing assumes that the drawer of the marbles regards some of the drawn light marbles as white and one as yellow, and this of course affects the posterior probabilities obtained.

Passing next to an application of the rule of succession, Bing considers the case in which 100 trials have yielded A, B and C respectively 49, 37 and 14 times. Then, it is claimed, the probability that the 101st trial will yield none of these three letters (but "something else") is 1/104. On the other hand, if we merely consider that a letter has been drawn in 100 trials, then the probability against drawing a letter on the next trial is 1/102. This position Bing finds paradoxical[12], but he notes that "the disparity originates exclusively from the differing application of Bayes's Theorem" [p.10]. He points out further that the two solutions are found from

$$\int_0^1 u^{100}(1-u)\,du \Big/ \int_0^1 u^{100}\,du = 1/102$$

and

$$\int x^{49}y^{37}z^{14}(1-x-y-z)\,d(x,y,z) \Big/ \int x^{49}y^{37}z^{14}\,d(x,y,z) = 1/104 \ ,$$

the integrals in the latter expression being taken over

$$\{(x,y,z): \ x \geq 0, \ y \geq 0, \ z \geq 0 \ \& \ x+y+z \leq 1\} \ .$$

A further example concerns the sampling of fruit from a large batch of 100,000 pieces, a sample of size 30 — all good — being taken. If the price of a good fruit is 10 Øre and x denotes the ratio of good to rotten fruit, then the total expectation (and hence the fair price to be paid) in Kronen is

$$10{,}000 \int_0^1 x^{31}\,dx \bigg/ \int_0^1 x^{30}\,dx = 9{,}687 \ .$$

If it is now discovered that each of the fruits sampled was of a different type, then the answer is given by considering the ratio of two thirty-fold integrals, in which case the value 9,836 is obtained. Bing seems to find the disparity unacceptable, for he writes

> most people will certainly regard it as absurd that the buyer should pay more for the merchandise because he has sorted the samples, in spite of the fact that the individual pieces before and after the sorting are assumed to be worth 10 Øre. [p.15]

The same theme is pursued in the next section, where an example relating to mortality statistics is presented. If of $l+d$ persons alive at the beginning of a specified time period (say a year), d died during that period, and if X denotes the actual probability of not dying in that year, then

$$\Pr[x < X < x + dx] = x^l(1-x)^d\,dx \bigg/ \int_0^1 x^l(1-x)^d\,dx \ . \qquad (1)$$

However

> a contradiction immediately arises as soon as allowance is made for the fact that it is possible to apply different subdivisions of time. [p.16]

To illustrate this assertion Bing supposes that a population of individuals initially aged 40 is considered, and that d_1 and d_2 are the numbers of individuals who die in the first and second half-year respectively. If X and Y are the (true) probabilities of not dying in the first and second half-years respectively, then "the correct *a posteriori* probabilities" [p.16] will be

$$x^{l+d_2}(1-x)^{d_1}\,dx \bigg/ \int_0^1 x^{l+d_2}(1-x)^{d_1}\,dx$$

and

$$y^l(1-y)^{d_2}\,dy \Big/ \int_0^1 y^l(1-y)^{d_2}\,dy \ .$$

The probability that X and Y "samtidigt ere rigtige" (are both correct) is then

$$\frac{\int_R\int x^{l+d_2}(1-x)^{d_1}y^l(1-y)^{d_2}\,dx\,dy}{\int_0^1 x^{l+d_2}(1-x)^{d_1}\,dx\,\int_0^1 y^l(1-y)^{d_2}\,dy} \ ,$$

where $R = \{(x,y): \ x \geq 0,\, y \geq 0,\, xy \leq \alpha\}$ and where the probability of the survival ratio for ages 40-41 is known to lie in the interval $[0,\alpha]$. On defining $y = z/x$ and $x = 1 - v(1-z)$, one finds that

$$\Pr[\alpha < A < \alpha{+}d\alpha] = \frac{\alpha^l(1-\alpha)^{d_1+d_2+1}\,d\alpha\,\int_0^1 v^{d_1}(1-v)^{d_2}[1-v(1-\alpha)]^{-1}\,dv}{\int_0^1 x^{l+d_2}(1-x)^{d_1}\,dx\,\int_0^1 y^l(1-y)^{d_2}\,dy} \qquad (2)$$

where A denotes the survival ratio. Bing notes that this formula differs from (1) above, the latter yielding $(l+1)/(l+2)$ when $d_1 = 0 = d_2$, while multiplication of (2) by α and integration from 0 to 1 yields, for the same d values, $[(l+1)/(l+2)]^2$. The extension of this situation to the division of the year into n parts, rather than two, shows that, in the limit as $n \to \infty$, each individual aged 40 must die within the specified time period.

Bing suggests that this is perhaps the first time that Bayes's Theorem has been applied to a subdivided year, and furthermore queries why it should be more logical to suppose a uniform distribution of probability for the 40-41 group rather than for the 40-$40\frac{1}{2}$. Claiming that "Bayes's Theorem is entirely unreliable for all cases in which no *a priori* information is available as to necessary causes" [p.18], Bing turns his attention to the following situation.

Suppose that of a population of n living individuals, d_1 die in the first year and d_2 in the second: suppose further that X_1 and X_2, the "relationships" between the numbers dying in the first or second year and the numbers living at the start of those years, are bound by an unknown function φ which satisfies

$$\varphi(x_1, x_2, d_1, d_2, n)\,dx_1\,dx_2 = \varphi(x_2, x_1, d_2, d_1, n)\,dx_2\,dx_1 \ . \qquad (3)$$

If Y_1 and Y_2 are the probabilities of surviving the first and second years, then

$$\Pr[y_1 < Y_1 < y_1 + dy_1] = \psi(y_1, d_1, n)\,dy_1 \ ,$$

$$\Pr[y_2 < Y_2 < y_2 + dy_2] = \psi(y_2, d_2, n - d_1)\,dy_2 \ .$$

Under the assumption that Y_1 and Y_2 are independent, we have

$$\Pr\left[y_1 < Y_1 < y_1 + dy_1,\, y_2 < Y_2 < y_2 + dy_2\right]$$
$$= \psi(y_1, d_1, n)\, \psi(y_2, d_2, n - d_1)\, dy_1\, dy_2 \quad . \tag{4}$$

The transformation $x_1 = 1 - y_1, x_2 = y_1(1 - y_2)$ applied to (4) and the use of (3) now yield the equation

$$(1 - x_1)^{-1}\, \psi(1 - x_1, d_1, n)\, \psi((1 - x_1 - x_2)\,/(1 - x_1)\,, d_2, n - d_1)\, dx_1\, dx_2$$

$$= (1 - x_2)^{-1}\, \psi(1 - x_2, d_2, n)\, \psi((1 - x_1 - x_2)\,/(1 - x_2)\,, d_1, n - d_2)\, dx_2\, dx_1\,.$$

On taking (natural) logarithms, applying the operator $\partial^2/\partial x_2\, \partial x_1$, and setting $z = (1 - x_1 - x_2)\,/(1 - x_1)$, Bing finds on solving the resulting differential equation that

$$\psi(1 - x_1, d_1, n) = (1 - x_1)^{a(n-d_1)-k}\, x_1^{ad_1 - 1} M \tag{5}$$

where a and k are constants and M is chosen so that $\int_0^1 \psi\, dx = 1$. In a comment in a subsequent paper (which will be discussed in due course), Bing states that k should be set at -1, because we ought to have

$$\psi(y_1, d_1, n) = \psi(1 - y_1, n - d_1, n) \quad ,$$

though he also says that one may keep k arbitrary.

Attention is drawn to the correspondence between this result and Bayes's Theorem. Furthermore, in order that the expression in (5) define a density, one must clearly have a positive (if $k = 1$), and even this choice proves unreasonable if $d_1 = 0$. (Whether the derivation leading to (4) is in fact valid if $d_1 = 0$ is not mentioned.) Bing then concludes[13]

> Den eneste mulige Funktionsform, som ikke giver Strid, har altsaa vist sig ubrugelig, og dermed mener jeg, at *det er bevist, at der aldeles ikke existerer nogen aposteriorisk Sandsynlighed,* naar der er Tale om Problemer, hvor man forud er absolut uvidende om de virkende Aarsager. [p.21]

Whether all this is really necessary to conclude that posterior probabilities cannot exist when there are no priors is moot.

Bing's work did not pass unnoticed: in a paper entitled "Bemærkninger til Hr. Bing's Afhandling 'Om aposteriorisk Sandsynlighed' ", also published in 1879, Lorenz raised several criticisms, the main thrust of which was Bing's misunderstanding of the practical applications of Bayes's Theorem.

Lorenz comments firstly on Bing's "bag and balls" example, and on that concerned with the fruit shipment, stressing that the evaluation of posterior probabilities is dependent on a clear statement of the initial information at one's disposal. When it comes to considering the example on mortality statistics, however, Lorenz's criticism is sharpened. He asserts firstly that Bing has proved more than was concluded in the preceding quotation: he has in fact shown that "der aldeles ikke existerer nogen aposteriorisk Sandsynlighed" [p.61] (there is no such thing as *a posteriori* probability[14]). For if $f(\cdot)$ denotes the prior probability density function of x_1 (say), then Bing's $\psi(1-x_1, d_1, n)$ becomes $f(x_1)\,(1-x_1)^{n-d_1}\,x_1^{d_1}$, and hence if ψ has no usable form, Bayes's Theorem must be false.

Consideration of the two cases $d_1 = 0, d_2 = n$ and $d_1 = n, d_2 = 0$ in turn persuades Lorenz that Bing's basic assumption that

$$\varphi(x_1, x_2, d_1, d_2, n)\,dx_1\,dx_2 = \varphi(x_2, x_1, d_2, d_1, n)\,dx_2\,dx_1$$

is wrong, and further investigation leads him to conclude that Bing's paradox illustrates the unreasonableness of the assumption that X_1 and X_2 are independent.

Unconvinced by Lorenz's comments, Bing eagerly seized the opportunity offered him by the editors of the *Tidsskrift for Mathematik* to reply; and this reply in turn was followed by a rejoinder from Lorenz, in which Bing's expression (4) for the posterior probability was used to show the equivalence of

$$x^{al-1}(1-x)^{ad-1}\,dx \Big/ \int_0^1 x^{al-1}(1-x)^{ad-1}\,dx$$

(an analogue of (1)) and

$$\frac{\alpha^{al-1}(1-\alpha)^{a(d_1+d_2)-1}\,d\alpha \int_0^1 v^{ad_1-1}(1-v)^{ad_2-1}\,dv}{\int_0^1 x^{a(l+d_2)-1}(1-x)^{ad_1-1}\,dx \int_0^1 y^{al-1}(1-y)^{ad_2-1}\,dy}$$

(the probability that the proportion between 40 and 41 years lies between α and $\alpha + d\alpha$).

Satisfied with his exposition, Lorenz stated at the conclusion of this paper that he regarded the matter as closed, a view which was not shared by Bing, who forcefully reiterated his argument. We shall not pursue the controversy further here[15].

8.6 A question of antisepticism

In December 1881 the following problem was proposed by Donald MacAlister[16] (1854-1934) in the columns of *The Educational Times*[17]:

> Of 10 cases treated by Lister's method, 7 did well and 3 suffered from blood-poisoning; of 14 cases treated with ordinary dressings, 9 did well and 5 had blood-poisoning; what are the odds that the success of Lister's method was due to chance? [Problem 6929]

This seemingly simple question occasioned much controversy in subsequent issues of the journal, and we shall take a brief look at some of the opinions (and hackles!) raised.

The first solution proposed, in the issue for 1st February 1882, was by Alexander MacFarlane (1851-1913), and since it gave rise to much comment, we shall present it in full.

> Let p denote the chance of a case treated by Lister's method doing well, and q the chance of a case treated with ordinary dressings doing well, then $p = 7/10$ and $q = 9/14$. But Lister's method consists in the ordinary dressings with the additional use of an antiseptic; hence the effect of the antiseptic is $p - q$, that is $2/15$ [sic]. Hence the odds that in a given case the success of Lister's method is not due to the characteristic part of it is $q/(p - q)$, that is $45/4$. [p.77]

Commenting on this solution, the proposer describes it as "very inadequate." This is followed by the sentiment

> It is a good rule in Probabilities to refrain from introducing any datum of your own into the conditions of the question [loc. cit.],

a suggestion as to which there might well be some debate. But be that as it may: after some further criticism of MacFarlane's solution, MacAlister proceeds to his own. To this end he rephrases the question in terms of balls and urns: suppose that two urns A and B each contain a large number of white and black balls. From A, $a + b$ balls are drawn in succession (a being white), while $p + q$ are similarly drawn from B (p being white). If $a/(a + b)$ is found to be greater than $p/(p + q)$, what are the odds that the proportion of white balls in B is actually less than that in A? (One might well ask whether MacAlister is obeying his own Good Rule in thus styling the problem.) If P denotes the probability that there are fewer white balls in B than in A, then

$$\frac{P}{1 - P} = \frac{\sum_{s=1}^{n-1} \sum_{t=1}^{s-1} s^a (n - s)^b t^p (n - t)^q}{\sum_{s=1}^{n-1} \sum_{t=s}^{n-1} s^a (n - s)^b t^p (n - t)^q}$$

or

$$P = \left[\binom{a+b+1}{a} \binom{p+q+1}{p+1} \Big/ \binom{a+b+p+q+2}{a+p+1} \right] \mathcal{S}$$

where $\mathcal{S}$ denotes the (terminating) series

$$\mathcal{S} = 1 + \frac{a}{b+2} \frac{q}{p+2} + \frac{a}{b+2} \frac{a-1}{b+3} \frac{q}{p+2} \frac{q-1}{p+3} + \&\text{c.}$$

with a similar expression for $1 - P$.

Using this result with $a = 5$, $b = 9$, $p = 3$, $q = 7$ MacAlister finds that $P = 0.59825$, whence $P/(1-P) = 1.49$. (We shall comment on this solution later.)

In the issue of the first of March 1882 of *The Educational Times* practically an entire page is taken up with this problem, the discussion being opened by the Editor who describes MacFarlane's solution as "a brief but somewhat obscure process" [p.103]. MacFarlane's defence is the first to be given: he re-affirms his earlier solution, but this time adduces more reasoning. The Scottish theme is further embellished by Hugh MacColl (1837-1909) who charges MacAlister with a violation of his own principle, inasmuch as

> He seems to me to have 'introduced into the conditions of the question' the datum that, *independently of the experiments*, it is an even chance (or 1/2) whether Lister's method has any advantage over the ordinary methods. [p.103]

Elizabeth Blackwood makes some general comments: these in turn are followed by a lengthy reply from MacAlister, in which he defends himself against MacColl by saying (perhaps rather weakly) that his assumption "is surely not merely a just assertion, but the only one possible" [loc. cit.]. His comment on the immediately preceding remarks is rather unkind, viz. "Miss Elizabeth Blackwood has perhaps not read my solution" [loc. cit.].

A new contestant enters the lists in the issue of the first of April. William Whitworth (1840-1905) proposes to denote by p the probability of success under the old treatment, and by μp the probability of success under Lister's treatment. The *a priori* probability of success is then

$$P = p^9(1-p)^5(\mu p)^7(1-\mu p)^3$$

where $0 < \mu < 1/p$. The probability that Lister's method is of no advantage is then

$$\begin{aligned} \Pr[\mu < 1] &= \int_0^1 \mu^7(1-\mu p)^3 \, d\mu \Big/ \int_0^{1/p} \mu^7(1-\mu p)^3 \, d\mu \\ &= 165p^8 - 440p^9 + 396p^{10} - 120p^{11} \; . \end{aligned}$$

If all we know is given in the statement of the question, we must set $p = 9/14$, and we then find the chance as 0.0078. The odds in favour of Lister's method being advantageous are then about 229 to 2. Whitworth notes further that P is maximal for $\mu = 7/10p$, which is thus the *most likely* value of μ, and he summarizes his results as follows:

> (α) The chance (from the observed cases only) that Lister's method should precisely make no difference, is less than any assignable chance. The question as stated can only mean this, and the only true answer is zero.
> (β) The odds that Lister's method is beneficial, as against the position that it is either useless or injurious, are about 114:1.
> (γ) The odds in favour of the statement that Lister's treatment will succeed in 7 cases out of 10, as against the position that his treatment makes no difference, are about 21:20. [p.127]

MacColl next re-enters the field, pointing out essentially that, in his view, the only correct interpretation of probability is in terms of long-run frequencies, and suggesting that it might be wiser to denote the unknown initial probability by x (say) and obtain the answer in terms of this quantity, rather than to make the (probably false) assumption that the *a priori* chance is 1/2.

Blackwood has the last word with some general remarks, and in a final sally on the first of May she takes exception to Whitworth's solution on account of his assuming that μ has a uniform distribution: "For this assumption I can discern no warrant whatever in the data of the question" [p.153].

In his solution to the problem MacAlister acknowledged the work of Carl Liebermeister (1833-1901), who had "published several tracts in which problems similar to mine are very clearly discussed" [p.78]. The pertinent problem, given in Liebermeister [1877], is reported by Winsor [1948] as follows:

> A sample from population 1 has given a failures and b successes; a sample from population 2 has given p failures and q successes. We have
>
> $$\frac{p}{p+q} < \frac{a}{a+b} \, .$$
>
> Required, the probability that the true proportion of failures in population 2 is less than that in population 1. [p.166]

Letting α and β be the probabilities of failure in the first and second populations respectively, we find that

$$\Pr\left[\text{observed result} \mid \alpha, \beta\right] = \binom{a+b}{a}\binom{p+q}{p}\alpha^a(1-\alpha)^b\beta^p(1-\beta)^q \; .$$

If one supposes that α and β are independently uniformly distributed over [0,1], then by Bayes's Theorem

$$\begin{aligned} P &= \int_0^1\int_0^\alpha \alpha^a(1-\alpha)^b\beta^p(1-\beta)^q\,d\beta\,d\alpha \Big/ \int_0^1\int_0^1 \alpha^a(1-\alpha)^b\beta^p(1-\beta)^q\,d\beta\,d\alpha \\ &= \sum_{j=p+1}^{p+q+1}\binom{a+j}{a}\binom{b+p+q+1-j}{b}\Big/\binom{a+b+p+q+2}{a+b+1} \; . \end{aligned}$$

Winsor [1948] compares the expression for $1-P$ derived from this formula with that given by Fisher's analysis of the 2×2 table, and concludes that "Liebermeister's probability is the same that Fisher would calculate from a table with the frequencies on one diagonal increased by unity" [p.167]. He concludes too that Liebermeister's method yields, for small samples, smaller values of $1-P$ ("and hence apparently stronger indications of significance" [p.168]) than Fisher's.

8.7 Francis Ysidro Edgeworth (1845-1926)

As Stigler [1986a, p.305] has noted, Edgeworth[18] stands out as a curiosity among nineteenth-century statisticians. His formal training in classical literature was followed by a deep personal study of mathematics, the fruit of which is abundantly evident in his various writings on ethics, economics and statistics[19]. Of those in the last group, the earliest falling within the scope of our study is a paper, published in 1883, on the method of least squares.

The following passage (I. A(3) in Edgeworth's paper) will serve as an example of the use made here of inverse probability:

> Given a set of observations x_1, x_2, &c., and given that they have been generated by divergence according to one and the same probability-curve from a single point, but given neither that point nor the modulus, to find both. [p.366]

The actual finding of the mean and the variance need not concern us here (the method is that of maximum likelihood). Germane however is the following discussion. Let us define

$$\begin{aligned} P &\equiv f(x_1, x_2, \ldots, x_n \mid \xi, c) \\ &= \left(\frac{1}{c\sqrt{\pi}}\right)^n \exp\left(-\frac{1}{c^2}\sum_1^n (x_i - \xi)^2\right) \quad . \end{aligned}$$

Then

$$f(\xi, c \mid x_1, \ldots, x_n)\, d\xi\, dc = \frac{P\, f(\xi, c)\, d\xi\, dc}{\int_0^\infty \int_{-\infty}^\infty P\, f(\xi, c)\, d\xi\, dc} \quad .$$

Edgeworth now supposes that the prior distribution is of the form kc^{-2} (where k is constant), an expression obtained by assuming that ξ and c are independent and that $h = 1/c$ has a uniform distribution. With this prior, and on integrating the joint density, Edgeworth finds that

$$f(\xi \mid x_1, \ldots, x_n) \propto \left[1 + n(\overline{x} - \xi)^2 \Big/ \sum (x_i - \overline{x})^2\right]^{-(n+1)/2} \quad .$$

This, as Welch [1958, p.779] has noted, reduces to[20]

$$f(t \mid x_1, \ldots, x_n) \propto \left[1 + t^2/(n-1)\right]^{-(n+1)/2}$$

on one's putting $t = \sqrt{n(n-1)}(\overline{x} - \xi) \Big/ \left[\sum (x_i - \overline{x})^2\right]^{1/2}$.

In 1884 Edgeworth published in the *Philosophical Magazine* a paper entitled "*À priori* probabilities." After having pointed out that

> In the measurement of a physical quantity it is generally assumed that, prior to observation, one value of the quaesitum is as likely as another [p.204],

he illustrates this point by considering a problem he had raised in an earlier paper [Edgeworth 1883], viz. consider a set of observations $\{x_1, x_2, \ldots\}$ "diverging according to a given probability-curve" from a point x. This point x is found by solving

$$\frac{d}{dx}\rho h \frac{1}{\sqrt{\pi}} \exp\left[-h^2 \sum (x - x_i)^2\right] = 0 \ ,$$

"where ρ is the *à priori* probability that the real value of the quaesitum is between x and $x + \Delta x$" [p.204]. He points out the modification required if ρ, rather than being constant, is equal to $\Delta x\, \chi(x)$. Mention is also made of more complicated problems, and the general remark is made that

> In so far as these methods are applications of Inverse Probabilities they involve à priori assumptions [p.206],

this remark encompassing the rule of succession.

Edgeworth points out further that when calculations *a posteriori* of the probability that some phenomenon is not due to chance are made, some assumption as to the *a priori* probability of the existence of chance is needed, and concludes that the theory advocated by Boole and Donkin is, in respect of such *a priori* probabilities, more correct than the practice of Laplace and Herschel. Supposing, then, that *a priori* probabilities are in fact needed, Edgeworth finds this need to be so far satisfied as

> to allow of a mathematical, though not a numerical, inference in cases where the *à posteriori* probability has a limiting value, provided that an involved *à priori* probability is not extreme. [p.207]

The correction introduced in the earlier example may safely be ignored when n is indefinitely large provided that $\chi'(x)/\chi(x)$ is finite. The general argument appears to be that the effect of the prior diminishes with increasing experience.

It is also pointed out that if X has a uniform distribution over (0,1), such will not be the case for X^2, and that when the form of a function is completely unknown one may assume that one which makes for most ease of calculation.

Edgeworth devotes another paper[21] of 1884, published in *Mind* and entitled "The philosophy of chance", chiefly to criticism of Venn's *The Logic of Chance.* He agrees with Venn on the essential similarity between inverse (*quâ* inverse) and direct probability, and suggests further that

> the much decried method of Bayes may be employed to deduce from the frequently experienced occurrence of a phenomenon the large probability of its recurrence [p.228],

a remark which might perhaps be interpreted as support for the rule of succession.

The assignment of equal probability-constants in the case in which nothing is known is founded upon the "rough but solid experience" [p.230] that such constants do, in practice, tend to have one value as often as another. Moreover,

> The ridicule which has been heaped upon Bayes's theorem and the inverse method will be found only applicable to the pretence, here deprecated, of eliciting knowledge out of ignorance, something out of nothing. The most formidable objection is that

> which was made by Boole, and is repeated by Mr. Venn, Mr. Peirce, and others with approbation. Our procedure in treating one value as *à priori* not less likely than another is, it is said, of a quite arbitrary character, and apt to lead to different conclusions from the plausible one which we have reached by accident. [p.230]

A parody of Boole's argument is given, the conclusion being that an appeal to experience is of prime importance: an appeal Pearson [1920a] notes "from which Bayes and Laplace ought to have started" [p.4], though it might in fact be argued that both Bayes and Laplace did in fact base their arguments upon some sort of prior experiment.

Following a discussion of some examples, Edgeworth states

> The preceding examples ... may show that the assumptions connected with 'Inverse Probability', far from being arbitrary, constitute a very good working hypothesis. They suggest that the particular species of inverse probability called the 'rule of succession' may not be so inane as Mr. Venn would have us believe. [p.234]

A useful summary of the state of the art was provided by Edgeworth in his article "Probability" in the eleventh edition of the *Encyclopædia Britannica* in 1911. Here the marriage of what Bowley [1928, p.6] terms Edgeworth's "metaphysical conception of probability" and his statistical investigations is clearly visible: indeed, Edgeworth is a realization of Dickens's Nicholas Tulrumble, who "contracted a relish for statistics, and got philosophical".

Arguing against Laplace's definition of probability, Edgeworth urges here that "merely psychological facts can at best afford a measure of belief, not of credibility" [¶2, p.377], but he nevertheless finds the frequency view "not so diametrically opposed as may at first appear" [¶3, p.377]. Again the question of the invariance of the prior distribution is raised, and it is suggested that, when values are constrained to lie in a small interval, any (reasonable) function "of a quantity which assumes equivalent values with equal probability" [¶8, p.377] will have approximately the same probability distribution as any similar function. Indeed,

> It may further be replied that in general the reasoning does not require the a priori probabilities of the different values to be very nearly equal; it suffices that they should not be very unequal; and this much seems to be given by experience. [¶8, p.377]

Passing, in Section II ("Calculation of Probability"), to the probability of causes deduced from observed events, Edgeworth points out firstly that the principal difference between problems to which these methods are applicable and others

> consists in the need of evidence, other than that which is afforded by the observed event, as to the probability of the alternative causes existing and operating. [¶44, p.382]

Three examples follow, the first being concerned with digits drawn at random from mathematical tables, the second being taken from Laplace's *Théorie analytique des probabilités* (Book II, chap.1, N<u>o</u>1), and the third coming from Bertrand's *Calcul des Probabilités* (art.134).

Paragraph 48 sees the start of Edgeworth's discussion of the probability of testimony, two basic assumptions in which are the following:

> (1) that to each witness there pertains a coefficient of probability representing the average frequency with which he speaks the truth or untruth, (2) that the statements of witnesses are *independent* in the sense proper to probabilities. [¶48, p.383]

(These assumptions Edgeworth finds open to serious criticism.) It is shown that for r witnesses of credibilities (or average truthfulness) $p_1, p_2, \ldots, p_r$, the probability that a statement is true is[22]

$$\prod_1^r p_i \Bigg/ \left[\prod_1^r p_i + \prod_1^r (1-p_i)\right] \quad .$$

Division of both numerator and denominator by $\prod_1^r p_i$ shows that this probability increases with r, provided that each $p_i > 1/2$.

The rule of succession comes under scrutiny in Paragraph 54, illustration being provided by the drawing of one further white ball from a mixture of an immense number of white and black balls, when it is known that n draws have all yielded white. Under the assumption of a uniform prior, Edgeworth obtains in the limit

$$\int_0^1 p^{n+1}\,dp \Bigg/ \int_0^1 p^n\,dp = (n+1)/(n+2) \quad .$$

In Part II of his article, "Averages and Laws of Error", Edgeworth turns in the first section to the law of error. He makes the perhaps somewhat unusual observation that

> there is a characteristic more essential to the statistician than the existence of an objective *quaesitum*, namely, the use of that method which is primarily, but not exclusively, proper to that sort of *quaesitum* — *inverse probability.* [¶123, p.395]

Inverse probability has a two-fold use here: (a) to determine the best values of the coefficients appearing in the law of error, and (b) to test the worth of the results obtained by using any values of these coefficients. As an example[23] of the procedure Edgeworth considers the case of n observations $x_1, x_2, \ldots, x_n$ from a Normal distribution with given modulus c. The probability P that the observations should have resulted from measurement of an object whose real position was between x and $x + \Delta x$ is then

$$P = \Delta x \, J \exp\Big\{ - \left[(x - x_1)^2 + \cdots + (x - x_n)^2 \right] / c^2 \Big\} \quad ,$$

where J is a constant of proportionality. The *most probable* value of x is found, by maximization of P, to be $\overline{x}$ (the arithmetic mean[24] of the n observations), a statistic with modulus $c/\sqrt{n}$. It is further pointed out that the same reasoning is applicable to the case in which data and quaesitum are proportions rather than absolute quantities,

> for instance, given the percentage of white balls in several large batches drawn at random from an immense urn containing black and white balls, to find the percentage of white balls in the urn — the inverse problem associated with the name of Bayes. [¶130, p.397]

Laplace, Edgeworth notes, did not adopt this approach. He saw the quaesitum not as the *most probable* value, but rather as "that point which may *most advantageously* be put for the real one" [¶131, p.397]. This necessitated calculation of "la valeur moyenne de l'erreur à craindre", that is, "the mean first power of the errors taken positively on each side of the real point." Gauss, on the other hand, took as the appropriate criterion the mean square of errors; and Edgeworth notes further that

> *Any* mean power indeed, the integral of any function which increases in absolute magnitude with the increase of its variable, taken as the measure of the detriment, will lead to the same conclusion, if the normal law prevails. [¶131, p.397]

Attention is also drawn to the modifications necessary if (i) different values of x are not equally probable prior to observation, (ii) the x_i come from distributions with different moduli, (iii) the modulus is also unknown, and (iv) the observations come from a bivariate Normal distribution with unknown correlation coefficient.

Some forty years after the paper in which he had considered Venn's book, Edgeworth returned to the topic in a review, also entitled "The philosophy of chance", of Keynes's *A Treatise on Probability*, a review published in *Mind* in 1922. In Keynes's dialectic Edgeworth finds support for his earlier contention that Venn had gone too far in his scepticism as regards *a priori* probabilities based on the principle of sufficient reason (or indifference).

Following on the recollection of some examples from Venn and some economic applications, Edgeworth notes that

> It may be observed that in general, for instance in all the applications which have just been noticed, the use of *à priori* probabilities has no connexion with inverse probability. That conjunction does occur in one very important branch of Probabilities — that which deals with errors-of-observation. [p.262]

This assertion is illustrated by an example involving several observations of the measure of some magnitude whose ascertainment is required, a special case being the estimation of a ratio (e.g. of black balls drawn from an urn) rather than an absolute magnitude. One requires that combination of the observations which yields the best value of the quaesitum.

Further reference is made to Keynes's views on *a priori* probabilities and the rule of succession, and on the latter Edgeworth remarks

> when the relevant *à priori* probabilities ... are overruled by the number of the observations, as may be shown by the reasoning above cited, the rule of succession is by no means so absurd. [p.265]

Moreover

> *à priori* probability is generally negligible in comparison with the evidence of repeated observations [p.266]

which strengthens a remark made earlier.

8.8 Morgan William Crofton (1826-1915)

In the ninth edition of the *Encyclopædia Britannica* in 1885 Crofton devotes twenty-one pages to the subject of probability[25], a concept which he understands in the following sense:

> The probability, or amount of conviction accorded to any fact or statement, is thus essentially subjective, and varies with the degree of knowledge of the mind to which the fact is presented (it

> is often indeed also influenced by passion and prejudice, which act powerfully in warping the judgement), — so that, as Laplace observes, it is affected partly by our ignorance partly by our knowledge. [p.768]

The determination of such probability, however, is to be accomplished *via* frequencies, viz.

> In fact we may say, considering how seldom we know *a priori* the probability of any event, that the knowledge we have of such probability in any case is entirely derived from this principle, viz., that the proportion which holds in a large number of trials will be found to hold in the total number, even when this may be infinite, — the deviation or error being less and less as the trials are multiplied. [p.769]

The second section, after a general introduction, of this article is entitled "Probability of future events deduced from experience". Here Crofton illustrates, in several simple examples, the following general principle: suppose that P_i denotes the antecedent probability of the ith cause[26] C_i, that p_i denotes the probability of an event E given C_i, and that a large number N of trials have been made. Then

> out of these the number in which the first cause exists is P_1N, and out of this number the cases in which the event follows are p_1P_1N. [p.773]

Continuing in this (frequentist) way Crofton finds the *a posteriori* probability π_i of C_i to be

$$\pi_i = p_iP_i \,/\textstyle\sum p_iP_i \quad .$$

It then follows that the probability of a further occurrence of the event is $\sum p_i\pi_i$.

A further illustration is concerned with sampling from an urn containing n white or black balls. If r drawings have resulted in white balls, the probability that the $(r+1)$th draw will also yield a white ball is given by

$$\frac{1}{n}\sum_{k=1}^{n} r^{k+1} \Big/ \sum_{k=1}^{n} r^k$$

if sampling is carried out with replacement, and

$$(r+1)/(r+2)$$

(independent of n) in the case of sampling without replacement.

Crofton next focuses attention on

> the important theorem of Bayes ... the object of which is to deduce from the experience of a given number of trials, as to an event which must happen or fail on each trial, the information thus afforded as to the real facility of the event in any one trial, which facility is identical with the proportion of successes out of an infinite number of trials, were it possible to make them. [p.774]

(Note the emphasis on a limiting frequency approach.) This problem[27] Crofton phrases in terms of a "balls and urn" example, and he derives

$$p = \int_\alpha^\beta x^m(1-x)^n\,dx \Big/ \int_0^1 x^m(1-x)^n\,dx \tag{6}$$

as "the probability that the ratio of the white balls in the urn to the whole number lies between any two given limits α, β" [p.774].

The usual extension to $p+q$ further draws following $m+n$ draws results in the probability

$$\binom{p+q}{p}\int_0^1 x^{m+p}(1-x)^{n+q}\,dx \Big/ \int_0^1 x^m(1-x)^n\,dx$$
$$= \binom{m+p}{p}\binom{n+q}{q} \Big/ \binom{m+n+p+q+1}{p+q} \ .$$

Lidstone [1920, §18] extends this result by an analogous argument to the case of an urn containing balls of i different colours. However he in fact questions

> whether Crofton's process ... can be considered strictly demonstrative; but it is at least simple, elegant, suggestive, and easily carried in the mind. [p.191]

Crofton deduces further that, when $(m+n)$ is large, (6) reduces to

$$\Pr\left[|p - m/(m+n)| < \delta\right] \sim \frac{2}{\sqrt{\pi}}\int_0^\lambda e^{-t^2}\,dt \ ,$$

where $\lambda = \delta(m+n)^{3/2}\big/\sqrt{2mn}$.

Some six and a third columns are devoted in this article to the probability of testimony. Examples involving one or more witnesses, of various credibilities, and one or more events, of various probabilities, are considered, the extension to "the verdicts of juries, the decisions of courts, and the results of elections" [p.779] being also mentioned. We shall not pursue the matter here.

8.9 Johannes von Kries (1853-1928)

In 1886 von Kries published his thought-provoking book *Die Principien der Wahrscheinlichkeits-Rechnung. Eine logische Untersuchung*, in which detailed discussion of the subjective theory may be found. We shall restrict our attention here, inasmuch as it is possible, to directly relevant matters.

Von Kries provides a precise definition of equiprobable events, viz. †

> als gleich möglich zwei oder mehrere Fälle anzusehen sind, wenn in dem jeweiligen Stande unserer Kenntnisse sich kein Grund findet, unter ihnen einen für wahrscheinlicher als irgend einen anderen zu halten. [p.6]

Here we clearly see the subjective basis on which von Kries's work rests[28], though that he himself viewed it in a logical sense is evinced by the following passage:

> Diese Deutung — wir wollen sie kurz als die logische Deutung bezeichnen, und das Princip, auf welches sie die Wahrscheinlichkeits-Rechnung basirt, als Princip des mangelnden Grundes — scheint auf den ersten Blick völlig zu befriedigen. [p.6]

In the second chapter, Article 3, we find a simple example of the rule of succession: one of two playing-cards lying face-down on a table is turned over and found to be black. The probability that the second card is black is, according to Poisson, 2/3, although one might ingenuously expect it to be 1/2. In a footnote von Kries mentions that the method illustrated here is based on the so-called Bayes's principle.

In Article 1 of the fourth chapter we find a further statement in favour of a subjective interpretation of probability, viz.

> jede Wahrscheinlichkeit ist subjectiv, der Ausdruck und die Folge unseres ungenauen oder unvollständigen Wissens; eine „objective Wahrscheinlichkeit" dagegen ist ein Unding, eine contradictio in adjecto. [p.77]

The next pertinent comment appears in Chapter V, "Die Arten der numerischen Wahrscheinlichkeit." Here von Kries considers the case of six dice, with die i having i faces marked with a "+" and $6 - i$ faces marked with a "0". If a die is drawn at random and tossed three times yielding the sequence +, 0, +, what is the probability that the die chosen was the first, second, &c.? The problem is solved in the usual way using

†References throughout this section are to the second edition of 1927.

> das sogenannte Bayes'sche Princip, welches mit Recht als einer der wichtigsten Sätze der Wahrscheinlichkeits-Theorie angesehen wird. [p.118]

The more general expression $p_1\alpha_1/\sum p\alpha$ is also given, and is described as the rule for deducing, as a consequence of Bayes's Theorem, "der ‚Ursache beobachteter Ereignisse' ". It is also stated that the use of Bayes's Principle is not uncontroversial. The continuous analogue of the discrete expression given above, viz.

$$\int_a^b \varphi(x-x_0)\,dx \bigg/ \int_{-\infty}^{\infty} \varphi(x-x_0)\,dx \ ,$$

is found in Article 5.

In Chapter VI, "Die Gewinnung und Begründung von Wahrscheinlichkeits-Sätzen", we find the assertion that, if a large number $(n+m)$ of draws from an urn have yielded n black and m white balls, the probability that the next draw will yield a black ball is approximately $n/(n+m)$.

> Man bezeichnet diese Verfahrungsweise als eine a posteriorische Wahrscheinlichkeits-Bestimmung. [p.133]

A work of this nature would of course be incomplete without mention of earlier work on the probability of testimony, and von Kries accordingly turns his attention to this problem, one which among all applications of probability is "vielleicht die merkwürdigste" [p.253], in the twelfth section of his ninth chapter "Weitere Anwendungen der Wahrscheinlichkeits-Rechnung." He argues that Laplace and Poisson reached their results from erroneous assumptions, an argument whose details we shall not give here, apart from mentioning that von Kries finds the absence of independence a major drawback.

The tenth and final chapter is devoted to the history of probability theory. Von Kries comments on the similarity between the "etwas schwerfällige" [p.267] definition of probability given by Bayes and that given a century earlier by Huygens in his *De Ratiociniis in Ludo Aleae*, the first proposition in which work reads

> Si a vel b expectem, quorum utrumvis æquè facilè mihi obtingere possit, expectatio mea dicenda est valere $(a+b)/2$. [Bernoulli 1713, p.4]

Further reference to Bayes's essay occurs later in this chapter, where von Kries brings in objective probability with the words

> Nach der von Bayes aufgestellten Regle pflegte man anzunehmen, dass ehe eine Erfahrung vorliegt, jeder Wert einer (objectiven) Wahrscheinlichkeit gleich wahrscheinlich ist. [p.277]

This rule becomes a method which may be applied, without any further thought, as soon as we have equally probable cases. Thus if n trials result in m known outcomes, one can give a determinate ("bestimmte") probability that the probability of the outcome in question lies between $(m/n) - \delta$ and $(m/n) + \delta$, a probability which may well be large even for moderately small values of δ; and moreover one may, with some degree of accuracy, ascertain any probability (this latter phrase, "jene Wahrscheinlichkeit bestimmen", is given within quotation marks in the original). And this rule can be used to determine the probability that

> bei einer Anzahl neuer Fälle wieder die relative Häufigkeit des betreffenden Verlaufes in irgend welchen Grenzen liegen werde. [p.278]

8.10 George Francis Hardy (1855-1914)

In 1889 some remarks by Hardy were published in volume 227 of the *Insurance Record*. The substance of these comments was republished in an editorial note to Whittaker [1920], and it is to this note that reference is made in the present discussion.

The correspondence in the *Insurance Record* arose in connexion with the following problem from Ackland and Hardy's *Graduated Exercises and Examples*:

> If the experience of a given mortality table indicates that, out of 2000 persons alive at age 30, 29 die before attaining age 31, is it theoretically correct to say that the probability of a person age 30 dying before 31 $= 29/2000$? [p.174]

The answer given runs as follows:

> If out of $(m+n)$ trials the result A has happened m times and the result B n times, then the probability that the next trial will produce the result A is strictly $\frac{(m+1)}{(m+1)+(n+1)} = \frac{(m+1)}{m+n+2}$, or in the present case $30/2002$ (*De Morgan on Probabilities*, chap.iii. p.65). This result is, however, based upon the assumption that all values of the required probability are *a priori* equally likely, which cannot be said to be true with regard to the probabilities of death. [p.175]

Commenting on this, a reviewer finds the solution[29] $(m+1)/(m+n+2)$ preferable to Bernoulli's $m/(m+n)$, but he finds the equally probable assumption a very obvious requirement in a mortality situation like this. To this latter comment Hardy takes exception, stating that

> As regards the probabilities of dying in a year, however, we know that the assumption is entirely incorrect at nearly all ages, and I fail to see how in a practical problem such as the constructing of a mortality table our results are to be improved by introducing an assumption known to be erroneous. [p.176]

As an illustration Hardy supposes that, of 1000 lives exposed to risk at age 70, 900 survive to age 71 and 800 to age 72. Using Laplace's formula one finds that

$$p_{70} = 901/1002 \ , \quad p_{71} = 801/902 \ ,$$

and hence ${}_2p_{70} = (901 \times 801)/(1002 \times 902)$, which is not equal to the ratio 801/1002 obtained from the original data,

> a different result, but one which has just the same claims to acceptance as the former, as there is no special sacredness in the year as a measure of time. [p.176]

Hardy concludes further that, while the usual formula (viz. $m/(m+n)$) may be open to a theoretical objection, no better formula can, in his opinion, be found.

The reviewer replied promptly to Hardy's letter, saying that, in his opinion, Hardy's comments showed that the deduction of $(m+1)/(m+n+2)$ from the formula $P_r p_r / \sum P_r p_r$ was wrong. He therefore rehearsed the usual deduction of the first of these formulae (in a mortality context), noting on the way that the most probable value, viz. $x = m/(m+n)$, of the facility of the event (or the probability of surviving a year), was given by maximizing $x^m(1-x)^n$.

Attention was then focused on Hardy's more general problem, which the reviewer framed as follows: $(m+n+p)$ lives, each aged k, are to be observed for 2 years: of these, $(m+n)$ survive the first year and m the second. From these observations it is required to find, for another person aged k years, the probabilities [p.178]

(i) that he will die in the first year;

(ii) that he will die in the second year;

(iii) that he will survive the first year;

(iv) that he will survive the second year.

Denoting by π_x and π_y the facilities of surviving a year at ages k and $(k+1)$ respectively, the reviewer shows that

$$\Pr\left[x < \pi_x < x + dx \;\&\; y < \pi_y < y + dy\right]$$

$$= \frac{x^{m+n}(1-x)^p y^m (1-y)^n \, dx\, dy}{\int_0^1 \int_0^1 x^{m+n}(1-x)^p y^m (1-y)^n \, dx\, dy}$$

$$= \frac{(m+n+p+1)!}{(m+n)!\,p!}\;\frac{(m+n+1)!}{m!\,n!} x^{m+n}(1-x)^p y^m (1-y)^n \, dx\, dy \ .$$

Multiplication of this result by $(1-x)$, $x(1-y)$, x and xy in turn and integration then yield the desired probabilities, viz.

(i) $(p+1)/(m+n+p+2)$;

(ii) $(m+n+1)(n+1)/(m+n+p+2)(m+n+2)$;

(iii) $(m+n+1)/(m+n+p+2)$;

(iv) $(m+n+1)(m+1)/(m+n+p+2)(m+n+2)$.

Hardy was unimpressed: he remarked firstly that the discrepant results he himself had given were based upon incompatible assumptions

> in the first case all values of p_k and p_{k+1} and in the second case all values of ${}_2p_k$ being assumed equally likely *a priori*. [p.180]

Further, there was no reason for regarding one of these assumptions as better than the other. He stressed further the importance of a suitable choice of the P_i, and suggested that they be chosen to[30]

> form a series which may be fairly represented by a curve of the form $x^r(1-x)^s$, where the relative values of r and s will depend on the most probable value of p_k (or x), and their absolute values on the extent of our prior knowledge of that function. [p.181]

This results in

$$\int_0^1 x^{m+r+1}(1-x)^{n+s}\,dx \bigg/ \int_0^1 x^{m+r}(1-x)^{n+s}\,dx$$

$$= (m+r+1)/(m+n+r+s+2) \ .$$

If one excludes from consideration the observations at age k, m and n become zero and one is left with the *a priori* best estimate of p_k as $p_k' = (r+1)/(r+s+2)$.

Although this concluded the discussion in the *Insurance Record*, the question was reopened by Whittaker in 1920, the sixth section of his paper being entitled *Hardy's paradox.* Whittaker finds that Hardy's paradox arises from "a misapplication of the Bayes-Laplace theory" [p.171], the correct application of which, in his opinion, runs as follows. Let H denote the hypothesis that the probability of a man aged 70 dying in his 71st year lies between x and $x + dx$, and that the probability of a man aged 70 dying in his 72nd year lies between y and $y + dy$. Then

$$\Pr[H \mid \text{Hardy's data}] = \frac{x^{100}y^{100}(1-x-y)^{800}\,dx\,dy}{\int\int x^{100}y^{100}(1-x-y)^{800}\,dx\,dy} ,$$

the integral being taken over the set $\{(x,y):\ x \geq 0,\ y \geq 0,\ x+y \leq 1\}$. The probability that "in a subsequent experience" [p.172] a 70 year old man will die in his 71st year is found, in the usual manner, to be 101/1003, while the probability (also in a subsequent experience) that a man aged 70 will die in his 72nd year is 101/1003. Thus

$$p_{70} = 902/1003\ ,\quad {}_2p_{70} = 1 - 202/1003 = 801/1003\ .$$

And since, in the usual manner, $p_{71} = 801/902$, it is clear that

$${}_2p_{70} = p_{70}p_{71}$$

as required. Whittaker emphasizes that the same prior knowledge is to be used in determining these three quantities, and that this is the cause of Hardy's paradox.

The editor in fact has noticed that the solution presented by Whittaker differs from that of the reviewer:

> The difference is due to their formulating in two different ways (both legitimate) the state of complete *a priori* ignorance. [p.179]

Discussion of a more general problem than that considered by Whittaker follows the latter's paper.

In his comments on Whittaker's paper [1920] Lidstone states that he finds Hardy's suggested prior "a highly valuable one" [p.196]. Moreover, he regards Hardy's solution as correct since the Bayes-Laplace theory is inapplicable "because there is no fixed basis on which it can be applied" [loc. cit.]. Stress is further laid on the fact that

> The change in the unit of time radically changes the resulting probability; yet mathematically one unit of time is as good as another. [p.198]

Thus an indeterminate constant (i.e. the time unit) is involved in the formula, and hence the formula itself yields an indeterminate value. This Lidstone believes to be

> essentially the argument of Bing and Hardy, and I must confess I do not think that Professor Whittaker has made any serious attempt to meet it. [p.198]

Furthermore,

> The conclusion I reach is ... that the formula is inapplicable where the "event" is capable of division in point of time or any other measurement. [p.198]

In his reply to the discussion Whittaker suggested essentially that common sense would dictate the choice of time unit, a suggestion which Lidstone was loth to accept in view of its appeal to experience and its contradiction of the assumption of *a priori* ignorance. The same sentiment was echoed in a letter[31] by Nicholl.

8.11 Joseph Louis François Bertrand (1822-1900)

In 1889 Bertrand published[32] his *Calcul des Probabilités*: the third edition of 1972, to which reference will be made here, is a textually unaltered reprint[33] of the second edition of 1907. In this edition also appears Bertrand's "Les lois du hasard" of 1884, a general essay on chance of no little interest.

Turning to the book itself, we find in Chapter I, entitled "Énumération des chances", the following definition:

> La probabilité d'un événement est le rapport du nombre des cas favorables au nombre total des cas possibles. Une condition est sous-entendue: tous les cas doivent être également possibles. [p.2]

Many illustrative examples, on which we shall not spend time, follow (indeed the whole work is amply illustrated).

Let us pass on immediately to Chapter VII, "Probabilité des causes". The term "causes" is defined at the outset thus:

> Les causes sont pour nous des accidents qui ont accompagné ou précédé un événement observé. Let mot n'implique pas qu'au sens philosophique l'événement soit un effet produit par la cause. [pp.142-143]

In §115 we find the following statement of the general problem (there is no mention of Bayes):

> Diverses causes $E_1, E_2, \ldots, E_n$ ont pu produire un événement observé. Les probabilités de ces causes, lorsque le résultat n'était pas encore connu, étaient $\omega_1, \omega_2, \ldots, \omega_n$. L'événement se produit; la cause E_i, lorsqu'on est certain que c'est elle qui agit, donne à l'événement la probabilité p_i. Quelle est la probabilité de chacune des causes qui sont, on l'admet, les seules possibles? [p.144]

Bertrand shows that the solution is given by

$$\Pr\left[E_i \mid \text{event}\right] = p_i\omega_i \Bigg/ \sum_1^n p_i\omega_i \quad .$$

Several applications follow. The first of these is concerned with the composition of an urn of μ balls (white or black in unknown proportion). If k draws, made with replacement, have all resulted in white balls, Bertrand finds the probability that the urn contains only white balls to be

$$\mu^k \Bigg/ \sum_{r=1}^{\mu} r^k$$

under the assumption that all compositions of the urn are *a priori* equally possible. This latter assumption is then relaxed in a particular numerical example, a further example illustrating the case of sampling without replacement.

The second problem is concerned with the same situation, except that now the μ draws result in m white and n black balls. In this case, and again under an equally-likely assumption, Bertrand finds that the most probable composition of the urn is that which makes the probabilities of the drawing of white or black balls proportional to the number of times they appear.

Denoting, in the above problem, the probability of drawing a white ball by x, Bertrand notes that "Chaque hypothèse sur la valeur de x a une probabilité" [p.149]. Starting with the function $x^m(1-x)^n$ he deduces, in the usual sort of way, that the limiting probability is proportional to

$$\exp\left(-\epsilon^2(m+n)/2pq\right)$$

where $p = m/(m+n) = 1-q$ and $x = p - \epsilon$. He also notes the difference between this result and the similar expression obtained in his study of Bernoulli's Theorem, this difference being given by what is known (i.e. p or n) in the two cases.

Bertrand observes immediately that

> La formule précédente est déduite d'une hypothèse qui se réalisera rarement. Toutes les probabilités désignées par x ont, en général, *a priori*, des valeurs inégales [p.151],

and follows this up with the following problem: from an urn containing N balls μ drawings have resulted in m white and n black balls, where initially the probability of drawing either of these colours is 1/2. What is the most probable composition of the urn? Under the assumption that N is large it is shown that the solution is given by $(N+2m)/2(N+m+n)$. Numerical variations on this theoretical theme follow[34].

Attention is next turned to the regularity in the ratio of male to female births, reference being made to work by Nicolas Bernoulli, Buffon and Laplace, and also to some miscellaneous problems.

In Article 136 Bertrand turns his attention to the probability of future events. As an example he considers the drawing of balls from an urn under the assumption[35] that "Toutes les suppositions sont également possibles" [p.172]. If μ draws have resulted in m white and n black balls then 'tis found, in the usual way, that the probability that the $(\mu+1)$th draw will yield a white ball is

$$(m+1)/(m+n+2) \ .$$

Turning to applications of this rule, Bertrand writes

> Les applications faites de cette formule ont été presque toutes sans fondement [p.173],

a sentiment which he illustrates by the example of the sun's rising tomorrow, given that it has risen daily for 6,000 years. Assimilating this to the repeated drawing of white balls from an urn, he finds that the probability of one further white given 2191500 white is 0.999999543: "Est-il besoin d'insister sur l'insignificance d'un tel calcul?" [p.174]. In the introduction we in fact find the further comment on the equating of these two cases:

> L'assimilation n'est pas permise: l'une des probabilités est objective, l'autre subjective. [p.xix]

Chapter XIII, "Probabilités des décisions", contains only one article, entitled "Résumé critique des tentatives faites pour appliquer le Calcul des probabilités aux décisions judiciaires." The description is accurate: Condorcet, Laplace, Poisson and Cournot all come under the spotlight. Thus, writing of Condorcet's *Essai* Bertrand says

> Aucun de ses principes n'est acceptable, aucune de ses conclusions n'approche de la vérité. [p.319]

Successors to Condorcet, while recognizing the insufficiency of his formulae, were not able to provide anything better: indeed

> Laplace a rejeté les résultats de Condorcet, Poisson n'a pas accepté ceux de Laplace; ni l'un ni l'autre n'a pu soumettre au calcul ce qui y échappe essentiellement: les chances d'erreur d'un esprit plus ou moins éclairé, devant des faits mal connus et des droits imparfaitement définis. [pp.319-320]

Further,

> Ni Cournot ni Poisson n'ont commis la plus petite faute comme géomètres; ils traduisent rigoureusement leurs hypothèses. Mais les hypothèses n'ont pas le moindre rapport avec la situation d'un accusé devant les juges. [p.326]

The criticism is just and reasonable, and the conclusion may perhaps be drawn that such matters are perhaps not completely suited to probabilistic examination.

8.12 George Chrystal (1851-1911)

The substance of an address delivered by Chrystal before the Actuarial Society of Edinburgh was published in 1891 in the *Transactions* of that body under the title "On some fundamental principles in the theory of probability". Following on the pioneering work of Venn, Chrystal proposed in this paper merely

> to state a little more clearly, from the mathematical point of view, the *reductio ad absurdum* of the rules of Inverse Probability. [p.421]

On "the view of Probability which has been gaining ground of recent years" [p.422], by which is no doubt meant the frequency theory espoused by Venn, Chrystal finds Laplace erring (if not sinning) in basing probability

> ultimately on a mere condition of the human mind, instead of resting it ultimately upon human experience of the objective world [p.422],

a position adopted by some of his (i.e. Laplace's) followers, especially de Morgan. While Boole seemed to be attempting to break this stranglehold[36], the grip of the past was perhaps too strong, and it was left to Venn, with his concept of the probability of a series, to fill the lacuna in Laplace's theory.

Chrystal defines the probability (or chance) of an event as follows:

> If, on taking any very large number N out of a series of cases in which an event A is in question, A happens on pN occasions, the probability of the event A is said to be p. [p.426]

He stresses that "probability is not an attribute of any particular event happening on any particular occasion" [p.426], and adds to this *caveat* the corollary that

> no information of any value regarding the probability of an event can be gathered from one or from a small number of observations. [p.426]

The sixth and seventh principles of Laplace's *Essai*, those concerned with inverse probability, are next recalled, and the following example, also due to Laplace, considered: two drawings are made, with replacement, from an urn containing two balls, each of which may be either black or white. If these two draws both yield white, what is the probability that the next ball to be drawn will also be white? Chrystal draws attention to Laplace's assumption that the two possible hypotheses as to the composition of the urn are equally likely, pointing out that this is not necessarily the case. A reading of Laplace's solution in the *Essai* shows that the desideratum $\Pr[W_3 \mid W_1, W_2]$ is obtained from

$$\Pr[W_3 \mid W_1, W_2] = \sum_{i=1}^{2} \Pr[H_i \mid W_1, W_2] \Pr[W_3 \mid H_i] \quad ,$$

a result which obtains under the assumption that, for each $i \in \{1, 2\}$,

$$\Pr[W_1 W_2 W_3 \mid H_i] = \Pr[W_1 W_2 \mid H_i] \Pr[W_3 \mid H_i] \quad .$$

The answer is $9/10$.

One of what Chrystal terms "the grand results of this method" [p.428] is the rule of succession. Reference is made to Laplace's use of this result in connexion with the sun's rising (comparison with Buffon's treatment being drawn) and to a simple example from Crofton [1885, p.774], which receives scornful treatment at Chrystal's hands. The same hands now turn to the manipulation of several problems: we shall consider them *seriatim*.

Problem I. Given a bag containing three balls, each of which may be black or white, to find the probability of drawing a black ball. [p.429]

Chrystal notes that the problem, as stated, is quite indeterminate[37], and stresses the need for the definition of an appropriate "*series*" for its solution. Two hypotheses are suggested, viz.

(A) all numbers of white balls will occur equally often in the long-run;

(B) each ball will be black or white equally often in the long-run.

Under these assumptions the four possible constitutions of the bag

$$\{(W,B)\} = \{(0,3),(1,2),(2,1),(3,0)\}$$

will occur in a large number N of trials with frequencies

(A) $\frac{1}{4}N$, $\frac{1}{4}N$, $\frac{1}{4}N$, $\frac{1}{4}N$, and

(B) $\frac{1}{8}N$, $\frac{3}{8}N$, $\frac{3}{8}N$, $\frac{1}{8}N$

respectively. Under either hypothesis the desired probability is $1/2$.

Problem II. Given a bag which contains one white ball and two others, each of which may be either white or black, what is the probability of drawing a white ball? [p.430]

In this case the possible constitutions

$$\{(W,B)\} = \{(1,2),(2,1),(3,0)\}$$

of the urn are considered subject to the hypotheses

(A) of the unknown balls 0, 1 or 2 white are equally likely, and

(B) each ball in the bag is equally likely to be black or white.

In a large number N of trials the possible constitutions will then occur with frequencies

(A) $\frac{1}{3}N$, $\frac{1}{3}N$, $\frac{1}{3}N$, and

(B) $\frac{3}{7}N$, $\frac{3}{7}N$, $\frac{1}{7}N$,

the required probability being $2/3$ or $4/7$ respectively.

Problem III. Given a bag which contains three balls. A ball is drawn, found to be white, and returned to the bag: calculate the probability of drawing a white ball on another trial. [p.431]

This is merely Problem II in an alternative form.

Problem IV. A white ball having been drawn from a bag containing three, required the probabilities that the bag from which it was drawn contained —

$$\{(W,B)\} = \{(3,0),(2,1),(1,2)\} \equiv \{1^{\circ},2^{\circ},3^{\circ}\}$$

respectively. [p.431] (notation altered)

Here some assumption as to the series is again required; but Chrystal first finds it necessary (at this stage!) to explain the meaning of the word "probability":

> Let a large number M of bags, each of which is filled with one white ball and two others, the occurrence of which is regulated in some given or supposed way, say on Hypothesis (A) or Hypothesis (B) as above, required the numbers pM, qM, rM of these cases in which *when a white ball was drawn* it came from bags having the constitutions $1^o, 2^o, 3^o$, respectively. [pp.431-432]

He also emphasizes that *conditional* probabilities are required. Under the respective assumptions of initial frequencies

(A) $\frac{1}{3}N$, $\frac{1}{3}N$, $\frac{1}{3}N$ and

(B) $\frac{1}{7}N$, $\frac{3}{7}N$, $\frac{3}{7}N$,

Chrystal finds that $(p, q, r) = (3/6, 2/6, 1/6)$ and $(1/4, 2/4, 1/4)$.

Problem V. From a bag containing three balls, each of which is white or black, two are drawn in succession, the first being replaced, to calculate the probability that whenever the first is white the second is white also. [p.432]

Following on from Problem IV the solutions 7/9 and 2/3 emerge under (A) and (B).

With these results as background material Chrystal turns his attention to Crofton's demonstration of Laplace's principles of inverse probability with particular reference to the following question:

> suppose an urn to contain three balls which are white or black; one is drawn and found to be white. It is replaced in the urn and a fresh drawing made; find the chance that the ball drawn is white. [p.434]

Crofton's solution of $\pi_1 = 7/9$ is stated by Chrystal to be the solution of (one case of) Problem V, rather than of the problem initially posed, inasmuch as Crofton

> *deceives himself into believing that he has solved his problem by the merely arbitrary statement*, that the probability π_1 is the *a posteriori* (or modified probability) of the cause C_1. It is, in reality, merely the probability that, when the event has happened, it happened from the cause C_1, which is a totally different thing. [p.434]

While one must agree that the probability found is in fact a conditional one, it might well be queried whether Crofton thought he had found anything else.

As a variation of the three-ball problem, and to illustrate the absurdity of the rules of inverse probability, Chrystal considers the following example:

> A bag contains three balls, each of which is either white or black, all possible numbers of white being equally likely. Two at once are drawn at random and prove to be white: what is the chance that all the balls are white? [p.435]

Chrystal's "common sense" solution runs as follows:

> Any one who knows the definition of mathematical probability, and who considers this question apart from the Inverse Rule, will not hesitate for a moment to say that the chance is 1/2; that is to say, that the third ball is just as likely to be white as black. For there are four possible constitutions of the bag:-
>
> | | 1° | 2° | 3° | 4° |
> |---|---|---|---|---|
> | W | 3 | 2 | 1 | 0 |
> | B | 0 | 1 | 2 | 3 |
>
> each of which, we are told, occurs equally often in the long-run, and among those cases there are two (1° and 2°) in which there are two white balls, and among these the case in which there are three white occurs in the long-run just as often as the case in which there are only two. [p.435]

Now this is a very curious solution: since there are initially more white balls in 1° than in 2°, one might well expect the answer to reflect this, and indeed that is just what emerges when one applies the inverse rules. For under these rules, argues Chrystal, there are only two possible constitutions of the bag, viz. 1° and 2°, each having *a priori* probability 1/2. The event consisting of the drawing of two white balls has for its probability under these hypotheses the values 1 and 1/3, and hence the *a posteriori* probabilities of 1° and 2° are 3/4 and 1/4, a result which Chrystal finds ridiculous [38].

If we look at the argument more closely, we find that Chrystal is suggesting the use of the hypothesis of an initial uniform distribution

$$\Pr[X = k] = 1/4, \quad k \in \{0, 1, 2, 3\}$$

(where X denotes the number of white balls in the bag) rather than the hypothesis (B) he used before, in which $X \sim b(3, 1/2)$. If we denote by C_i

the ith constitution and by E the drawing of two white balls, then

$$\Pr[E \mid C_1] = 1 \quad , \quad \Pr[E \mid C_2] = 1/3$$

$$\Pr[C_1 \mid E] = 3/4 \quad , \quad \Pr[C_2 \mid E] = 1/4 \ .$$

Where Chrystal errs is in supposing that, after E, the constitutions 1° and 2° are equally probable with chance $1/2$.

Chrystal argues further that the fallacy embodied in the inverse rules consists in the confusion of what we might write as $\Pr[C_i]$ with $\Pr[C_i \mid E]$, a confusion which in turn arises

> from neglect of the consideration that a probability is not unambiguously defined until the "series" of the "event" to which it relates has been given. [p.436]

He suggests further that Laplace's two principles can be written in the form

$$\Pr[C_i \mid E] = \Pr[E \mid C_i]\Pr[C_i] \Big/ \sum_1^n \Pr[E \mid C_i]\Pr[C_i]$$

and

$$\Pi = \sum_1^n (\Pr[E \mid C_i])^2 \Pr[C_i] \Big/ \sum_1^n \Pr[E \mid C_i]\Pr[C_i] \quad ,$$

where Π is the probability of one further occurrence of E after it has occurred once. To these formulations no exception can of course be taken, and one may be sure that Chrystal's interpretation is indeed that intended by Laplace.

As a further example of the unreasonableness of inverse probability Chrystal considers the following situation:

> A bag contains five balls which are known to be either all black or all white — and both these are equally probable. A white ball is dropped into the bag, and then a ball is drawn out at random and found to be white. What is now the chance that the original balls were all white? [p.437]

Chrystal's answer is that the chance is still $1/2$, unlike the solution obtained by Whitworth [1878, p.151] of $6/7$. This latter answer is interpreted by Chrystal as follows:

> if you were to drop a ball among the five a great many times, and draw one out again, then in about $6/7$ths of the times that you got a white ball you would get it from a bag in which all the balls are white. About this there is nothing mysterious whatever; but it is not the meaning of the question as it stands. [p.437]

The distinction is clear: Chrystal is concerned with an absolute and Whitworth with a conditional probability.

The theory of inverse probability is finally dismissed as follows:

> both from the point of view of practical common-sense, and from the point of view of logic, the two so-called laws of Inverse Probability are a useless appendage to the first principles of the Theory of Probability, if indeed they be not a flat contradiction of those very principles. [p.438]

Chrystal's attack[39] on inverse probability (one might even refer to it as a diatribe) did not pass unchallenged. In 1920, in a paper entitled "On some disputed questions of probability", E.T. Whittaker (1873-1956) considers the variation of the three-ball problem discussed by Chrystal, changing it, to intensify the effects, to a bag containing 1,000,001 balls, each either white or black, and all possible numbers of white balls equally likely *a priori.* If 1,000,000 balls are drawn, and all are found to be white, there is clearly an overwhelming probability that the remaining ball is also white. Whittaker presents both a "common-sense" argument and a frequency one to confute Chrystal, and argues further that considerations analogous to those presented by the latter are correctly applied in the following instance:

> An urn A contains a very large number of white balls, and the same number of black balls; from it n balls are drawn at random and placed in a second urn B without being examined. From B $(n-p)$ balls are drawn (without being replaced) and are found to be all white. What is the probability that the next ball drawn from B will be white? [p.167]

Arguing from the assumption that all constitutions of B are equally likely, Whittaker deduces from "Bayes's formula" that the required probability is 1/2.

He also deduces, in the usual manner, the formula

$$\int_0^1 x^{m+1}(1-x)^n v(x)\,dx \Big/ \int_0^1 x^m(1-x)^n v(x)\,dx \quad . \tag{7}$$

An unusual facet of his derivation is the interpretation of this as the probability that a person aged s will die before attaining age $(s+1)$, given that of $(m+n)$ persons alive at age s, $(m+1)$ die before attaining age $(s+1)$, with $v(x)\,dx$ denoting the probability that the facility lies between x and $x+dx$. As a limiting case it is supposed that $v(x)=1$, in which case (7) reduces to

$$(m+1)/(m+n+2) \quad . \tag{8}$$

> Since, however, it is almost inconceivable that anybody could be in the position of having no *a priori* knowledge whatever regarding mortality, the formula [8] has no practical value; the really important formula is [7]. [pp.169-170]

He suggests further that, as an approximation, one might well use $m/(m+n)$.

In the discussion of Whittaker's paper, J.R. Armstrong suggests that Chrystal's paper should not be viewed merely as an attack on the Bayes-Laplace theory. Rather, its aims are threefold: (i) a reiteration of Venn's criticism of mathematical probability as a calculus of belief, (ii) a criticism of certain (then current) interpretations of results obtained by a cavalier application of Bayes's formula, and (iii) a protest against the use of the formula where such use is illegitimate. As regards (i) Armstrong sides with Venn and Chrystal; as far as (iii) is concerned he notes that such enlivening problems only become amenable to the Bayes-Laplace theory "by a process of abstraction that deprives them of all their specific content" [p.199], while in connexion with (ii) he in the main stresses the importance of a clear distinction between absolute and conditional probabilities.

This last point is also stressed by W.L. Thomson, in the discussion, while the president, A.E. Sprague, in his concluding speech said

> I speak as an old pupil of the late Professor Chrystal, and with great diffidence and great respect, but I am sorry to say that I cannot make out from his paper precisely what his meaning was, and I think that his arguments as stated therein are open to criticism in various directions. [p.202]
> My own view is that Professor Whittaker's guns in the contest have outclassed Professor Chrystal's and Mr. Thomson's. [p.203]

In his reply to the discussion Whittaker defends his opinions against Armstrong and Thomson's defence of Chrystal, stressing that if an event E can occur only as a result of one and only one of the causes $A_1, A_2, \ldots,$ then to say that "when E happens, it happens as a result of A_1" is surely equivalent to saying that A_1 exists.

Hard on the heels of Whittaker's paper followed one by John Govan, entitled "The theory of inverse probability, with special reference to Professor Chrystal's paper 'On some fundamental principles in the theory of probability.' ". This paper, although not published until 1920, had in fact been read before the Actuarial Society of Edinburgh in 1893: it was apparently published at Whittaker's suggestion.

Govan first considers the variation on the three-ball problem discussed by Chrystal. Under a long-run frequency interpretation it is argued that the desired answer is indeed 3/4. Chrystal's five-ball problem is examined, and Whitworth's solution of 6/7 is confirmed. Furthermore, the usual form of the rule of succession (i.e. $(m+1)/(m+n+2)$) is derived in the case of sampling from an urn of indefinitely large size when the proportion p of white to black balls in the urn is unknown, but is uniformly distributed. Govan extends this example to the case in which $(m+n)$ draws which resulted in m white and n black balls were preceded by $(m' + n')$ draws yielding m' white and n' black balls. In this case the *a priori* probability of p (before the $(m+n)$ draws) is no longer dp but

$$\frac{\left(m' + n' + 1\right)!}{m'!\,n'!} p^{m'} (1-p)^{n'} \, dp \ ,$$

and the probability that the next draw will yield a white ball is found, as expected, to be

$$\frac{m' + m + 1}{m' + m + n' + n + 2} \ .$$

Exception is in fact taken to most, if not all, of Chrystal's arguments. Thus, for example, in discussing Chrystal's Problem II Govan criticizes the assumption of hypothesis (B) that each ball is equally likely to be white or black: for how, he says, "can we suppose that, when we are told that one ball is white?" [p.220]. If one supposes rather that one ball is white and each of the remaining two is equally likely to be white or black, then the possible constitutions arise with relative frequencies 1:2:1 rather than Chrystal's 3:3:1. A generalization of this problem is also provided.

> The fundamental error which vitiates nearly every conclusion in Professor Chrystal's paper, is his denial of the fact that (in the class of problems here discussed) the result of every trial modifies our data, or series, to use his own term. ... In Problem III. for instance (Hypothesis (A)), the series as at first given puts the four possible constitutions on an equal footing. The result of the first trial makes the constitution three black impossible, but Professor Chrystal will not admit that, just as three black has become impossible, so three white has become more probable than, say, one white and two black. [p.223]

Govan next turns his attention to the following general problem:

> p is the ratio of white, q of black ($p+q=1$), in an urn containing an indefinitely large number M of balls. N balls are drawn at random, N being a number very great in itself, but insignificant as compared with M. The proportion of white among the balls drawn will be p. [p.223]

To prove this Govan proceeds as follows: since M is large and N negligible as compared with M, the probability that the sample contains r white and $N-r$ black balls, viz.

$$\binom{pM}{r}\binom{qM}{N-r}\Big/\binom{M}{N} \quad ,$$

reduces to

$$\binom{N}{r}p^r q^{N-r} \quad .$$

This expression being maximized by the setting of $r = pN$ (approximately), the probability of the most probable ratio, p, in the drawing is

$$\binom{N}{pN}p^{pN}q^{qN} \quad ,$$

an expression which use of the Stirling-de Moivre formula reduces to

$$P \equiv 1\Big/\sqrt{2\pi pqN} \quad .$$

It follows further that the probability of a deviation of $\pm x$ in the number of white balls drawn is

$$\psi(x) = P\exp\left(-x^2/\left(2pq\,N\right)\right) \quad ,$$

and hence the expectation of the deviation from the most probable number, pN, of white balls will approximately be

$$\int_0^{pN} x\,\psi(x)\,dx + \int_0^{qN} x\,\psi(x)\,dx \quad ,$$

which is easily found to be

$$\sqrt{pqN/2\pi}\,[2 - \exp(-pN/2q) - \exp(-qN/2p)] \quad .$$

For large N this behaves like $\sqrt{2pqN/\pi}$, and it follows that the ratio of this to N tends to zero as N tends to infinity, as asserted in the proposition.

8.13 William Matthew Makeham (1826-1891)

In 1892 Makeham published, in volume 29 (1891) of the *Journal of the Institute of Actuaries*, a paper entitled "On the Theory of Inverse Probabilities." The paper consists of five sections.

In the first section Makeham declares his intent to use the word "chance" as signifying "a way of happening", a meaning which he finds in Lubbock and Drinkwater-Bethune [1830, ¶5]. The term is to be distinguished from *probability*, about which the following is recorded:

> We cannot be said to be ignorant of the *probability* of a given event, for the term "probability" has no reference to the chances (for and against) actually *existing*, but only to our *knowledge* of them. The *probability*, therefore, can always be determined by calculation, provided, of course, that we possess the skill necessary for the purpose. [p.243]

In this same section Makeham cites Laplace's "well-known formula in inverse probabilities", viz.

$$(m+1)/(m+n+2) \ ,$$

a formula which is deduced under the following fundamental assumptions:

> first, that the ratio of chances, for and against, may have any value from 0 to 1; and, secondly, that all values within those limits are *a priori* equally probable. [p.245]

In an attempt to counter objections raised by G.F. Hardy as to the applicability of this formula to assurance, Makeham proposes to generalize the result. This generalization is undertaken in Section 2, the following situation being considered: suppose that several urns are filled by withdrawing balls randomly from an urn containing a large number of white and black balls, the (known) ratio of white to total number being p and that of black to total number being q. Suppose further that in a particular filled urn the ratio of white to black balls is as $p' : q'$. Makeham now gives the following definition:

> The quantity denoted by p is the limit towards which the unknown ratio p' (in any particular urn) necessarily tends more and more to approximate as the number of balls contained in the urn is increased. [p.246]

The ratio p is then the antecedent, or *a priori*, probability of drawing a white ball from any urn; moreover, it is what Laplace terms "le milieu de probabilité" not only of all *possible* values of p' in a specific urn, but also of the several values of p' actually existing in the different urns[40].

Now to the problem in hand: suppose that $(m+n)$ draws (with replacement) have been made from a specific urn, m balls being white. What is the probability $p_{m,n}$ of obtaining a white ball on the next draw? Makeham states further that

> p represents the *a priori* probability (before any drawings have yet been made); while $p_{m,n}$ represents the *a posteriori* probability (after the fact that m white and n black balls have been drawn *has become known to the observer*). [pp.246-247]

This may seem slightly in conflict with the earlier definition of $p_{m,n}$ (after all, is a predictive probability the same thing as a posterior probability?): it seems, however, from what follows that $p_{m,n}$ *is* intended in a predictive sense.

Two postulates are established for the solution of this question [p.247], viz.

Postulate 1. If $p = m/(m+n)$, then $p_{m,n}$ is also equal to $m/(m+n)$, and to p.

Postulate 2. In all other cases $p_{m,n}$ will necessarily lie between $m/(m+n)$ and p.

In defence of the first postulate, Makeham argues that if $p = m/(m+n)$, the result of the $(m+n)$ trials provides no reason for altering the estimate of the probability. As regards the second, since p is the *milieu de probabilité* of the possible values of p' in the urn concerned, if $m/(m+n) < p$ it is *probably* less than p', and hence $m/(m+n) < p_{m,n}$. Further, since p is the *milieu de probabilité* of the values of p' in the different urns, p' is *probably* less than p (in the urn in question) if $m/(m+n) < p$, and so $p_{m,n} < p$. A similar argument may be applied if $m/(m+n) > p$, in which case it follows that $p < p_{m,n} < m/(m+n)$.

It now follows that $p_{m,n}$ may be supposed to be given by[41]

$$p_{m,n} = (m + rp)/(m + n + r) \tag{9}$$

for some $r > 0$. This may alternatively be written

$$p_{m,n} = [m/(m+n) + \alpha p]/(1+\alpha) \ ,$$

where $\alpha = r/(m+n)$. Now r may be shown to be independent of $m+n$, though it may well be a function of p — say $\varphi(p)$. On interchanging m and n, and replacing p by q, we find that the probability $q_{n,m}$ of drawing a black ball on the next draw is

$$q_{n,m} = [n + \varphi(q)q]/[m + n + \varphi(q)] \ .$$

But since $q_{n,m} = 1 - p_{m,n}$ it follows that r either must be constant or must be symmetric in p and q.

In Section 3 the general formula (9) is compared with Laplace's formula, the expressions of course coinciding for $p = 1/2$ and $r = 2$. Then, says Makeham,

> For convenience of calculation I propose, for the present, to assume a mean value in all cases for the function r, whatever may be the value of p, for which purpose it is evident that the mean value in question must be taken = 2. [p.249]

His choice of r yields the (approximate) *generalized* formula[42]

$$(m + 2p)/(m + n + 2) \ ,$$

and the values thus obtained for $n = 0, p = 0.01$ and 0.99 are compared with those given by Laplace's formula and by the "ordinary formula" $m/(m+n)$ (a result Makeham attributes to James Bernoulli).

In his fourth section Makeham turns to the four "Principes" given by Laplace in his *Théorie analytique des probabilités*, each of which is illustrated by a coin-tossing example. Of note is the following remark:

> we shall find that, without exception, the application of each one of Laplace's four fundamental principles (covering, as they do, the whole field of the doctrine of probabilities) involves an application of the *inverse* theory. [p.445]

The first principle may be formulated as follows:

$$\Pr[E \ \& \ F] = \Pr[E] \Pr[F \mid E] \ .$$

Applying this principle (or more accurately an extension of it) to repeated tosses of a coin, Makeham deduces that the probability of getting "heads" m times in succession is $p \prod_{j=1}^{m-1} p_j$, where p is the probability of "heads" on the first throw and

$$p_j = \Pr[\text{"heads" on } (j+1)\text{th trial} \mid \text{"heads" on preceding } j \text{ trials}] \ .$$

Makeham now asserts that the form of p_j will vary

> according to *the observer's knowledge* of the "actual ratio of chances", that is, of the inherent tendency of the coin to fall head or tail [p.446],

and three cases are considered in support of this assertion:

(i) $C_H : C_T :: (1 + w) : (1 - w)$, where C_x denotes the chances for x;

(ii) $p_x = (x + rp)/(x + r)$;

(iii) either $C_H : C_T :: (1+w) : (1-w)$ or $C_T : C_H :: (1+w) : (1-w)$, and these two suppositions are equally likely *a priori*.

The value of $p \prod_{j=1}^{m-1} p_j$ is found in these cases to be

$$\left(\frac{1+w}{2}\right)^m \quad ; \quad \frac{1}{m+1} \quad ; \quad \frac{1}{2}\left[\left(\frac{1+w}{2}\right)^m + \left(\frac{1-w}{2}\right)^m\right]$$

respectively.

The second principle may be given thus: if F denotes a future event and E an observed event, and if $\Pr[E \,\&\, F]$ and $\Pr[E]$ are determined *a priori*, then

$$\Pr[F \mid E] = \Pr[E \,\&\, F] / \Pr[E] \quad .$$

In the notation adopted in the discussion of the first principle, one has

$$\Pr\left[m' \text{ heads} \mid m \text{ heads}\right] = p_m p_{m+1} \cdots p_{m+m'-1} \quad .$$

For $m' = 1$ this reduces to $(m+rp)/(m+r)$, if the general formula is used, or $(m+1)/(m+2)$, if Laplace's formula is used.

The third principle runs as follows: suppose that E, an observed event, can occur in conjunction with one (and only one) of n different causes $C_1, \ldots, C_n$. Then[43]

$$\Pr[C_i \mid E] : \Pr[C_j \mid E] :: \Pr[E \mid C_i] : \Pr[E \mid C_j]$$

and

$$\Pr[C_i \mid E] = \Pr[E \mid C_i] / \sum_j \Pr[E \mid C_j] \quad .$$

The example used to illustrate this principle shows that an equi-probable assumption is needed.

The fourth principle states that, for a future event F,

$$\Pr[F \mid E] = \sum_i \Pr[C_i \mid E] \Pr[F \mid C_i] \quad .$$

Here Makeham shows that the probability of a head given m heads in succession is

$$\tfrac{1}{2}\left\{\left[(1+w)^{m+1} + (1-w)^{m+1}\right] / \left[(1+w)^m + (1-w)^m\right]\right\} \quad .$$

Some criticism of an example given by Laplace then follows ‡, together with the astute observation that

‡This example is concerned with the drawing, with replacement, of balls from an urn of three balls, each of which is either white or black: m of these drawings yield m white balls. To determine the *a posteriori* probabilities of the possible constitutions of the urn, Makeham suggests that one should consider the prior probabilities as $1/8, 3/8, 3/8$ and $1/8$.

> for the correct solution of these inverse problems, everything depends upon the proper determination of the elementary values, that is, upon the correct analysis of the elementary hypotheses of equal antecedent probability. [p.456]

The final section of the paper is devoted to a brief application of some of the preceding results to the problem of the use of observed mortality rates in assurance.

Some comment on Makeham's work seems necessary. Firstly, Laplace's formula, viz. $(m+1)/(m+n+2)$, applies only if the number of balls in the urn is infinite, a condition not made explicit by Makeham. Secondly, the equating of Laplace's expression to (9) results in

$$r \equiv r(p) = (m-n)/[q(m+1) - p(n+1)] \ . \tag{10}$$

Note that $r(p) = r(q)$ implies that $p = 1/2$, in which case $r = 2$. But while the pair $(r = 2, p = 1/2)$ certainly reduces Makeham's formula to Laplace's, so will many others: the condition $r(p) = r(q)$ ensures uniqueness.

It follows from (10) that, since $0 < p < 1$, r must satisfy

$$0 < (n-m)/(n+1) < r \ ,$$

which not only sets some extra condition on the permissible values of r, but also requires that n exceed m (this requirement is in fact not met in the numerical example used by Makeham, in which $n = 0$).

One might also note that Laplace's formula may be written in the form

$$\alpha[m/(m+n)] + (1-\alpha)1/2$$

where $\alpha = (m+n)/(m+n+2)$. Now Makeham states that

> the supposed value of p, or the antecedent probability, is 1/2 in Laplace's investigation. [p.249]

It is perhaps difficult to reconcile this with the fundamental assumptions stated earlier in this section. Of course, if p is uniformly distributed over the unit interval, a number of statistics of p (e.g. its mean, median and mode) take the value 1/2; and in view of Makeham's assertion that p is "le milieu de probabilité" of the possible values of p' in a given urn,

> that is to say, p is a quantity such that the true value of p' in any particular urn is just as likely to be *above* as *below* it [p.246],

we might well consider the median as the appropriate statistic. In this case p should be replaced by the median value m obtained from tables of the incomplete beta-function $I_m(\alpha, \beta)$.

A further short note by Makeham, "On a problem in probabilities", appeared in the same volume of the *Journal of the Institute of Actuaries.* This note was devoted to the following problem: consider four urns of respective composition three white balls, two white and one black, one white and two black, three black balls. One of these urns having been chosen at random, m draws (with replacement) are made, and all result in white balls. Find the *a posteriori* probability that urn $1, 2, 3$ or 4 was chosen. The problem is solved in the usual way: all we might note here is Makeham's justification of the choice of a uniform prior, viz.

> As the urn chosen may be any one of the four, it is evident that, *a priori*, there is precisely the same chance in favour of each of the four hypotheses in question. We have here, then, *necessarily*, the identical condition *gratuitously* assumed by Laplace in the solution of his well-known problem. [p.475]

(The reference is to p.183 of the *Théorie analytique des probabilités* (first edition): it may be found on p.185 of the *Œuvres complètes* edition of 1886.)

Makeham's theory did not go unchallenged: in 1892 Edward L. Stabler published a paper in which he presented

> some considerations which I think will show that this formula is not suitable for any application. [p.240]

He agrees with Makeham's formula

$$p_{m,n} = (m + rp)/(m + n + r) \quad ,$$

but states that

> in this form the formula gives no more information as to the probability desired than was already evident from the nature of the case. [p.240]

Stabler takes exception to Makeham's "proof" that

> r is "some undetermined constant independent of $m + n$", and either independent of p or "not affected by the substitution of q, or $1 - p$, for p." [p.240]

his counter-examples showing (i) that r may be affected by the replacement of q by p, and (ii) that r may depend on $m + n$.

Turning his attention to the general case of sampling from an urn of N (finite) white or black balls, in which urn the initial probability of drawing a white ball is known to be p, Stabler shows that the probability $p_{m,n}$ that the $(m+n+1)$th draw will yield a white ball, after $(m+n)$ draws have resulted in m white and n black balls, is given by

$$p_{m,n} = \frac{\sum_{s=0}^{N} \binom{N}{s} p^{N-s}(1-p)^s (N-s)^{m+1} s^n}{N \sum_{s=0}^{N} \binom{N}{s} p^{N-s}(1-p)^s (N-s)^m s^n} ,$$

which is a generalization of that given earlier by Terrot.

Further criticism was raised by John Govan [1920]. Govan concentrated on Makeham's "urn and balls" example, agreeing with Chrystal's assertion that the problem, as stated, is simply indefinite, and he argued that Makeham had made a serious error in his solution. For, *contra* Makeham,

> if we know that each individual ball is equally likely to be white or black, we cannot know *in addition* that one ball is certainly black (unless we know further that one ball is certainly white), inasmuch as the one condition is incompatible with the other. [p.228]

Govan suggested that the following meaning might be attached to Makeham's problem: suppose we have $\binom{3}{r}$ bags, each containing r white and $(3-r)$ black balls. Each bag is taken N times, and, on each of these occasions, $(s+t)$ draws are made (with replacement) from that bag. Then the number of times that we get s white and t black balls is approximately

$$N\binom{3}{r}\binom{s+t}{s}\left(\frac{r}{3}\right)^s\left(\frac{3-r}{3}\right)^t .$$

As r takes on the values $0, 1, 2, 3$ in turn, we find the relative frequencies (ignoring constants)

$$0^s.3^t, \quad 3.1^s.2^t, \quad 3.2^s.1^t, \quad 3^s.0^t .$$

The *a posteriori* relative frequencies, after the drawing of the m white balls, are of the same form, with s replaced by $s+m$. If now $s=0$ and $t \neq 0$ (Makeham's first hypothesis), we obtain the sequence

$$0^m.3^t, \quad 3.1^m 2^t, \quad 3.2^m.1^t,$$

which does not agree with Makeham's solution. Further, if $s \neq 0$ and $t \neq 0$, the sequence yields Makeham's result only if $s = t$.

8.14 Karl Pearson (1857-1936)

In a life so richly productive of statistical innovation as that of Karl Pearson[44], particularly in the biometrical field, one might well expect to find little time devoted to matters of historical or philosophical concern. Pearson's interest in statistics (and science) in general, however, was such as to lead him to not inconsiderable speculation on these matters[45], and among his voluminous writings[46] eight have been singled out as bearing on the present investigations.

The first of Pearson's works which is pertinent is his justly celebrated *The Grammar of Science* (first published[47] in 1892), a work which Haldane [1957] regards as Pearson's "main contribution to philosophy". In Chapter 4, entitled "Cause and Effect. Probability", we find in Section 13, headed "Probable and Provable", a discussion of the rule of succession phrased in the following words:

> A certain order of perceptions has been experienced in the past, what is the probability that the perceptions will repeat themselves in the same order in the future? [p.168]

Pearson's belief in the frequency interpretation of probability is born out by his further statement

> The probability is conditioned by two factors, namely: (1) In most cases the order has previously been very often repeated, and (2) past experience shows us that sequences of perceptions are things which have hitherto repeated themselves without fail. [p.168]

He states further Laplace's assertion that the probability of the further occurrence of an event which has already occurred p times and has not been known to fail, is $(p+1)/(p+2)$, and illustrates this result by considering (a) the further solidification of hydrogen after one such success, and (b) the further rising of the sun after a million dawns. Believing that the numbers obtained in these two cases "do not in the least represent the degrees of belief of the scientist regarding the repetition of the two phenomena" [p.169], Pearson[48] argues that the problem ought rather to be posed as follows:

> p different sequences of perception have been found to follow the same routine, however often repeated, and none have been found to fail, what is the probability that the $(p+1)$th sequence of perceptions will have a routine? Laplace's theorem shows us that the odds are $(p+1)$ to one in favour of the new sequence having a routine. [p.169]

In Section 14, "Probability as to Breaches in the Routine of Perceptions", Pearson points out that Laplace's result permits one to take account of "possible 'miracles', anomies, or breaches of routine in the sequence of perceptions" [p.170] (perhaps all of these are covered by the second term). He concludes that one is justified in saying that miracles have been *proved* incredible, where "proved" is interpreted as the establishment of an overwhelming probability in favour of.

In Section 15, "The Bases of Laplace's Theory lie in an Experience as to Ignorance", Pearson turns his attention more closely to Laplace's result, drawing an analogy between the world of perceptions (divided into routine-order and anomy) and a bag containing white and black balls. Talking further of a coin-tossing set-up, Pearson mentions the following Laplacean principle:

> "If a result might flow from any one of a certain number of different constitutions, all equally probable before experience, then the several probabilities of each constitution after experience being the real constitution, are proportional to the probabilities that the result would flow from each of these constitutions." [pp.173-174]

and in expanding further on its use he emphasizes the rôle played by experience in the determination of *a priori* probabilities.

In Section 16, "Nature of Laplace's Investigation", Pearson returns to his "nature bag" example, supposing no longer that routine and breach of routine are equally probable, but rather that every possible ratio of black to white balls is equally likely[49]. He then deduces an expression of the form

$$\Pr[\text{white}] = \sum_i \Pr[\text{white} \mid \text{constitution } i] \Pr[\text{constitution } i] \ ,$$

and points out that Laplace's result follows. A particular case is discussed in the following section, "The Permanency of Routine for the Future".

In some measure *The Grammar of Science* is still pertinent to modern science[50], but one must agree with Haldane [1957] that "the discussion of probability and statistical method in the first edition of *The Grammar of Science* is superficial".

We now come to Pearson's papers, the first relevant one of which was written with Filon and published in 1898 (read on the 25th of November 1897). This paper, entitled "On the Probable Errors of Frequency Constants and on the Influence of Random Selection on Variation and Correlation" formed the fourth part of "Mathematical Contributions to the Theory of Evolution." Commenting on this paper, E.S. Pearson [1967] writes

> The basis of the approach used here is a little obscure and there seems to be implicit in it the classical concept of inverse probability. [p.347]

A similar comment[51] has been expressed by MacKenzie [1981, p.241].

The main result of this paper (to be found in the second article) has been reformulated by MacKenzie [1981, pp.241-243] and Welch [1958] in terms of inverse probability: we shall present a similar (but more general) interpretation[52]. Pearson and Filon show that if $z = f(x_1, \ldots, x_n; \eta_1, \eta_2, \ldots)$ is a frequency surface, where the η_i are frequency constants (i.e. means, standard deviations, &c.), then, on neglecting cubic and higher terms in the deviations $\Delta\eta_i$, "the frequency surface giving the distribution of the variations in the deviations" [p.236] is

$$P_\Delta = P_0 \exp\left\{-\tfrac{1}{2}\left[\sum a_{rr}\left(\Delta\eta_r\right)^2 - 2\sum a_{rs}\Delta\eta_r\Delta\eta_s\right]\right\} ,$$

where P_0 is a normalizing constant and

$$a_{rr} = -\int\cdots\int f\left[d^2(\log f)/d\eta_{rr}^2\right]dx_1\ldots dx_n$$

$$a_{rs} = \int\cdots\int f\left[d^2(\log f)/d\eta_r d\eta_s\right]dx_1\ldots dx_n .$$

The desideratum is

> Σ_r, the standard deviation of $\Delta\eta_r$, and R_{rs}, the coefficient of correlation between $\Delta\eta_r$ and $\Delta\eta_s$ [p.236]

which requires consideration of the (posterior) distribution of the $\Delta\eta_i$ and hence specification of a prior.

As a specific illustration Pearson and Filon consider a random sample $\{(X_i, Y_i)\}$ of size n drawn from the bivariate[53] Normal distribution $N\left(\mu_x, \mu_y, \sigma_x^2, \sigma_y^2, \rho\right)$. The joint density (of the data as a function of the parameters) is then viewed as a density of the parameters in order to determine things like the standard deviations of errors in σ_x, σ_y and ρ. If we denote the joint density by $f(S \mid \mu_x, \mu_y, \sigma_x, \sigma_y, \rho)$, where S denotes the data, then

$$f(S, \mu_x, \mu_y, \sigma_x, \sigma_y, \rho) = f(S \mid \mu_x, \mu_y, \sigma_x, \sigma_y, \rho)\, f(\mu_x, \mu_y, \sigma_x, \sigma_y, \rho)$$

where f is used indiscriminately to denote a density function. The posterior distribution of the parameters given the data is then found in the usual manner. The choice of a uniform prior distribution for the parameters yields a posterior distribution which is proportional to the likelihood, and it is this latter function with which Pearson and Filon are concerned. The standard deviations of the errors in the parameters given in this paper are today well known.

The next paper demanding our attention is entitled "On the influence of past experience on future expectation": it was published in the *Philosophical Magazine* in 1907 with the avowed aim of putting

> into a new form the mathematical process of applying the principle of the stability of statistical ratios, and to determine, on the basis of the generally accepted hypothesis, what is the extent of the influence which may be reasonably drawn from past experience. [p.365]

After pointing out inadequacies in common application of the principle, Pearson states[54]

> I start as most mathematical writers have done, with "the equal distribution of ignorance," or I assume the truth of Bayes' Theorem. [p.366]

If Pearson is equating "the equal distribution of ignorance" with Bayes's Theorem then he is simply wrong. He goes on further to say "I hold this theorem not as rigidly demonstrated" [p.366], and again he errs: granted the assumptions made by Bayes, the theorem is correct.

Pearson now passes on to the statement of Bayes's Theorem, which he gives as follows

$$\Pr\left[x < X < x + \delta x \mid p \text{ occurrences of } E \text{ and } q \text{ failures of } E\right]$$

$$= x^p(1-x)^q\,dx \Big/ \int_0^1 x^p(1-x)^q\,dx$$

"on the equal distribution of our ignorance" [p.366]. The chance that in a further m trials the given event E will occur r times and fail $s = m - r$ times is then

$$C_r = \binom{m}{r} \int_0^1 x^{p+r}(1-x)^{q+s}\,dx \Big/ \int_0^1 x^p(1-x)^q\,dx \quad ,$$

and "This is, with a slight correction, Laplace's extension of Bayes' Theorem" [p.367]. We have already commented on the correctness of this assertion.

Noting that the usual method of evaluation involves, via beta-functions and the Stirling-de Moivre approximation, the expression of C_r in terms of ordinates of the Normal distribution (an approach which later illustrations in the paper show to be sometimes unsatisfactory), Pearson proposes to use the hypergeometric series

$$C_0 \left\{ 1 + \frac{m(p+1)}{1!\,(q+m)} + \frac{m(m-1)}{2!} \frac{(p+1)(p+2)}{(q+m)(q+m-1)} \right.$$
$$\left. + \frac{m(m-1)(m-2)}{3!} \frac{(p+1)(p+2)(p+3)}{(q+m)(q+m-1)(q+m-2)} + \&c. \right\}$$

whose successive terms give C_r, $r \in \{0, 1, \ldots, m\}$, with

$$C_0 = \Gamma(q+m+1)\Gamma(n+2)/\Gamma(q+1)\Gamma(n+m+2) \ .$$

Note that the term in braces in this series is the hypergeometric function ${}_2F_1(p+1, -m; -(q+m); 1)$.

A detailed comparison of the moments of the hypergeometric series with those of the standard Normal distribution leads to the following conclusions:

> it is not possible in judging expectancy from past experience (i.) to neglect the relative sizes of the first and second samples, or (ii.) to neglect, even in characteristics which appear in 10 p.c. of the sample, the sensible deviation from the Gaussian distribution. [p.373]

A further conclusion drawn is the following:

> The frequency of future samples is given by a certain hypergeometrical series, which is not at all closely represented by the Gaussian curve except when the past experience is very large as compared with the proposed sample, and further the characteristic expected does not occur in either a very large or very small percentage of the population. [p.378]

Some thirteen years later, in 1920, we find Pearson returning to this question in his paper "The fundamental problem of practical statistics", a paper which in a sense is an amplification of that just discussed.

The question, stated to be "as ancient as Bayes" [p.1], explored in this paper runs as follows:

> An "event" has occurred p times out of $p + q = n$ trials, where we have no *a priori* knowledge of the frequency of the event in the total population of occurrences. What is the probability of its occurring r times in a further $r + s = m$ trials? [p.1]

Pearson briefly discusses the contributions made by Bayes, Price, Condorcet and Laplace to the solution of this problem, and before adding his own solution he comments on criticism by Boole and Venn of inverse probability, and notes also that

> Edgeworth returns to the appeal to experience from which Bayes and Laplace ought to have started. [p.4]

Pearson finds that those antagonistic to inverse probability generally attack two hypotheses used by Bayes, viz.

(i) the hypothesis that *a priori* we ought to distribute our ignorance of the chance of a marked individual occurring *equally*,

(ii) the hypothesis that earlier occurrences do not modify the chance of later trials [p.2],

and in an attempt to divert the assault on the first hypothesis (the one which is usually attacked), he proposes to investigate whether *any* continuous distribution of *a priori* chances would lead to the same result.

To this end let a stroke be made at random on a line of length a, the position of this stroke at distance x from one end being known. A further n strokes are now made at random on the line, p falling in the segment 0 to x and $q = n - p$ in x to a. Unlike Bayes, Pearson now supposes the probability density function for the strokes to be given by $y = \varphi(x)/a$, where φ is any continuous function. Denoting by X the chance of a stroke, we have

$$\Pr[x < X < x + \delta x] = \varphi(x)\delta x/a \quad .$$

Thus P_x, the chance of a stroke afterwards occurring between 0 and x, is given by

$$P_x = \int_0^x \varphi(x)\, dx/a \quad ,$$

and similarly

$$Q_x = 1 - P_x = \int_x^a \varphi(x)\, dx/a \quad ,$$

while $P_0 = 0$ and $P_a = 1$. The probability of the combined event will be

$$\delta P_x . (P_x)^p (Q_x)^q \binom{p+q}{p} \quad ,$$

and hence the probability that the unknown probability lies between P_b and P_c (i.e. X lies between b and c) will be

$$\int_{P_b}^{P_c} (P_x)^p (1 - P_x)^q \, dP_x \bigg/ \int_0^1 (P_x)^p (1 - P_x)^q \, dP_x \quad .$$

Similarly the chance that $m = r+s$ trials will yield r successes and s failures will be

$$\binom{r+s}{r} \int_0^1 (P_x)^{p+r} (1 - P_x)^{q+s} \, dP_x \bigg/ \int_0^1 (P_x)^p (1 - P_x)^q \, dP_x \quad .$$

This latter expression, like that given by Laplace, reduces to[55]

$$C_r = B(p+r+1, q+s+1)/B(p+1, q+1)B(r+1, s+1) \ .$$

Two methods are now proposed for the simplification of C_r. The first of these, a somewhat more complete development than that given by Laplace, requires the replacement of the beta-functions by gamma-functions and the latters' approximation by the Stirling-de Moivre formula for large values of p, q, r and s. The final result is

$$C_r = C\exp\left(-h^2 T_1/2\sigma^2\right)\exp\left(-hT_2/2\sigma + h^3 T_3/6\sigma^3\right) \ , \qquad (11)$$

where

$$T_1 = 1 - [(1+2\rho)/2m(1+\rho)][(1-2PQ)/PQ]$$

$$T_2 = (Q-P)\Big/\sqrt{m(1+\rho)PQ}$$

$$T_3 = (1+2\rho)(Q-P)\Big/\sqrt{m(1+\rho)PQ}$$

$$C = e^{-n}p^p q^q \Gamma(m+2)/B(p+1, q+1)\Gamma(n+m+2)$$

$$1+\rho = (n+m)/n \ ; \ P = p/n \ , \ Q = q/n \ ;$$

$$r = (mp/n) + h \ , \ s = (mq/n) - h \ .$$

Equation (11) clearly shows that unless $1/\sqrt{m}$ be small, the terms in h and h^3 cannot be neglected: in other words, a *skew* frequency curve is suggested, rather than a Normal one. After some further discussion of the effect of the magnitude of $1/\sqrt{m}$, Pearson notes that the Gauss-Laplace distribution fails

> (a) for small samples. Its whole method of deduction is then wrong for Stirling's Theorem is invalid;
> (b) when the sample is large, but the probability of occurrence is small, so that mP is finite and small. [p.8]

As a second method Pearson proposes to find a less rough approximation to the original hypergeometrical (sic) series for C_r. Just as the Normal density had been shown by Laplace to correspond to the symmetrical binomial histogram, so Pearson finds in the present case (after considerable manipulation and starting from the assumption that $C_0, C_1, \ldots, C_r, C_{r+1}, \ldots$ are plotted as a histogram of rectangles of base c and heights C_0/c, C_1/c, $\ldots$,

C_r/c, $C_{r+1}/c, \ldots$) that the curve corresponding as closely to the skew binomial histogram satisfies the differential equation

$$\frac{1}{y}\,\frac{dy}{dx} = -x \Big/ \left(\sigma_0^2 + \tfrac{1}{2}c(Q-P)x\right)$$

where $\sigma_0^2 = PQ(m+1)c^2$. Assuming rather more generally that

$$\frac{1}{y}\,\frac{dy}{dx} = -x \,/ \left(a_0 + a_1 x + a_2 x^2\right)$$

where $a_0 > 0$, $a_2 < 0$ and

$$a_0 = PQ(m+1)(1+(m+1)/n)c^2$$

$$a_1 = (Q-P)\left(\tfrac{1}{2} + (m+1)/n\right)c$$

$$a_2 = -1/n \quad ,$$

Pearson writes

$$\frac{1}{y}\,\frac{dy}{dx} = -x\,/b_0\,(b_1 - x)\,(b_2 + x) \quad ,$$

and hence obtains

$$y = y_0\,(1 + x/b_2)^{s_2}\,(1 - x/b_1)^{s_1}$$

where $s_1 = b_1\,/b_0\,(b_1 + b_2)$ and $s_2 = b_2\,/b_0\,(b_1 + b_2)$, and y_0 is the modal ordinate.

This result is then applied to the following problem: in a sample of size 1,000 , 20% of the individuals are found to possess a certain characteristic. What is the chance that such a percentage occurs in a further sample of size 100? This problem, of a type Pearson terms "the *fundamental* problem of statistics" [p.12], is explored by both the above methods, the skew curve giving a much better fit to the series than does the Normal curve. A further problem, in which an indefinitely large population contains 10% of a given character, is considered, similar conclusions once again obtaining.

Hard on the heels of this paper (indeed, in the same volume of *Biometrika*) came Pearson's "Note on the 'Fundamental Problem of Practical Statistics.' " The previous paper had apparently occasioned some misunderstanding:

> I believe it to be due to the critics not having read Bayes' original theorem as given by Price in the *Phil. Trans.*, Vol. LIII. [p.300]

Pearson repeats here Bayes's argument: a ball is placed at random on a table (of unit breadth, say), its distance from one side being x (a variate) and its chance of falling between x and $x + \delta x$ being δx. With Bayes's definition of "success" and "failure" it follows that the chance of p successes and q failures will be

$$\binom{p+q}{p} x^p (1-x)^q \, dx \ .$$

Pearson now sagely notes that

> It is solely the fact that all possible values of the variate x are made *a priori* equally likely that makes the chance of a success x, equal to the variate itself. [p.301]

He now repeats his argument concerning P_x of the earlier paper, showing once again that the same conclusion is reached in this case as that in which the "equal distribution of ignorance" is assumed. The final paragraph is worth noting:

> I believe that in most cases such a variate [as x] may be hypothecated and if it can the objection to Bayes that he made all positions of his balls on the table "equally likely" can be removed, and if removed one fundamental objection to his theorem as he stated it, i.e. in terms of excess or defect of a variate, disappears. [p.301]

In 1924 in his "Note on Bayes' Theorem", Pearson becomes more personal: instead of referring vaguely to "critics" he begins the present paper with a sharp attack:

> Dr. Burnside, I venture to think, does not realise either the method in which I approach Bayes' Theorem, or the method in which Bayes actually approached it himself. [p.190]

(Burnside's note, which immediately preceded this paper by Pearson, is discussed in an appendix to this chapter.) Pearson once more repeats his argument: suppose that an occurrence takes place if a certain variate X, known to lie between two values 0 and a (say), exceeds a certain value ξ, and suppose that the occurrence does not occur if X does not exceed ξ. The value ξ being unknown, let us suppose that the frequency curve of the *a priori* possible values of ξ is $y = \varphi(\xi)\, d\xi$. (We assume that $\varphi(\cdot)$ is a probability density function over $[0, a]$.) Suppose further that the frequency curve of X in the population (of size N) is $Nf(x)$. Then the chance of an occurrence is

$$P_\xi = \int_0^\xi f(x)\,dx \ ,$$

and hence

$$\Pr[p \text{ occurrences \& } q \text{ non-occurrences} \mid \xi] = \binom{p+q}{p} P_\xi^p (1-P_\xi)^q \varphi(\xi)\,d\xi \ .$$

Supposing next (as did Bayes) that $\Pr[\xi=\xi_0] \propto \Pr[\text{event} \mid \xi=\xi_0]$, we find that "the probability of the constitution being ξ" [p.190] is

$$P_\xi^p (1-P_\xi)^q \varphi(\xi)\,d\xi \Big/ \int_0^a P_\xi^p (1-P_\xi)^q \varphi(\xi)\,d\xi \ .$$

Hence the chance of an (r,s) sample following a (p,q) sample is

$$C_{(p,q)(r,s)}$$

$$= \binom{r+s}{r} \int_0^a P_\xi^{p+r} (1-P_\xi)^{q+s} \varphi(\xi)\,d\xi \Big/ \int_0^a P_\xi^p (1-P_\xi)^q \varphi(\xi)\,d\xi \ . \quad (12)$$

Pearson next points out that this result generalizes that of Bayes in two respects:

(i) Bayes assumes $\varphi(\xi) = 1/a$, i.e. all values of ξ are *a priori* equally likely;

(ii) Bayes takes $f(x) = 1/a$ also.

It was indeed to overcome Bayes's assumption that all values of X and all values of ξ are equally likely, says Pearson,

> that I wrote my paper of which Dr. Burnside, who does not seem to have read Bayes, disapproves. [p.191]

Since $dP_\xi = f(\xi)\,d\xi$, expression (12) can be rewritten in the form

$$C_{(p,q)(r,s)} = \binom{r+s}{r} \frac{\int_0^1 P_\xi^{p+r} (1-P_\xi)^{q+s} \varphi(\xi)\,/\,f(\xi)\,dP_\xi}{\int_0^1 P_\xi^p (1-P_\xi)^q \varphi(\xi)\,/\,f(\xi)\,dP_\xi} \ .$$

If we take $\varphi(\xi) = f(\xi)$ and let $P_\xi = z$, we obtain Bayes's Theorem — or so Pearson [1924b, p.191] asserts, though as we have already noted he errs in this conclusion. Commenting further on the choice of φ and f, Pearson notes that

> If Bayes' Theorem does not give us reasonable results, then we must select a better value of the ratio $\varphi(\xi)/f(\xi)$ than unity, but at present it has not been demonstrated to lead to results contrary to experience; it has been solely criticised on the ground that equal distribution of ignorance is not logical. [p.191]

Noting the difference between (12) and the formula cited by Burnside, Pearson says

> Dr. Burnside cites as Bayes' formula, what is only an element in Bayes' Theorem, and he does so on the strength of Poincaré, who in all probability had not studied Bayes' original work. [p.191]

This is followed by fairly extensive discussion of the applicability of the "equal distribution of ignorance" assumption, and Pearson stresses that

> it cannot be too generally recognised that it is the basis of Bayes' Theorem to assume no knowledge beyond the (p, q) observation. [p.192]

Pearson also adduces reasons for his preferring the use of φ and f to Bayes's assumptions, but notes that, in the preceding notation, the probability distribution function F of P_ξ satisfies

$$F(P_\xi) = \varphi(\xi)/f(\xi) \ .$$

It thus follows that if $\varphi(\xi)/f(\xi)$ is constant, then P_ξ has a uniform distribution, "or as in Bayes' Theorem all chances are equally likely" [p.192].

The four papers by Pearson discussed so far form a quartet on which some comment may well be made. We have already mentioned Burnside's criticism and Pearson's rebuttal thereof, and shall say no more on this point. Writing in 1921, F.Y. Edgeworth comments as follows on Pearson's "The fundamental problem of practical statistics":

> Apparently Professor Pearson does not withdraw the countenance which in an earlier writing [*The Grammar of Science*, 3rd edition, ch.iv, p.146] he had given to the doctrine upheld by the present writer (*Mind*, 1884), that the equal distribution of *a priori* probability (in the absence of specific knowledge) rests on a rough but solid basis of experience. Professor Pearson now seems to regard the doctrine, not indeed as untrue, but as unnecessary for the purposes of Inverse Probability. [1921, p.82]

Commenting further on Pearson's question "Is it not possible that any continuous distribution of a priori chances would lead equally well to the Bayes-Laplace result?" (loc. cit. p.4), Edgeworth notes that one may indeed answer this in the affirmative without rejecting his own remark on the equal distribution of *a priori* probability.

Further, although Pearson's question *is* generally answered in the affirmative, it is not for the reason advocated by its proposer:

> His reasoning seems to rest upon a very peculiar — not to say, hardly supposable — relation between the antecedent probability that a certain "possibility" (in Laplace's phrase) or constitution (e.g. of a coin or die) would have existed, and the *a posteriori* probability that, if it existed, such and such events (e.g. so many Heads or Aces in n trials) would be observed. [1921, p.83]

Assuming with Pearson that the *a priori* distribution of chances is $\varphi(x)\delta x/a$, one notes that φ should not appear in the *a posteriori* probability. The fact that the usual Bayes-Laplace result is obtained under almost any continuous initial distribution was in essence noted by Cournot [1843] and Mill [1843]. This idea of the "swamping" of prior knowledge by experience is of course well known to modern Bayesians.

In a remarkably controlled passage R.A. Fisher (1890-1962) remarks that "The fundamental problem of practical statistics" is a paper

> in which one of the most eminent of modern statisticians presents what purports to be a general proof of Bayes' postulate. [1922, p.311]

This is of course a totally inaccurate observation: no attempt at a "proof" of the postulate was essayed.

More recently A.W.F. Edwards has commented in two papers on Pearson's early work involving Bayes's Theorem. In the first of these papers, entitled "A problem in the doctrine of chances", Edwards takes exception to the way in which Pearson framed his question, since

> to speak of 'the probability of its occurring r times' is to beg part of the question, for probability may not be the proper calculus for prediction. [1974, p.44]

After noting Pearson's expression

$$C_r = B(p+r+1, q+s+1)/B(p+1, q+1)B(r+1, s+1) \ ,$$

a result which is independent of any non-uniform prior, Edwards writes

> Pearson's capacity for not seeing the wood for the trees was exceptional. Instead of commenting on this remarkable independence, he thought he had solved the fundamental problem, and busied himself in the rest of the paper with evaluating beta-integrals. [1974, p.46][56]

Edwards notes further in this paper that Edgeworth [1921] and Burnside [1924] showed the unsuitability of the Bayes model inasmuch as it implies a relation between the prior and the likelihood, and "Pearson ... took the point eventually" [1974, p.46].

In his second paper, "Commentary on the arguments of Thomas Bayes", published in 1978, Edwards essentially repeats his earlier arguments, though in a more concise form. He makes the additional point that

> Pearson had made the mistake of identifying the distribution of the throws of the ball with the prior distribution of the probability. [1978, p.118]

In 1979 D. Hinkley took a fresh look at Pearson's "fundamental problem of practical statistics", providing a definition of predictive likelihood "which can produce a simple prediction analog of the Bayesian parametric result, posterior $\propto$ prior $\times$ likelihood" [p.718]. Hinkley errs however in asserting that "Pearson's purpose was to reexamine the general applicability of Bayes's earlier solution" (loc. cit.): at least, while that *might* have been Pearson's aim, Bayes's result, as we have already seen, is not concerned with future events (that extension is due to Price).

Now let us return to Pearson's work. In his paper "James Bernoulli's theorem", published in *Biometrika* in 1925, Pearson, in between his discussion of Bernoulli's proof and his own treatment of the problem, remarks

> Bernoulli then turns the problem round and says that if the observed value in nt trials be p, then the true value p_0 will lie between $p \pm 1/t$ with the given probability. This is rather stated than proved, but it is of course the kernel of much later developments of importance. Leibnitz raised objections to it. [p.205]

No further comments on this point are however made: we have already said something on this score elsewhere[57].

Pearson next turns his attention to a critical examination of a commonly used sampling method in his 1928 paper entitled "On a method of ascertaining limits to the actual number of marked members in a population of given size from a sample" published in *Biometrika*. The problem considered is the following: a population of size N contains p marked and $q = N - p$

unmarked individuals, a sample of size n from this population being found to contain r marked and $s = n - r$ unmarked members. It is usual to estimate the percentage of marked individuals as

$$100r/n \pm 67.449\sqrt{rs/n^3} \ ,$$

the probable error[58] thus found being taken as a rough measure of the possible deviation of the sample value $100r/n$ from the actual (though unknown) value $100p/N$. Pearson finds the reasoning leading to this result unsatisfactory, on the following grounds:

> first, because the result is independent of the size of the population sampled, and secondly because it really throws us back on the normal curve as representing the binomial, and this will only be correct if r or s be not small as compared with n. [1928, p.149]

Pearson sees in this general question two distinct problems:

(i) on the basis of a sample of size $n = r+s$, what will be the distribution of r' and s' in a further sample of size $n' = r' + s'$?

(ii) on the basis of a sample of size $n = r + s$, what is the distribution of p and q?

(Pearson also refers to the quaesitum in this second question as "the likelihood of various values of p and q in the actual population N" [1928, p.149].) The first of these problems Pearson views as involving an appeal to Bayes's Theorem, and it was discussed in his paper of 1907. It is hardly necessary once again to stress that it was Price who extended Bayes's result to the question of future observations: thus Pearson is slightly inaccurate in his present observation.

Proceeding to the second question, we find (sampling occurring without replacement), that

$$C_{r,p} \equiv \Pr[r \mid p] = \binom{p}{r}\binom{N-p}{n-r} \Big/ \binom{N}{n} \ .$$

Hence, by the theory of inverse probability — a theory which Pearson does not associate with Bayes, apparently — we have

$$C_{p,r} \equiv \Pr[p \mid r] \propto p!\,(N-p)!\,/(p-r)!\,(N-p-n+r)! \ .$$

Pearson determines the constant of proportionality by summing an appropriate hypergeometric series (a method advantageously used in his paper of 1907): recourse to the definition of conditional probability results more swiftly in the solution

$$C_{p,r} = \frac{n+1}{N+1} C_{r,p} \ ,$$

the prior probability of the population's containing p marked items and the probability of a sample of size n containing r marked individuals being $(N+1)^{-1}$ and $(n+1)^{-1}$ respectively.

Consideration is next given to the finding of various moments of the hypergeometric series; and in view of the labour that might be incurred in computing the successive terms of such a series, Pearson suggests that the series be replaced by its appropriate frequency curve, found to be

$$y = y_0(rb/n + x)^r(sb/n - x)^s \ ,$$

where $y_0 = M(n+1)!\,/b^{n+1}\, r!\, s!$, and M is "the total number of possible populations from which the sample may have been drawn" [p.157]. Four different determinations of the curve range b are suggested, and Pearson plumps eventually for $b = \sqrt{(N+2)(N-n)}$. Several examples follow.

The paper concludes with two appendices: in the first of these Venn's criticism of inverse probability is examined, while in the second it is a solution by Laplace that falls under the microscope. These appendices have been discussed in the present work in the appropriate chapters.

Before we leave Pearson we might note that he paid some attention to Bayes's Theorem and its applications in his lectures (see E.S. Pearson [1938] for further details). The interested reader might also consult Pearson [1978].

8.15 Miscellaneous

There are a few works which, although they fall in this period, have not been discussed here as I have been unable to examine them in detail. They are Fujisawa [1891] (in which the rule of succession as generalized to $(r+s)$ future events is discussed), Gosiewski [1886] (the inversion of Bernoulli's Theorem), and Nekrassoff [c.1890] (the inversion of Bernoulli's Theorem).

There are also pertinent, though slight, passages in Hagen [1837] and Sorel [1887]. In the first of these works we find a discussion of the rule of succession and of the sun's rising, while in the second it is noted that Bernoulli's Theorem is incomplete without an inverse. Neither contribution, however, is of sufficient depth to warrant detailed discussion in the present work.

8.16 Appendix 8.1

In 1924 William Burnside (1852-1927) published a note "On Bayes' formula" in *Biometrika.* Here, following Poincaré, he stated this formula as

$$\Pr[A_i \mid B] = \Pr[A_i]\Pr[B \mid A_i] \Big/ \sum_1^s \Pr[A_j]\Pr[B \mid A_j], \quad i \in \{1, 2, \ldots, s\} \ .$$

Burnside claimed that the argument given by Pearson [1920a, p.5] was unsatisfactory in that the numerical value of $\Pr[B \mid A_i]$ depended not only on the nature of A_i but also on $\Pr[A_i]$. However, an examination of the problem initially posed by Pearson had persuaded him that the value of $\Pr[A_i]$ had no relation to nor effect on the value of $\Pr[B \mid A_i]$, and he concluded

> There is therefore no reason for supposing that any conclusions drawn from the investigation on p.5 will hold with respect to the statistical problem stated at the beginning of Professor Pearson's paper. [p.189]

8.17 Appendix 8.2

Original text of extracts given in translation in §8.5.

1. *A blindfolded person ... black and white marbles.* Af en Pose er der udtrukken Kugler iblinde; det har vist sig, at et vist Antal af disse vare hvide, et andet Antal sorte, der spørges om Sandsynligheden for, at Posen har et bestemt Indhold, f. Ex. lige mange hvide og sorte; man ved, at Kuglerne ere enten sorte eller hvide.

2. *[This] above-mentioned disparity ... Bayes's theorem.* Den Strid, som i det foregaaende omtales, hidrører udelukkende fra den forskjellige Anvendelse af Bayes' Regel.

3. *Most people will ... assumed to be worth 10 Øre.* Men det vil vistnok af de fleste betragtes som absurd, at Kjøberen skal betale mere for sine Varer, fordi han har sorteret Prøverne, uagtet de enkelte Stykker, før og efter Sorteringen, antages at vaere 10 Øre vaerd.

4. *A contradiction immediately ... subdivisions of time.* Men alligevel kommer der strax Strid, saasnart man tager Hensyn til, at der kan anvendes forskjellige Inddelinger af Tiden.

5. *Bayes's theorem is ... as to necessary causes.* Bayes' Regel i alle saadanne Tilfaelde, hvor man intet véd a priori om de søgte Aarsager.

Notes

Neither indeed would I have put my selfe to the labour of writing any Notes at all, if the booke could as well have wanted them, as I could easilie have found as well, or better to my minde, how to bestow my time.

Marcus Aurelius Antoninus.

Chapter 1

1. This description is from Hacking's introduction to the English translation of Maistrov [1974, p.vii].
2. Forsaking, in this respect, what is described in the 14th edition [1939] of the *Encyclopaedia Britannica* as its "career of plain usefulness" [vol. 3, p.596]. A supplementary volume of the *Dictionary*, soon to be published, will contain a note on Thomas Bayes.
3. See Barnard [1958, p.293].
4. The note, by Thomas Fisher, LL.D., reads in its entirety as follows: "Bayes, Thomas, a presbyterian minister, for some time assistant to his father, Joshua Bayes, but afterwards settled as pastor of a congregation at Tunbridge Wells, where he died, April 17, 1761. He was F.R.S., and distinguished as a mathematician. He took part in the controversy on fluxions against Bishop Berkeley, by publishing an anonymous pamphlet, entitled "An Introduction to the Doctrine of Fluxions, and Defence of the Mathematicians against the Author of the Analyst," London, 1736, 8vo. He is the author of two mathematical papers in the Philosophical Transactions. An anonymous tract by him, under the title of "Divine Benevolence", in reply to one on Divine Rectitude, by John Balguy, likewise anonymous, attracted much attention." For further comments on this reference see Anderson [1941, p.161]. The *Winkler Prins Encyclopaedie* [1948], in the

entry under "Bayes, Thomas", has only this to say by way of biography: "Engels wiskundige (?-1763), omtrent wiens leven wij vrijwel niets weten" [vol. 3, p.396]. It does, however, give a clear and correct discussion of Bayes's Theorem.

5. Anderson [1941, p.160]. If Thomas was born before 25th March (New Year's Day in the old English calendar), the year could be 1701 (old style), though it would be 1702 (new style). The day of his death being the seventh, we find on subtracting the 11 days lost by the calendar reformation of 1752, that his death was on the 27th March 1761 (o.s.). Subtraction of his age at death – viz. 59 – shows that he was either born on or just after New Year's Day 1702 (o.s.), or in 1701 (o.s.). See Bellhouse [1988a].
6. Hacking [1970] and the 15th edition of the *Encyclopædia Britannica* [1980], (vol. I of the *Micropædia*).
7. Holland [1962, p.451].
8. Barnard [1964], Hacking [1970].
9. Holland [1962, p.452], Pearson [1978, p.355] and Wilson [1814]. Bogue and Bennett [vol. II, 1809] point out that "The necessities of the church may render it proper that men should be ministers, who have not enjoyed the advantages of an academical, or even a liberal education" [p.7].
10. Pearson [1978, p.356] states that this academy was founded by the Congregational Fund Board in 1695.
11. Writing of the course of instruction for the Christian ministry, Bogue and Bennett [vol. III, 1810] say "To mathematics and natural philosophy it has usually been judged proper to apply a portion of the student's time. As they tend to improve the mind, and peculiarly to exercise its powers, and call forth their energies, the general influence of both may be favourable to his future labours, and the hearers as well as the preacher experience their good effects" [p.270].
12. H.M. Walker, in her biographical notes appended to the 1967 reprint of de Moivre's *The Doctrine of Chances*, writes "...Thomas Bayes, with whom he [i.e. de Moivre] is not known to have been associated" [p.367]. An outrageous statement is made by Epstein [1967]: speaking of de Moivre, he writes "His mathematics classes were held at Slaughter's Coffee House in St. Martin's Lane: one successful student was Thomas Bayes" [p.5].
13. Holland [1962, p.452]. Proving how little times have changed since then, Kac [1985] in his autobiography says "The way Stan [Ulam] did mathematics was by talking, a work style which goes back to his young days in Lwów, which were spent largely in coffee houses

(mainly in Szocka, which is Polish for "Scottish Café") endlessly discussing problems, ideas and conjectures. Great stuff came of this highly unorthodox way of doing mathematics ..." [pp.xx-xxi]. See also Ciesielski [1987].

14. See Dale [1990].
15. Holland [1962, p.453]. See also James [1867].
16. James [1867].
17. Holland [1962, p.453] states that this move of Thomas's was made in 1731, while Barnard [1958, p.293] merely notes that "he was certainly there in 1731". See also Pearson [1978, p.357] and Timpson [1859, p.464]. However, in the *Minute Books of the Body of Protestant Dissenting Ministers of the Three Denominations in and about the Cities of London and Westminster* the following entry may be found: "Oct. 3rd. 1732. List of approved Ministers of the Presbyterian denomination. $\left.\begin{array}{l}\text{Mr. Bayes Sen}^{\text{r}}.\\ \text{Mr. Bayes Jun}^{\text{r}}.\end{array}\right\}$ Leather Lane." I would suggest therefore that Thomas Bayes moved to Tunbridge Wells somewhat later than the usually cited date of 1731.
18. Barnard [1958, p.293]. Burr [1766] points out further that "a Methodist meeting-house has also been erected at Tunbridge-Wells since the rise of that deluded sect" [p.104].
19. The following descriptions of Tunbridge Wells are from Burr [1766]: "Tunbridge-Wells is situated on the southern side of the county of Kent, just on the borders of Sussex, and about thirty-six miles from London. It is partly built in Tunbridge parish, partly in Frant parish, and partly in Speldhurst parish; and consists of four little villages, named Mount-Ephraim, Mount-Pleasant, Mount-Sion, and the Wells" [pp.98-99]. "An excellent bowling-green, the old assembly-room, and a capacious handsome Presbyterian meeting-house, are all situated upon Mount-Sion" [p.104]. For further details see Holland [1962, pp.453-454].
20. Some doubt as to Ditton's religious convictions exists. *The Imperial Dictionary of Universal Biography* says "The son [i.e. Humphrey], in opposition to the nonconformist wishes of the father [also Humphrey], entered the English church." On the other hand, in the *Dictionary of National Biography* we find "The younger Ditton afterwards became a dissenting preacher at his father's desire". *The New General Biographical Dictionary* is more cautious and merely states "he [i.e. Humphrey] at the desire of his father, although contrary to his own inclination, engaged in the profession of divinity".
21. Holland [1962, p.455].

22. Cajori [1919a] writes "the publication of Berkeley's *Analyst* was the most spectacular mathematical event of the eighteenth century in England" [p.219].
23. Pearson [1978, p.360] claims that Jurin was the secretary of the Royal Society, and gives his name in the list of secretaries on p.357. But this claim is not supported by the *Signatures in the First Journal-Book and the Charter-Book of the Royal Society* [1912].
24. For further details see Pearson [1978, p.360].
25. Extracts from this tract are given in Holland [1962, pp.455-456]: for a detailed discussion of the paper see Smith [1980]. *The Dictionary of Anonymous and Pseudonymous English Literature* cites de Morgan as its source of information on the authorship of this work.
26. Holland [1962] writes "His 1736 tract in defence of the mathematical art was published at an opportune time and was of such merit that it is likely that his election was unanimously agreed to by members of this distinguished body ... We do not believe he obtained election in the manner of some members who contributed nothing of merit but were wealthy enough to pay the high admittance fees and yearly dues" [p.459]. See also Hacking [1970, p.531], Maistrov [1974, p.88], Pearson [1978, p.356], Keynes [1973, p.192], Pearson [1978, p.350] and Timerding [1908, p.44]. The last three of these references incorrectly give Bayes's date of election as 1741. Elected at the same time as Bayes were Walter Bowman and Michel Fourmont: see *Signatures in the First Journal-Book and the Charter-Book of the Royal Society* [1912].
27. Quoted in Holland [1962, p.459].
28. One trusts that this last word is not used with the meaning that *Chambers Twentieth Century Dictionary* assures one it has in booksellers' catalogues!
29. Reprinted in Pearson [1978, p.357]: for details of the signatories see pp.357-358, op. cit.
30. Pearson [1978, pp.360-361]. That Bayes (and his clerical coevals) should have had time to indulge in mathematical and scientific pursuits is hardly surprising when one bears in mind the following passage from an editor of Derham's *Physico-Theology* (first issued in 1713): "The life of a country clergyman is in every respect more favourable to the cultivation of natural science, by experiment and observation, than any other professional employment. He has all the leisure that is requisite to philosophic researches; he can watch the success of his experiments from day to day, and institute long processes without interruption, or record his observations without chasm or discontinuation." Not that their professional duties were ignored, however:

Bogue and Bennett write "As to the quantity of labour performed by dissenting ministers of evangelical principles (the religious principles of the old nonconformists), they need not blush at a comparison with those of the preceding times. To the two public services of former times, a third has now been generally added, and evening lectures are become in most congregations the stated practice. In the course of the week too, there is a public season for worship in one of the evenings, so that the minister has to preach four times from Sabbath to Sabbath" [vol. IV, 1812, pp.343-344].

31. This notebook, although strictly speaking to be numbered among the adespota to be attributed to Bayes, bears on its first page the handwritten words "This book appears to be a mathematical notebook by Rev. Thomas Bayes, F.R.S.. The handwriting agrees very well with papers by him in the Canton papers of the Royal Society Vol. 2, p.32." This note is dated 21-1-1947 and is signed by M.E. Ogborn.
32. Holland [1962, pp.456-459]. Richard Price (1723-1791) was a dissenting minister, mathematician and political economist. His actuarial work led to his election as Fellow of the Royal Society in 1765, and the honorary degree of D.D. was awarded him on the 7th August 1767 by Marischal College, Aberdeen (*not* by the University of Glasgow, as sometimes stated). An LL.D. from Yale followed in 1781. William Morgan (1750-1833) was the son of Price's sister Sarah. His actuarial writings led to the award of the gold medal of the Royal Society and a fellowship.
33. The key is to Aulay Macaulay's system. The shorthand actually used by Bayes in the notebook has been identified as being basically that derived in the 17th century by Thomas Shelton and modified by Elisha Coles. See Holland [1962, p.458] and Home [1974-1975, p.83].
34. See Archibald [1926], Molina and Deming [1940], Pearson [1924a, p.404] and Pearson [1978, p.358]. For details of the relative contributions of de Moivre and Stirling see Tweedie [1922, pp.203-204], and, for a detailed examination of de Moivre's work, see Schneider [1968]. De Moivre wrote a short paper on the approximation of the greatest term of $(a+b)^n$ as a series. This paper, originally printed on 12th November, 1733, was later translated, with only minor alterations, in the second [1738] and third [1756] edition of de Moivre's *The Doctrine of Chances.* The first printing had only a limited circulation: indeed, de Moivre prefaced his later translation with the words "I shall here translate a Paper of mine which was printed November 12, 1733, and communicated to some Friends, but never yet made public, reserving to myself the right of enlarging my own Thoughts, as occasion shall require" [1756, p.242]. If Barnard's speculation that Thomas Bayes

might have learned mathematics from de Moivre is correct [Barnard 1958, p.293], it is tempting to conjecture further that Bayes might have been one of the privileged circle to see the 1733 *Approximatio ad Summam Terminorum Binomii* $(a+b)^n$ *in Seriem expansi.* However, several papers on infinite series had been published in the *Philosophical Transactions* before Bayes's paper on this subject (communicated 1761, published 1764), so one should perhaps not rely too much on the possible friendship between de Moivre and Bayes as an explanation of the latter's consideration of "Stirling's Theorem".

35. In Morgan's biography [1815] of Price may be found the following remarks: "it [i.e. Bayes's Essay] was presented by Mr. Canton to the Royal Society, and published in their Transactions in 1763. — Having sent a copy of his paper to Dr. Franklin, who was then in America, he [i.e. Price] had the satisfaction of witnessing its insertion the following year in the American Philosophical Transactions." The records of the American Philosophical Society for 1762-1766 have apparently been lost, and recent research has failed to find any record of this alleged communication by Price to Franklin. It seems that on this point Morgan erred. Walter Ashburner, a direct descendant of Price's sister, in a memorandum sent to the president of the Massachusetts Historical Society in 1903 (see *Letters to and from Richard Price*) in fact wrote "William Morgan was a distinguished mathematician ... but he was not a good biographer" [p.4]. Thomas [1924] describes Morgan's *Memoirs* of Price as "inadequate, and, unfortunately, often inaccurate" [p.iii]. For biographical notes on Price see Bogue and Bennett, [vol. IV, 1812, pp.421-425] and Holland [1968].
36. See Clay [1895] and Leader [1897].
37. See Pearson [1978, p.355] and Stephen [1885].
38. The saint's day seems to be a movable feast: the 14th edition [1939] of the *Encyclopædia Britannica*, the source of the quotation cited, gives it variously also as August 24th (1666) and 12th; W.B. Forbush's edition of *Fox's Book of Martyrs* [1926] gives it as August 22nd, while Charles Dickens, in his *A Child's History of England*, gives August 23rd.
39. *Encyclopædia Britannica*, 14th edition [1939, vol. 8, p.470]. Ejected at the same time from "a good living at Moreton, in Essex, near Chipping Ongar" was the father of Edmund Calamy (see Calamy [1830, vol. I, p.65]). For further details of this ejectment see the entry "Nonconformity" in Hastings [1967].
40. Enclosures in [...] are the present author's.
41. The description is from Wilson [1814]; for further details of Frankland see Holland [1962, p.452]. Bogue and Bennett [vol. I, 1808, p.225]

describe Frankland as "an eminent dissenting tutor, who taught university learning."

42. The distinction between such dissenting academies and the dissenting schools of that period is succinctly discussed in Holland [1962, p.452]: see also Dale [1907]. Parker [1914] notes that while the Dissenting Schools were charity foundations, the Dissenting Academies "were schools of university standing" [p.50].
43. Pearson [1978, p.355]. Parker [1914] points out further that, on Frankland's death in 1698, his academy at Rathmell declined: it was succeeded by one under Chorlton's tutorage. Parker [1914, p.121] finds no justification for the claim (see, for example, Holland [1962, p.452]) that Chorlton's, and thus Frankland's, Academy may be viewed as one of the forerunners of Manchester College, Oxford. For further details of Frankland's Academy see Bogue and Bennett [vol. I, 1808].
44. Wilson [1814, p.396].
45. The others were Joseph Bennett, Thomas Reynolds, Joseph Hill, William King, Ebenezer Bradshaw and Edmund Calamy. Bradshaw and Calamy were the sons of ejected ministers (see Calamy [1830]).
46. Pearson [1978, p.355]. See also Barnard [1964] and the more correct Barnard [1958, p.293]. The latter, however, mentions six, rather than seven, ordinees.
47. Stephen [1885]. The following details are from Calamy [1830]: the ordainers on this occasion were Dr. Samuel Annesley, Mr. Vincent Alsop, Mr. Daniel Williams, Mr. Richard Stretton, Mr. Matthew Sylvester, and Mr. Thomas Kentish. The proceedings opened with a prayer by Dr. Annesley, followed by Mr. Alsop's preaching from 1 Peter v.1, 2, 3. Mr. Williams then prayed, made a discourse concerning the nature of Ordination, and read the names and testimonials of those to be ordained. Confessions of faith on the part of the latter and prayers then followed, the whole concluding with a solemn charge, a psalm, and a prayer. "The whole," according to Calamy [1830, p.350], "took up all the day, from before ten to past six o'clock." Before being ordained, each candidate had to defend a thesis upon a theological question, the several ministers present warmly opposing it. Joshua Bayes's question was "An Deus sit Essentiâ suâ omnipresens?" *Aff.* See also Bogue and Bennett [vol. II, 1809, pp.121-122].
48. Calamy [1830].
49. In Southwark and Leather Lane, according to Pearson [1978, p.355], though Wilson [1814] is more restrained and merely writes "It does not appear where Mr. Bayes spent the first years of his ministry, but it was, most probably, in the neighbourhood of London" [p.397].

These peregrinations are not mentioned by Holland [1962] who writes "Joshua was ordained in 1694 ... at Little St. Helen's Meeting House and was the minister at Box Lane, Bovingdon, Herts., until 1706" [p.451].

50. Joshua succeeded Mr. Batson (Wilson [1814, p.312]).
51. Stephen [1885]. Sheffield died in 1726 (Calamy [1830, vol. II, p.487]).
52. Pearson [1978] refers to this gentleman as "Brook Taylor" [p.355]: this is clearly one of the slips which he would indubitably have corrected had he prepared his lectures for publication (see the preface to Pearson [1978]).
53. On early English presbyterianism see Anderson [1941, p.160].
54. Pearson [1978, p.355] and Stephen [1885]. For details of the other ministers involved in this work see Bogue and Bennett [vol. II, 1809, p.297].
55. James [1867], Stephen [1885] and Wilson [1814].
56. James [1867].
57. James [1867]. Joshua was deemed a Calvinist, "that is, such as agree with the Assembly's Catechism", and Thomas an Armenian, "or such as are far gone that way, by which are meant such as are against particular election and redemption, original sin at least the Imputation of it, for the power of man's will in opposition to efficatious Grace, and for Justification by sincere obedience in the room of Christ's righteousness &c." (The quotations are from Anon (b), pp.87 & 88.)
58. For details of the Salters' Company see the 14th edition [1939] of the *Encyclopædia Britannica* [vol. 14, pp.236-237]. The Merchants' Lecture was originally established in 1672 in Pinners' Hall, Broad Street, but after an attack on heresies by Daniel Williams in one of his lectures, a Presbyterian Lectureship was set up at Salters' Hall in 1694 (Dale [1907, p.481]).
59. Pearson [1978, p.355], Stephen [1885] and Wilson [1814].
60. Such a statement is made by Barnard [1958, p.293], [1964], Hacking [1970, p.531] and Maistrov [1974, p.88]. Holland [1962, p.452] points out the error, and his assertion is vindicated by the *Signatures in the First Journal-Book and the Charter-Book of the Royal Society* [1912].
61. Wilson [1814] writes "the inscription upon his tomb-stone says, in his 52nd year, but it is evidently a mistake" [p.398]. The vault, in which the mortal remains of Thomas and other members of the Bayes family were also interred, and which had fallen into disrepair, was restored in 1969, the erroneous phrase being omitted from the inscription. An engraving of Joshua Bayes may be found in Wilson [1814]: it is copied from the portrait in Dr Williams's Library.

62. Wilson [1814].
63. This cemetery is referred to by Calamy [1830] as " ... the new burial place for Dissenters, by Bunhill Fields, near London ...". Richard Price and his wife Sarah are also buried here: their tomb, like many others here, is sorely in need of restoration. Hicks [1887] has suggested that the original name of the burial ground was Bon- or Bone-hill Fields; this is disputed by others.
64. James [1867] describes the congregation as "mainly tradesmen" [p.670]. Bogue and Bennett [vol. III, 1810, p.495] write "Among the churches in London, the first rank of respectability was assigned to ... Joshua Bayes". Joshua played an active rôle in the nonconformist circles of his time, serving — sometimes as chairman — on the committee of the three denominations (Presbyterian, Congregational and Antipædobaptist) which saw to many matters, in and around London , pertaining to nonconformity (see *Minute Books of the Body...*). Further comment on Bayes's meeting house is given in Anon (b), as follows: "This meeting house is about 15 square of building, with 3 Galleries. In 1695, Mr. Buris was minister to this people, but not living many yeares after that time Mr. Christopher Taylor was chosen Pastor in his room. he was accounted a G^{t}man [≡ gentleman] of a bold spirit & a good preacher & about 1714 Mr. Bayes was chosen to assist him. Mr. Taylor dying about 1724 Mr. Bayes succeeded as pastor, & since that time Mr. Bayes Junr was chosen to assist his father. This congregation was never large. but were a people generally of substance. It does not certainly appear what difference there is between the congregation in 1695 & the present, tho it is apprehended to be somewhat less. Mr. Bayes is a judicious serious and exact preacher and his composures appear to be laboured. He is of a good temper & well esteemed by his brethren. Mr. Bayes is a lecturer at Salter's Hall" [p.35]. It is also recorded here that "his congregation collects £100 annually for a fund to assist country ministers" [p.89].
65. Holland [1962]: Rebecca's name does not appear on the restored Bayes-Cotton vault in Bunhill Fields, while the year of Ann's death, quoted here from Clay [1895], is given on the restored vault as 1758: the change might well have come about at the time of the restoration. The birth-years given here are found by subtracting the age at death from the year of death.
66. On the 22nd September, according to the Bunhill Fields tombstone; but the obituary in *The Gentleman's Magazine* ... 31 [1761, p.188] has "Aug. 16. At Brighthelmstone, Mrs. Bayes, wife of Sam. Bayes esq. of Clapham".

67. This tract is discussed in Pearson [1978, p.359 et seqq.]. Quotations are given in Barnard [1958, p.294] and Holland [1962, pp.454-455]. It was apparently unknown to de Morgan: see his [1860]. On probability as a guide in religious matters as in secular affairs see Shapiro [1983, p.80].
68. Writing of Henry Grove, Bogue and Bennett [vol. III, 1810] state that his "theological learning was considerable, and his attainments in polite literature were superior to those of most of his brethren.... Unhappily Mr. Grove was not sound in the faith; and as he advanced in years, he contracted a more keen and rooted aversion to evangelical doctrines" [p.275]. Bogue and Bennett (loc. cit. and vol. I) also mention that Grove was a tutor in pneumonology and ethics at an academy formed by Matthew Warren at Taunton, Somerset. On Robert Darch's resignation, mathematics and natural philosophy were added to Grove's department, and on Stephen James's resignation in 1725, he was appointed to the chair of divinity.
69. The *Dictionary of Anonymous and Pseudonymous English Literature* (Halkett and Laing [1926]) attributes *Divine Benevolence* to Thomas Bayes, citing as authority Darling's *Cyclopaedia Bibliographica* [1852-1854]. *The National Union Catalogue*, on the other hand, attributes it to Joshua Bayes. The definitive statement is perhaps however made by Price [1787], who, in a footnote on page 429 of his paper, states that "The author [of *Divine Benevolence*] was Mr. Bayes, one of the most ingenious men I ever knew, and for many years the minister of a dissenting congregation at Tunbridge Wells."
70. Holland [1962] describes Bayes's defence as "the most scathing reply" [p.455].
71. Anderson [1941, pp.160-161] and Wilson [1814, p.401].
72. For comments on Whiston's (curious) views on the universe see Gardner [1957, pp.33-34]; for details of Whiston see Anderson [1941, p.160], Holland [1962, p.454] and Pearson [1978]. Commenting on Whiston's rapture with the writings of the early fathers, Bogue and Bennett [vol. III, 1810, p.216] say "Nothing more is necessary to characterize the man."
73. Barnard [1958, p.294].
74. Pearson [1978] and Whiston [1753, pp.325-326].
75. Pearson [1978, pp.349-350].
76. On the amount a clergyman of that period could earn, see the entry "passing rich" in Brewer's *The Dictionary of Phrase and Fable* [1978, p.947].

77. The formation of the Independent church is reported by Timpson [1859, p.466] as follows: "Having heard of a faithful dissenting minister at Goudhurst, they [i.e. Thomas Baker, Edward Jarrett, an aged man, named Bunce, and Robert Jenner] went one Lord's day, May 21st, 1749, to hear him. Delighted with his sermon, they conversed with him, and he informed them that the Rev. Mr. Jenkins would be ordained pastor of the Independent church at Maidstone on the following Wednesday. They went to that service, and became thus acquainted with the Rev. Mordecai Andrews, of London; and he came with Mr. Booth, for the season, to the Wells, where he engaged the Presbyterian chapel, from the Rev. Mr. Bayes, its minister. They enjoyed the gospel preached by ministers sent from London for nearly a year, until Easter Sunday, in 1750, when Mr. Bayes resumed his pulpit, disliking the doctrine of the Independents, and they again attended at the Established Church, for the sake of the Lord's Supper." The time-scale has been differently recorded: see Anon (a), "Early Presbyterianism at Tunbridge Wells", where it is stated that the dissenters used Bayes's meeting-house from 1743 to 1750. Thomas [1924], quoting Drysdale's *History of the Presbyterians in England* writes "Presbyterian ministers were, as a whole, much more dignified and clerical in tone than their Independent brethren" [p.25].
78. See Holland [1962, p.456] and Timpson [1859, p.464]. The latter writes of Bayes "he was a gentleman of fortune; but though he was said by the Rev. Mr. Onely, a clergyman of Speldhurst, to have been the best Greek scholar he had ever met with, he was not a popular preacher, nor evangelical in his doctrine."
79. Was this the original "Disgusted, of Tunbridge Wells"?
80. Holland [1962, p.456]. The quotation is from the "Church Book" of the Independents, which, according to Miss J. Mauldon of the Tunbridge Wells Library, has disappeared.
81. Timpson [1859] writes that Bayes "bequeathed his valuable library to his successor, the Rev. William Johnson, M.A., who became minister of the chapel in 1752" [p.464]. This statement has been repeated by Barnard [1958, p.294], Holland [1962, p.459], and Strange [1949, p.17], but there is no mention of such a bequest in Bayes's last will and testament.
82. He directed, in his will, that his funeral expenses "may be as frugal as possible" [Holland 1962, p.459]. The date of his death is given variously as the 7th (*The Gentleman's Magazine and Historical Chronicle* XXI (1761), p.188), the 14th (*The London Magazine, or Gentleman's Monthly Intelligencer* XXX (1761), p.220) and the 17th (Rose [1848]). See also Anderson [1941, p.162], Jones [1849, p.8], Waller [1865] and

Wilson [1814]. According to Jones [1849], the original inscription read "The Rev. Thomas Bayes, son of the said Joshua, died April 7th, 1761, aged 59 years." What is reputed to be a portrait of Thomas Bayes may be found on page 335 of O'Donnell [1936], above the legend "Rev. T. Bayes Improver of the Columnar Method developed by Barrett." No reference to the source from which the portrait is taken is given, and O'Donnell is elsewhere unreliable (see Dale [1988a], and, for further comment, Bellhouse [1988a], O'Hagan [1988] and Stigler [1988]). The portrait was reprinted in Press [1989] and Stigler [1980a].

83. This vault, having fallen into disrepair, was restored "In recognition of Thomas Bayes's important work in probability ... in 1969 with contributions received from statisticians throughout the world." According to O'Hagan [1988], the tomb, after being sadly weathered, is once again in good condition.
84. According to Holland [1962, p.452], Coward's (later known as the Hoxton Academy) was the *only* academy in the London area from 1716 to 1730. For further details of the Hoxton Academy see McLachlan [1931, pp.18, 118, 120].
85. Writing of Chauncy, Bogue and Bennett [vol. II, 1809, p.35] say "Though a learned divine, he was not a popular preacher".
86. Although commenting with approval on Ridgeley's suitability as a theological tutor, Bogue and Bennett [vol. III, 1810] cannot forbear from noting that "had [Ridgeley's] style but possessed neatness, elegance, and force, what an additional value it would have imparted to his ample treasures of sacred truth" [p.283]. Ridgeley died in 1734, in his 67th year.
87. Dale [1907, p.501] writes "Eames, though distinguished as a scholar, was disabled for the ministry by a defect in the organs of speech, and by a pronunciation that was 'harsh, uncouth, and disagreeable'. He once attempted to preach, but broke down, and never repeated the experiment." Bogue and Bennett [vol. III, 1810] more sympathetically merely say "extreme diffidence and a defect in the powers of elocution deterred him from preaching more than one sermon" [p.284]. According to these authors (loc. cit.) Dr. Isaac Watts once remarked to one of his students "your tutor [i.e. Eames] is the most learned man I ever knew." McLachlan [1931] writes of Eames that "he was the only layman ever placed in charge of an academy, and, unlike most other tutors, published nothing. He was eminent alike in classics and mathematics, attracted to his lectures, despite a lack of oratorical gifts, some of the most promising pupils of other academies, and after his death his lectures continued to be used in manuscript by

tutors of academies other than his own " [p.18]. Eames died suddenly in June 1744.

88. The rules of Doddridge's Academy are listed in Appendix III in Parker [1914]. Bogue and Bennett [vol. III, 1810] regard this Academy as a revival of one established at Kibworth, Leicestershire, by John Jennings in 1715, and temporarily suspended on his death. Doddridge had in fact been a student at Jenning's Academy. Bogue and Bennett (op. cit., p.480) criticize Doddridge's lectures for having "a tendency to generate a controversial spirit", and add further [p.482] that "As a man, he [i.e. Doddridge] cannot be said to have been endued with genius in the highest sense, nor was his learning very profound, though it was extensive, rendering him respectable rather than eminent."

Chapter 2

1. See Brewer, *The Dictionary of Phrase and Fable* [1978, p.938].
2. This pudency (or prudency?) seems first to have been noticed by William Morgan [1815, p.24], who wrote "On the death of his friend Mr. Bayes of Tunbridge Wells in the year 1761 he [i.e. Price] was requested by the relatives of that truly ingenious man, to examine the papers which he had written on different subjects, and which his own modesty would never suffer him to make public." Hacking [1965, p.201] writes "Cautious Bayes refrained from publishing his paper; his wise executors made it known after his death. It is rather generally believed that he did not publish because he distrusted his postulate, and thought his scholium defective. If so he was correct." In 1971, however, and writing on this same point, Hacking says of Bayes that "His logic was too impeccable" [p.347]. Stigler [1986a, p.130] suggests that any reluctance Bayes might have felt towards publication could perhaps be attributed to difficulty in the evaluation of the integral of his eighth proposition. Good [1988] mentions three reasons for non-publication, viz. (i) the implicit assumption that a discrete uniform prior for r (the number of successes) implies a continuous uniform prior for p (the physical probability of a success at each trial), (ii) the essential equivalence of the assumptions as to the two priors in (i) when N (the number of trials) is large, and (iii) the first ball (by means of which p is determined) is essentially a red herring.
3. Canton is described in Pearson [1978] as "the Royal Society Secretary" [p.369]. However his name does not appear in Pearson's list of secretaries on p.369, nor in *Signatures in the First Journal-Book and the Charter-Book of the Royal Society*: nor is he listed as holding office in the Royal Society in *The Record of the Royal Society of London* [1912].

4. Commenting on this letter, Savage, in an unpublished note of 1960 (printed as the Appendix to the present work), wrote "this is apparently the first notice ever taken of asymptotic series". On this point see Appendix 1.1 to Chapter 1. Deming (see Molina and Deming [1940, p.xvi]) states that the manuscript was submitted to the Royal Society by Price.
5. On works attributed to Bayes see Pearson [1978, p.360-361].
6. For reprints and summaries of the Essay the following should be consulted: Barnard [1958] (reprinted in Pearson and Kendall [1970]: note also comment in Sheynin [1969]), Bru and Clero [1988], Dinges [1983], Edwards [1978], Fisher [1956/1959], Molina [1931], Molina and Deming [1940] (reviewed by Lidstone [1941]), Press [1989] and Timerding [1908] (see Pearson [1978, pp.366, 369]). The 1918 catalogue of the Printed Books in the Edinburgh University Library lists, as number 0*22.14/1, a work entitled "A Method of Calculating the Exact Probability of All Conclusions founded on Induction. By the late Rev. Mr. Thomas Bayes, F.R.S." I am indebted to Mrs. Jo Currie of the Special Collections section of that library for the information that this work is in fact merely a reprint of the Essay: with it is bound the supplement (listed as 0*22.14/2), both being reprinted in this edition in 1764.
7. Unlike all Gaul.
8. Price's nephew, William Morgan, writes: "Among these [i.e. Bayes's papers] Mr. Price found an imperfect solution of one of the most difficult problems in the doctrine of chances ..." [1815, p.24]. Later he speaks of Price as "completing Mr. Bayes's solution". It seems clear from Price's introductory remarks to the essay, however, that the major part of the latter was presented as Bayes had left it, though Price did expand on the Rules given by Bayes.
9. See de Finetti [1972, p.159].
10. See Savage [1960]. Hacking [1971] finds in Price's introduction "perhaps the most powerful statement ever, of the potential relations between probability and induction" [p.347].
11. Condorcet [1785, p.lxxxiii] traces the idea to Jacques Bernoulli & de Moivre.
12. Bernoulli's Law of Large Numbers, in modern terms, runs as follows: let $\{A_i\}$ be a sequence of independent events with $\Pr[A_i] = p$, where i is a natural number. For every $\epsilon > 0$,

$$\Pr[|S_n/n - p| \geq \epsilon] \to 0 \qquad \text{as } n \to \infty,$$

where $S_n = \sum_1^n I_k$. (Here I_k denotes the indicator function of A_k, i.e. that function taking on the values 1 on A_k and 0 off A_k.) Baker [1975, p.162] does not find it surprising that the foundations for the inversion of Bernoulli's Theorem were laid in England; he traces this to some aspects of Newtonian philosophy. Further he notes (op. cit. p.166) the relationship between Bayes's passage from a physical model of probability to an epistemological interpretation, and Price's appendix showing clear evidence of the logic of Hume's *Treatise*. For further details on this last point see Gillies [1987].

13. Price also comments [p.373] on the defects of the asymptotic nature of de Moivre's results. The respect in which Price held de Moivre is commented on by Morgan [1815, p.39] as follows: "In the first of these papers [the two published in the *Philosophical Transactions* in 1770] he corrected an error into which M. De Moivre had fallen; ... From the high opinion he entertained of the accuracy of De Moivre, he conceived the error to be his own rather than that of so eminent a mathematician, and in consequence puzzled himself so much in the correction of it, that the colour of his hair, which was naturally black, became changed in different parts of his head into spots of perfect white." See Dale [1988b] for a discussion of the relationship between Bayes's theorem and the inverse Bernoulli theorem. In the twelfth volume (1763-1769) of the *Philosophical Transactions (Abridged)* we find the following comment on Bayes's problem: "In its full extent and perfect mathematical solution, this problem is much too long and intricate, to be at all materially and practically useful, and such as to authorize the reprinting it here; especially as the solution of a kindred problem in Demoivre's Doctrine of Chances, p.243, and the rules there given, may furnish a shorter way of solving this problem. See also the demonstration of these rules at the end of Mr. Simpson's treatise on 'The Nature and Laws of Chance'." [p.41]. The reference to de Moivre being to his *Approximatio ad Summam Terminorum Binomii* $(a+b)^n$ *in Seriem expansi*, it appears that there was also some confusion here between Bayes's result and the inverse Bernoulli theorem.
14. The existence of this supplement to the Essay is not mentioned by Barnard [1958] (see Sheynin [1969, p.40]), although note is taken of it in the note at the end of the reprint of his article in Pearson & Kendall [1970].
15. See Sheynin [1969] for further comments and discussion.
16. Bayes's formulation of the problem is viewed by de Finetti, a leading subjectivist, as unsatisfactory — see his [1972, p.158].
17. This note is printed as the Appendix to the present work.
18. Seven definitions and seven propositions — is their purpose analogous to that of Carroll's maids and mops?

19. But see Price's introduction to the Essay, foot of p.372.
20. See de Moivre [1756].
21. Savage [1960] calls it "of course most interesting". See also Bernoulli's *Ars Conjectandi* and de Moivre's *The Doctrine of Chances.*
22. See Savage [1960].
23. Edwards [1974, pp.44-45].
24. See Fine [1973, pp.60-61], Hacking [1975, pp.152-153] and Shafer [1976].
25. Perhaps this supports de Finetti's [1937] view that the idea of "repeated trials" is meaningless for subjective probability — see Kyburg and Smokler [1964, p.102].
26. See Edwards [1978, p.116] for references.
27. More correctly, one postulate in two parts.
28. Some writers, including Pearson [1920a] and Fisher [1959], have referred to Bayes's table as a billiard table (which of course is not square). One might wonder whether such referral, occurring as it does in connexion with matters of chance, perhaps embodies a pun, as the word "hazard" was formerly used for a pocket of a billiard table.
29. This specifies a uniform distribution in the *plane*: the deduction of a uniform distribution over the *side* of the table is tacit.
30. Note also Edward's [1978, p.116] reformulation.
31. "A deliberately extramathematical argument in defense of Bayes' postulate", Savage [1960].
32. This seems to imply exchangeability: for example, if a coin is tossed three times, the scholium says that

$$\Pr[3 \text{ heads}] = \Pr[2 \text{ heads}] = \Pr[1 \text{ head}] = \Pr[0 \text{ heads}]$$

($= \frac{1}{4}$, presumably!), and hence, for example

$$\Pr[HHT] = \Pr[HTH] = \Pr[THH] = \frac{1}{12} \ .$$

See Edwards [1974, p.48] and Zabell [1982].
33. See Savage [1960].
34. See Dale [1982] and, for a contrary assertion, Edwards [1978, p.117].
35. For comment on Bayes's evaluation of the incomplete beta-integral see Lidstone [1941, pp.178-179], Molina and Deming [1940, pp.xi-xii], Sheynin [1969, p.4], [1971a, p.235], Timerding [1908, pp.50-51] and Wishart [1927]. The last of these authors points out [p.10] an erroneous value given by Bayes and undetected by Timerding. For a detailed study of the incomplete beta-function see Dutka [1981].

36. See Pearson [1978, p.369].
37. Notice that this is also framed in terms of ratios of causes — see p.406 of the Essay.
38. Note the comment by Waring in Todhunter [1865, art.839]. See also Savage [1960], and Pearson [1978, pp.365-366]. It seems that Price, and not Bayes, was perhaps the first to frame a sort of "rule of succession" argument. See Keynes [1921/1973, chap. XXX] for commentary on this rule.
39. Dinges [1983, p.95] is one of the few authors to acknowledge this problem as being posed by Bayes. The mentioning of events occurring *under the same circumstances* as they have in the past can perhaps be traced back to G. Cardano (1501-1576), in whose *Liber de Ludo Aleae*, caput VI, we read "Est autem, omnium in Alea principalissimum, aequalitas, ut pote collusoris, astantium, pecuniarum, loci, fritilli, Aleae ipsius."

Chapter 3

1. In his review of Molina and Deming [1940], Lidstone [1941] says that Todhunter's criticism is "rather harsh, and it is in any case based on a particularly high standard of comparison" [p.179]. He also notes that "De Morgan, who was no bad judge, was much more appreciative" (loc. cit.).
2. *The Doctrine of Chances* [1756, pp.1-3].
3. Savage [1960].
4. Savage [1960] remarks that the derivations of the propositions " . . . are beclouded by the idea that numbers are a little more shameful than ratios".
5. See Shafer [1978, p.345]. Shafer also points out (loc. cit.) that de Moivre was apparently the first to give a statement of a rule of additivity for probabilities, viz. $\Pr[E] + \Pr[\overline{E}] = 1$.
6. Hartigan [1983, p.6] has drawn attention to the fact that Bayes's definition "describes how a person *ought* to bet, not how he *does* bet." See also Stigler [1982a, p.250]. For a detailed discussion of whether Bayes's concept of probability places him in the subjective or objective school see Dinges [1983, §6]. Note also Shafer [1985].
7. According to Savage [1960], this section of the Essay contains but a *germ* of a theorem about the probability of causes (so often wrongly attributed to Bayes). The present discussion owes much to Edwards [1978]. For further comment see Dinges [1983, §4]: the latter asserts that Bayes shines in this section "als ein Experte der damaligen Integralrechnung" [p.80].

8. These are phrased respectively by Edwards [1978, p.117] as "the event 'that the first ball thrown lies in a particular interval on the table' " and "the event 'that the probability at each of the subsequent trials lies in a particular interval'."
9. Edwards [1978, p.117] states "If, therefore, all Bayes' propositions are interpreted in terms of the event 'that the probability lies in a particular interval' rather than 'that the ball lies in a particular interval', the first postulate, of a uniform table, is redundant." The well-known birthday problem (see Feller [1968, p.33]) is usually stated under the assumption that birthdays are uniformly distributed throughout the year. Bloom [1973], in solving a problem posed by Knight, shows that the solution $1 - (365)_n / 365^n$ obtained under (1) independence and (2) uniform distribution of birthdays, is a lower bound attained only when all days are equally probable. The point is further explored in Munford [1977].
10. Molina [1930, p.383] points out that this important fact was omitted by Todhunter: "Failure to appreciate this point kills the significance of Bayes' scholium". (Todhunter also failed to discuss the scholium.)
11. On the equivalence stated in this proposition see my preceding remarks and Edwards [1978, p.117].
12. See Maistrov [1974, p.92].
13. I have throughout interpreted integrals as areas and vice versa.
14. The argument advanced by Bayes here has been well summarized by Edwards [1978, p.111]. For further discussion of the Scholium see Dinges [1983, §8] and Gillies [1987, §5].
15. See de Finetti [1932] and Feller [1966, p.224].
16. Stigler [1982a, p.250] cites Karl Pearson, R.A. Fisher, Ian Hacking and Harold Jeffreys as misinterpreters of the argument.
17. Bayes of course does not qualify the noun.
18. Compare this with the discussion in §2.6.
19. Notation altered.
20. Stigler [1982a, p.253] finds it "tempting to speculate that it was Reverend Thomas Bayes's experience as a minister that made this approach more congenial than his original postulate of an *a priori* uniform distribution for θ [my x]: All men may know the works of God, and through these works know God, but only men of great faith know God directly."
21. "Unlike the marvellously flexible principle of insufficient reason, which is immediately (if dubiously) adaptable to any parametric model" [Stigler 1982a, p.253].

22. In a footnote on p.405 Price writes "There can, I suppose, be no reason for observing that on this subject unity is always made to stand for certainty, and $\frac{1}{2}$ for an even chance."
23. See Dale [1982].
24. For good discussions of the rule of succession see Keynes [1921/1973, chap. XXX] and Zabell [1989b]. Gillies [1987, §3] distinguishes between "Price's rule of succession", viz.

$$\int_a^b x^n\,dx \Big/ \int_0^1 x^n\,dx = b^{n+1} - a^{n+1} \ ,$$

and "Laplace's rule of succession", viz.

$$\int_0^1 x^{n+1}\,dx \Big/ \int_0^1 x^n\,dx = (n+1)/(n+2) \ .$$

By the unappellational, or uneponymised, term we shall always mean the latter of these rules. Herschel, in his 1850 review of Quetelet on probabilities, showed a clear understanding of the two rules. He wrote "the expectation that the sun will rise tomorrow, grounded on the sole observation of the fact of its having risen a million times in unbroken succession, has a million to one in its favour. But to estimate the probability, drawn from that observation, of the existence of an influential cause for the phenomenon of a daily sunrise, we have to raise the number 2 to the millionth power ... and the ratio of this enormous number to unity, is that of the probability of the phenomenon having happened *by cause*, to that of its having happened *by chance*" [Herschel 1857, p.415].
25. Price was well acquainted with Condorcet: see Pearson [1978, p.375].
26. For further comment on the solution see Lidstone [1941, p.179] and Pearson [1978, pp.368-369]. Pearson (loc. cit.) finds neither Todhunter nor Timerding illuminating on this point. Note also the discussions in Gillies [1987, pp.332-333] and Zabell [1988a], [1989b].
27. Dinges [1983, pp.67-68] points out that, while Price's intervals should not be interpreted as confidence intervals, they could perhaps be considered in a fiducial context. See Barnard [1987] for a justification of R.A. Fisher's claim that the term "probability" is used in the same sense both in the fiducial argument and in Bayes's Essay.
28. An example described by Lidstone [1941, p.179] as "now notorious".
29. See, for example, §XI, "Of the probability of chances", of his *Treatise of Human Nature*, where we find the words "One wou'd appear

ridiculous, who wou'd say, that 'tis only probable the sun will rise to-morrow, or that all men must dye; tho' 'tis plain we have no further assurance of these facts, than what experience affords us." Further, in his *Essays Literary, Moral and Political*, we find in §4, "Sceptical doubts concerning the operations of the understanding" of the *Inquiry concerning Human Understanding*, the sentence "*That the sun will not rise to morrow*, is no less intelligible a proposition, and implies no more contradiction, than the affirmation, *that it will rise*." And as a footnote to §6, "Of probability", in the same essay, we have "... we must say, that it is only probable all men must die, or that the sun will rise tomorrow."

30. No doubt our observer is "an agéd, agéd man" by now!
31. Lidstone [1941, p.179] emphasizes Price's recognition of the distinction between *casual* and *causal*.
32. For a discussion of (part of) the Essay from a decision-theoretic viewpoint see Dinges [1983, §2]. While noting that Bayes was not decision orientated, Ferguson [1976, p.338] states that in the Essay "Not even the probability of the occurrence of the event on the next trial is calculated", though such a calculation is given in the Appendix. He ascribes the first such calculation to Laplace in 1774.

Chapter 4

1. The spelling is from his own hand, rather than as given by Todhunter [1865]. The date of his death is incorrectly given as 1796 by Lancaster [1968].
2. Keynes [1921/1973], Pearson [1978] and Todhunter [1865] all give the date 1771: this seems to be wrong (the first edition was printed in 1761).
3. For further discussion see Molina and Deming [1940, p.xv]: this paper is reviewed by Lidstone [1941]. A general discussion of Bayes's work on infinite series may be found in Dale [1991].
4. See Chapter 1, Note 35.
5. Deming in fact gives a fourth reason, which I do not find very compelling.
6. For a form in which Bernoulli numbers are used see Archibald [1926, p.675].
7. See Dale [1991] and Molina and Deming [1940, p.xvi].
8. One must concur with Todhunter [1865, art.553] that "these investigations are very laborious, especially Price's."
9. Perhaps even Price found these calculations wearisome!

10. See Sheynin [1969].
11. This value is arrived at by finding the second derivative of the beta density (or the "Bayessche Kurve", as Timerding [1908, p.51] calls it) $k\, x^p(1-x)^q$.
12. For further details of this approximation see Sheynin [1969, §5] and Timerding [1908, pp.58-59].
13. Shelton published two shorthand books: *Tachygraphy* [1641] and *Zeiglographia* [1654]. It is the latter of these that is most closely related to Coles's work of 1674.
14. This notation is of course not used by Bayes.
15. References in this section are to the second edition of 1768. The *Four Dissertations* was reviewed in *The Monthly Review* 36 (1767), pp.51-66 & 80-93.
16. For his nephew's views on Price's use of Bayes's results see Morgan [1815].
17. For comment on Price's criticism of Hume's views see Gillies [1987] and Sobel [1987, p.169]. The relationship between miracles and statistics is explored in Kruskal [1988]. See also Zabell [1988a], [1988b].
18. According to Whittaker [1951], in the century between Newton's death and Green's scientific activity, "the only natural philosopher of distinction who lived and taught at Cambridge was Michell" [p.153].
19. Arbuthnott's name is spelled variously with one or two t's.
20. Michell gives the ratio $(60'/6875.5')^2$, the denominator in fact being 2 radians.
21. A modern calculation replaces Michell's figure of "somewhat more than 496000" by 476189.
22. The figures are very little changed if $\sin 3\frac{1}{3}'$ is used instead of $3\frac{1}{3}'$.
23. As noted by Todhunter [1865, art.622], the numerical results quoted by Herschel as being Struve's do not agree with the latter's work. The results are however correctly given in the 1873 edition of *Outlines of Astronomy*.
24. According to the *Dictionary of National Biography*, Forbes, whose mother had been "the first love of Sir Walter Scott", was elected F.R.S. "at the unprecedented early age of nineteen".
25. Forbes (op. cit.) points out that this value is approximately $n^2/2p$.
26. For biographical details of Beguelin see Netto [1908, p.227], and for a general discussion of his work on probability see Todhunter [1865, arts 603-616].
27. Sur les suites ou séquences dans la lotterie de Genes, *Histoire de l'Académie ... Berlin*, 1765 (published 1767).
28. The year of publication is uncertain.

29. The problems are numbered I to XI, but there is no Problem IX.
30. See Pearson [1978, pp.599, 601].
31. All references are to Serret's edition of the *Œuvres de Lagrange.*
32. Pearson [1978, pp.598-599].
33. "Lagrange ... gave such a cloudy discussion of a problem in inverse probability that it is doubtful whether he had read Bayes" [Stigler 1975, p.505]. Stigler [1986a, p.118] concludes also that "I think it is fair to say that his work was untouched by any real sense of inverse probability."
34. Pearson [1900].
35. On the naming of this distribution see Patel and Read [1982].
36. Pearson [1978, p.156].
37. Quoted here from Pearson [1978, pp.600-601].
38. Pearson [1978, p.601].
39. Mis-spelt in the bibliography in Keynes [1921/1973]. Hacking [1971] finds Emerson "a curious example" of "the lesser minds of the period that take the conceptual matters seriously instead of ploughing on with the mathematics" [p.350]. Cajori [1919b] writes of Emerson, a self-taught mathematician, that "he wrote many mathematical texts which indicate a good grasp of existing knowledge, but not great originality" [p.192].
40. This refusal is given in Emerson [1793] as follows: "It was a d—n'd hard thing that a man should burn so many farthing candles as he had done, and then have to pay so much a year for the honor of F.R.S. after his name. D—n them and their F.R.S. too."
41. Compare the axioms and definitions given by Bayes and de Moivre. Dinges [1983, p.88] finds evidence of both the aleatory and the epistemic notions of probability in Emerson's work; Hacking [1971] sees merely a "groping for the idea of probability as 'judgement' or credibility." [p.351].
42. Part of the text of this essay is Buffon's *Mémoire sur le jeu de franc-carreau* of 1733, an early attempt at geometrical probability (see Roger [1978, p.29]).
43. It is clear from other passages in this memoir that Buffon did not regard probability as normed. For further discussion of his work see Coolidge [1949, chap. XIII] and Zabell [1988a].
44. In the original, p.64 is followed immediately by p.85.
45. A rough translation runs as follows: Distinguished mathematicians have studied this matter, especially the famous Laplace in the Notes of the Paris Academy. Since however in the solving of problems of this type advanced and hard analysis may have been applied, I have

considered it worth the effort to address the same questions by an elementary method and appropriate use of a knowledge of series. By that theory this changed part of the probability calculus might be reduced to the theory of combinations, as I first derived in a dissertation transmitted to the Royal Society. I shall undertake to touch upon these questions briefly here, by a lucid, especially rigorous method.

46. Todhunter [1865, arts 766, 767 & 774]. For a similar opinion see Cantor [1908, p.243].
47. In his papers Trembley abbreviates his first name to "Io."
48. Trembley's word is "schedulas": our translation is more convenient than the literal "small strips of papyrus" or "small leaves of paper".
49. Todhunter [1865, art.773] is slightly inaccurate in noting that "Trembley remarks that problems in Probability consist of two parts": what Trembley in fact wrote was "E supradictis sequitur Probabilitatem causarum ab effectibus oriundam, methodam requirere quae duabus constat partibus."
50. A particular case was later considered by Terrot (see §7.16).
51. Todhunter's reference here is in fact to Condorcet rather than Laplace.
52. Further details may be found in Todhunter [1865, art.851]. For a discussion of Prevost's work on testimony see Zabell [1988b].
53. Notice a curious inversion of the "editorial we" in the last clause of this quotation.
54. For a general discussion of Gauss's contributions to statistics see Sprott [1978]. Fisher [1970, pp.21-22] may be consulted for an opinion on Gauss's appreciation of the method of maximum likelihood.
55. According to Bühler's biography of 1981, Gauss's christian names were Johann Friedrich Carl.
56. This passage is translated by Davis [1857, p.255] as follows: "If, any hypothesis H being made, the probability of any determinate event E is h, and if, another hypothesis H' being made excluding the former and equally probable in itself, the probability of the same event is h': then I say, when the event E has actually occurred, that the probability that H was the true hypothesis, is to the probability that H' was the true hypothesis, as h to h'." Le Cam [1986] states that "The 'proof' of Bayes formula by Gauss cannot even be considered adequate by the standards of his time or earlier ones" [p.79].
57. For a discussion of the reasons for the qualification $\mu > \nu$ see p.254 of Davis's translation of the *Theoria Motus Corporum Coelestium.*
58. Sprott [1978] sees here a special case of Bayes's Theorem, a result which he describes as "merely an expression of the addition and multiplication rules of probability" [p.190]. He notes further [p.191] that Gauss's interpretation of his results was given in a frequency rather than a Bayesian sense.

59. Morgan's mother Sarah, Price's sister, married William Morgan, a surgeon in Bridgend, Glamorganshire. For biographies of Price (1723-1791) see Holland [1968] and Thomas [1924].
60. For further details of the Morgans see Pearson [1978, pp.395-396, 408].
61. This reference of Morgan's is mysterious. From correspondence with the Library of Congress (Science and Technology Division) and the American Philosophical Society I learn that (i) the Society was not in existence between 1745 or 46 and 1767-68, and (ii) the minutes or records of an earlier society, with which Franklin was associated and which preceded the American Philosophical Society, for 1762-66 have been lost. It can only be assumed, from lack of corroborative detail, that Morgan erred here. For (similar) remarks on Morgan's accuracy see Holland [1968, pp.45-46].
62. The date is mistakenly given as 1806 in Laurent [1873]. Crepel [1988a] says of this work "Lacroix n'est que partiellement disciple de Condorcet: son objectif, beaucoup plus pédagogique, est différent, il s'agit de rendre accessible à un public suffisamment nombreux non seulement certaines idées de Condorcet, mais surtout la 'Théorie Analytique des Probabilités' de Laplace. La traité de Lacroix, qui constituera le manuel de référence en français jusqu'aux trois-quarts du 19e siècle, va en fait gommer les aspects qui nous semblent aujourd'hui les plus novateurs dans l'oeuvre de Condorcet" [§7(c)].
63. The reference is probably to Laplace's *Mémoire sur les probabilités.*
64. See Todhunter [1865, art.1057].

Chapter 5

1. For a general discussion of Condorcet's work see Gouraud [1848, pp.89-104], Maistrov [1974], Pearson [1978] and Todhunter [1865]. For a brief discussion of his work on probability see Baker [1975, p.81].
2. Gillispie [1972, p.15], writing of the memoir, says "It will hardly be worth while to follow him in these writings obscurely expounding the reasonings and procedures of probability itself in relation to causality and epistemology." Recent work by Crepel, however, has been devoted to denying, if not indeed refuting, the existence of such obfuscation (see in particular his [1988a], [1988b], [1989a] and [1989b], and Bru and Crepel [1989]).
3. See Laplace's *Mémoire sur les probabilités*, and also Todhunter [1865, art.773] and Trembley [1795-1798].
4. Condorcet's rebarbative notation has been altered and some obvious misprints have been corrected.

5. Translated by Pearson [1978, p.456] as "between two contingent events becoming actual".
6. Various other "multiple Bayes's integrals" are given, but this illustration is sufficient.
7. The reference is to Laplace's *Sur les approximations des formules qui sont fonctions de très grands nombres* of 1782.
8. See Pearson [1978, p.457].
9. For further comments on Condorcet's failure to mention Bayes see Stigler [1975, p.505]. Note also Pearson [1978, p.181].
10. The persistence of this habit to this day is remarked on by Neveu [1965, p.ix].
11. Todhunter [1865, art.734] has "the next $p+q$ trials", as does Pearson [1978, p.458]: the adjective is not present in the original, though this was probably the intent.
12. For further details see the preface to Pearson [1978].
13. See Dale [1982].
14. For some comments on Condorcet's work on testimony see Zabell [1988b].
15. For further examination of this formula see Owen [1987, §3], Sobel [1987, §2] (where the Bayes-Laplace rule is recast as "The Hume-Condorcet Rule for the Evidence of Testimony") and Todhunter [1865, art.735].
16. This sixth part is discussed in some detail in Todhunter [1865, arts 737-751].
17. For an opinion in turn on Gouraud's exuberance see Todhunter [1865, art.753]. The awkwardness of Condorcet's expression seems to have manifested itself early in his career. According to Baker [1975, p.6] the first paper submitted to the French Academy of Sciences was rejected by Clairaut and Fontaine, who had been charged with its examination, on account of "its sloppiness and its lack of clarity".
18. See Todhunter [1865, art.467].
19. Hacking [1971, p.351] considers no phrase in our subject "less felicitous" than Condorcet's *probabilité moyenne.*
20. Similarly harsh sentiments have been expressed by Bertrand, who, in commenting on Condorcet's *Essai*, wrote "Aucun de ses principes n'est acceptable, aucune de ses conclusions n'approche de la vérité" [1972, p.319]. Gillispie [1972, p.12], on the other hand, describes Todhunter's judgement as "harsh", and he provides some comments by Condorcet's contemporaries as evidence of the esteem in which he was held.
21. See Hacking [1971, p.351].

22. Hacking [1971, p.351] considers Condorcet as the first to render explicit the "groping for the idea of probability as 'judgement' or credibility." For comments on the distinction between "logical" and "physical" probabilities to be found in the works of D'Alembert, Condorcet and Laplace, see Baker [1975, pp.177-178].
23. Condorcet describes the third part of this work as an "Ouvrage plein de génie & l'un de ceux qui sont le plus regretter que ce grand homme ait commencé si tard sa carrière mathématique, & que la mort l'ait si-tôt interrompue" [p.viij].
24. See Gillispie [1972, p.15].
25. Writing of the use of Bayes's Theorem in the probability of judgements, Poisson [1837, p.2] says "il est juste de dire que c'est à Condorcet qu'est due l'idée ingénieuse de faire dépendre la solution, du principe de Blayes [sic], en considérant successivement la culpabilité et l'innocence de l'accusé, comme une cause inconnue du jugement prononcé, qui est alors le fait observé, duquel il s'agit de déduire la probabilité de cette cause."
26. Perhaps one sees here an adumbration of the Principle of Irrelevant Alternatives.
27. For a summary of the *Essai* see Cantor [1908, pp.253-257].
28. As precursors in the search for a method for the determination of the probability of future events from the law of past events Condorcet cites Bernoulli, de Moivre, Bayes, Price and Laplace [p.lxxxiij].
29. In Condorcet's notation, $\binom{m+n}{n}$ is written $\frac{m+n}{n}$. In Problem 3, however, $\frac{1}{0}$ is used to mean "infinity". This is a prime example of what Todhunter [1865, art.660] describes as Condorcet's "repulsive peculiarities". Pearson [1978, p.480] argues that the curve of judgements should be of the form $y_0(x-1/2)^p(1-x)^q$.
30. See Dinges [1983, pp.68, 95] for comment on the occurrence of this result in Condorcet's work.
31. The integrand in the second integral is given by Condorcet as $(1-x)^n$.
32. Todhunter [1865, art.698].
33. Here we have another example of Condorcet's awkward notation: the integral $\int_{1/2}^{1} x^m(1-x)^n\,dx$ is written as $\int \frac{\frac{1}{2}}{x^m(1-x)^n}\,dx$ in the original.
34. As Todhunter [1865, art.701] points out, Condorcet *ought* to say "let the probability not be assumed constant".
35. It is this result that Pearson [1978, p.366] describes as "really Condorcet's and Laplace's extension of Bayes."
36. The factor $\binom{q}{q'}$ is missing in the original.
37. For a general discussion of Condorcet's application of probability to the voting problem see Gillispie [1972, p.16]: Auguste Comte's opinion on the matter is discussed in Porter [1986, p.155].

38. See Pearson [1978, pp.482-489] for a discussion of Parts 4 and 5. A general discussion of the probability of decisions, with special reference to the work of Condorcet, Laplace and Poisson, may be found in Chapter XIII of Bertrand [1972].
39. For a general discussion of this paper see Pearson [1978, pp.501-505].
40. Two of which are numbered VI.
41. "Qui a pour objet l'application du calcul aux sciences politiques et morales" [p.171].
42. These two articles are respectively entitled "De l'intérêt de l'argent" [pp.2-31] (the first page is an introduction) and "Sur une méthode de former des tables" [pp.31-56]. See Crepel [1988a] for a discussion of these articles.
43. For reference to earlier work on testimony by John Craig (d.1731) see Pearson [1978, p.465] and Stigler [1986c].
44. On this point see Pearson [1978, p.502].
45. The first Article VI (see Note 40 above) is entitled "Application du calcul des probabilités aux questions où la probabilité est determinée" [pp.121-145]; the second is "De la manière d'établir des termes de comparaison entre les différens risques auxquels on peut se livrer avec prudence, dans l'espoir d'obtenir des avantages d'une valeur donnée" [pp.145-150], while the seventh article is "De l'application du calcul des probabilités aux jeux de hasard" [pp.150-170].

Chapter 6

1. For biographical details of Laplace see Cantor [1908, p.228], David [1965], Maistrov [1974, pp.135-138], Pearson [1929], Pearson [1978, pp.637-650] and Whittaker [1949].
2. Some of these memoirs are cited here only for general definitions, and not for any Bayesian results: Gillispie [1972, p.3] in fact finds only nine memoirs relevant to probability. For a general discussion of Laplace's early work see Gillispie [1979] and Stigler [1978]. Also useful are Sheynin [1977] and Stigler [1975].
3. This definition of probability as a ratio of numbers of cases occurs in the second edition of de Moivre's *Doctrine of Chances* of 1738 (see Schneider [1968, p.279]). However the idea is also evident in the fourteenth chapter, "De punctis geminatis" of Cardano's *Liber de Ludo Aleae* (written c.1564), where we find the words "Una est ergo ratio generalis, ut consideremus totum circuitum, & ictus illos, quot modis contingere possunt, eorumque numerum, & ad residuum circuitus,

eum numerum comparentur, & iuxta proportionem erit commutatio pignorum, ut aequali conditione certent." (See Boldrini [1972, p.125], David [1962, chap. 6] and Ore [1953].) For a discussion of equipossibility as it arises in probability see Hacking [1971], [1975] and van Rooijen [1942] (the latter contains an illuminating contrast between the Dutch terms "gelijkwaardig" and "even waarschijnlijk"). Laplace's approach to his "definition" of probability was not uncommon. Robinson [1966, p.265] notes that "in the approach of Euclid and Archimedes, which is also the approach of de l'Hospital, a definition frequently is an explication of a previously given and intuitively understood concept, and an axiom is a true statement from which later results are obtained deductively". See Gini [1949] for a discussion of the difference between the *concept* and the *measure* of probability.

4. For a general discussion of (parts of) this memoir see Cantor [1908, pp.241-242] and Gillispie [1972, pp.4-5]. In addition to the passages considered here, this memoir is noteworthy for its discussion of the Normal probability density function (see Keynes [1921, chap. XVII, §5]). Stigler [1986b] provides a general discussion and a translation of the memoir.
5. When this memoir was *written* is uncertain: Baker [1975, p.433], acting on a suggestion by Hahn, suggests that it might have been written in 1774. Stigler [1978, p.253], however, is not convinced by this suggestion, and his investigations lead him to a date of 1773.
6. Molina [1930] finds Todhunter's discussion of the work of Bayes and Laplace on the probability of causes "most inadequate".
7. We might also point out, as Porter [1986, p.93] has noted, that Laplace should be viewed as an independent developer of inverse probability.
8. The word used in the original is "établirons", which may be translated in terms of "assert", "prove" or "establish": since no proof is given, I have chosen the first.
9. A discussion of this principle may be found in Keynes [1921, chap. XVI, §§11-14]. Van Dantzig [1955, p.36] seems to regard Laplace's elaboration of the theory of the probability of causes as a youthful aberration, while Hacking [1971, p.348] suggests that Laplace had "fewer philosophical scruples" than Bayes — an opinion which seems to be shared by Dinges [1983, p.67]. Gouraud [1848], in commenting on Condorcet's *Essai*, writes of the "principe récemment entrevu par Bayes et démontré par Laplace" [pp.95-96]: however he finds in Bayes's Essay both a direct determination of the probability "que les possibilités indiqueés par les expériences déjà faites sont comprises

dans des limites données" and "la première idée d'une théorie encore inconnue, la théorie de la probabilité des causes et de leur action future conclue de la simple observation des événements passés" [p.62], and it is not clear to which of these he is referring.

10. It should be noted that Laplace nowhere bestows on it this appellation, despite what Maistrov [1974, p.100] says.
11. Catalan [1888], in his discussion of this problem, finds it necessary to draw attention to the wording of the "*futur* admirable écrivain" [p.256].
12. See also Molina [1930, p.382].
13. The reason for the "dx" in the numerator in (1) is nowhere explained. However, it was not an uncommon practice in the nineteenth century to assign infinitesimal masses to points (rather than infinitesimal volumes) in the case of continuous distributions. Thus, while we would today interpret the numerator in (1) as $\Pr[x \leq X \leq x + dx]$, Laplace had little choice in arriving at (1) as he did.
14. For comment on Laplace's and Gauss's introductions of the Normal probability density function see Stigler [1980b, p.153].
15. The problem of appropriate division of the accumulated pot as it occurs in the game "primero" is discussed in Cardano's *Liber de Ludo Aleae*: see Ore [1953, p.117].
16. For further comment see Keynes [1921, chap. XVII, §§5-7], Sheynin [1977] and Stigler [1986a, pp.105-117].
17. See Laplace's *Mémoire sur les probabilités* [1778, pp.476-477] for discussion of the case of different φ's.
18. See de Morgan [1837, Part II, p.247]. The *milieu de probabilité* is, as Stigler [1986a, p.109] notes, just the posterior median.
19. As Wilson [1922-1923, p.841] has noted, "the first two laws of error that were proposed both originated with Laplace." The first of these laws is the one discussed here: the second was given in the memoir of 1778. For the place of this work in the theory of least squares see Harter [1974].
20. My discussion here owes much to Stigler [1986a, pp.113-117].
21. This formulation seems to get round a difficulty seen by Sheynin [1977, p.7] in the defining of the integral.
22. For further comment on the St Petersburg paradox see Note 10 to Chapter 7. The game known as "cross or pile" is discussed in Brewer's *The Dictionary of Phrase and Fable.*
23. Hacking [1971] traces the publicization of the principle of indifference to Bernoulli's *Ars Conjectandi.*

24. The published version mistakenly gives the date of reading as "10 Février 1773" instead of "10 Mars 1773" (and continued on the 17th): see Baker [1975, p.432] and Stigler [1978, p.252]. The memoir is described by Gillispie [1972, p.5] as "astonishing".
25. For comment on the precise use of the term *hazard* in the *Encyclopédie ou Dictionnaire raisonné des sciences, des arts et des métiers* [1751-1765], see Nový [1980, p.29].
26. See Baker [1975, p.434] and Gillispie [1972, p.8].
27. This abstract is known to be by Condorcet — see Baker [1975, p.169] and Stigler [1975, p.252].
28. For a general summary of the memoir see Gillispie [1972, pp.8-10].
29. For further discussion of this problem see Netto [1908, p.232] and Makeham [1891a, pp.242-243]. See also Stout and Warren [1984, p.212], where it is stated "We are concerned with efficiently using flips of a coin of unknown bias to simulate a flip of an unbiased coin. This problem is quite natural in that when given an arbitrary coin one should assume that it has some unknown bias."
30. For the case in which $\alpha = 0$, see p.390 of the memoir discussed and also Todhunter [1865, art.891]. L'Hopital's rule is required. (According to Boyer [1968, p.460], this rule is in fact due to Jean Bernoulli.)
31. For details see Todhunter [1865, art.891].
32. See Edwards [1978, p.116].
33. See Netto [1908, pp.243-244]. It should be noted that Laplace stresses the time-order of the events, and on this topic Shafer [1982] may be consulted.
34. See Todhunter [1865, art.893].
35. There is no mention here of either de Moivre or Stirling, though in Article XXIII Laplace writes of the "beau théorème de M. Stirling sur la valeur du produit $1.2.3\ldots u$, lorsque u est un très grand nombre".
36. A generalization of this result is given in Article XXIII: see also Article XXV.
37. For some discussion of this matter see Gillispie [1972, pp.8-10]: it received further consideration by Laplace in a memoir discussed in §6.7 of the present work.
38. Sheynin [1971b, p.235] has pointed out that Laplace often used the Bayesian conception of supposing that a constant but unknown parameter had a prior distribution. This was in fact not done by Bayes himself.
39. The problem is again considered in the *Théorie analytique des probabilités*, but for a period of 40 years rather than 26. According to Boldrini [1972, p.184], W. Lexis, at the end of the nineteenth century, showed "that the probability of masculine births varies with time and place."

40. Legendre attributes this result to Euler — see Plackett [1972, p.243].
41. See Todhunter [1865, art.902] for details of other discussions of this problem.
42. In defining u and $u - x$ Laplace uses "probabilité" and "possibilité" respectively: this seems to suggest that he did not always find it necessary to observe a distinction between these two terms. On D'Alembert's observation of the difference see Daston [1979, p.266].
43. An error in Laplace's evaluation is given by Todhunter [1865, art.902].
44. Todhunter [1865, art.902] finds the solution "very obscure", and indicates where a better solution may be found.
45. De Morgan [1838, pp.87-88] provides a discussion of the separate advantages conferred by the terms *probability, chance, presumption, possibility, facility* and *expectation.* Lagrange preferred " facilité" for the physical, objective concept of probability (see Hacking [1971, p.350]): Laplace was not always so careful.
46. Compare the quotation from p.419 of this memoir given earlier in this chapter.
47. Todhunter [1865, art.904] writes "The theory does not seem, however, to have any great value."
48. For more details on Laplace's theory of errors see Sheynin [1977].
49. A discussion of Laplace's *milieu de probabilité* may be found in Makeham [1891a, p.246] and Stigler [1986a, p.109]: the interpretation is as a posterior median.
50. This is effected by setting $f(z+b) = f(z) + bf'(z) + b^2 f''(z)/2! + \cdots$, where $z = \alpha x$ and $b = \alpha p$.
51. Gillispie [1972, p.10] states that it was here that Laplace "first employed phrasing famous from his later popularization" (i.e. the *Essai philosophique sur les probabilités*), but I suggest that the *sentiment* is already patent in the memoir discussed in §6.4.
52. There is some slight discussion of this formula in Netto [1908, pp.244-245].
53. He in fact finds "la probabilité que la valeur de x est comprise entre les deux limites $a - \theta$ et $a + \theta'$" [p.305].
54. See Netto [1908, p.246].
55. Laplace omits this phrase.
56. For comment on this example and related work see Todhunter [1865, art.909].
57. There is some superficial discussion of this memoir in Westergaard [1968, p.82]: a more detailed treatment may be found in Gillispie [1972, pp.10-11].
58. The memoir was in fact read on 30th November 1785.

59. Chang [1976] notes that Laplace was the only one of the mathematicians of his time who examined this estimation problem, to see the need, and to find an expression, for the prevision of the estimate.
60. Compare Stigler [1986b, p.361].
61. See Pearson [1928, pp.170-171] for an alternative solution.
62. There had probably been an earlier publication of these lectures, for a footnote on p.169 reads "Depuis la première publication de ses leçons ..." The reference might be to the 1810 publication "Notice sur les probabilités".
63. For a discussion see Sheynin [1977].
64. The memoir also contains an example of Laplace's procedure for finding the probability distribution of the sum of a number of identically distributed random variables: see Seal [1949, pp.225-226]. Moreover, it is here that Laplace "first developed the characteristic function as a tool for large-sample theory and proved the first general Central Limit Theorem" [Stigler 1975, p.506].
65. These limits are incorrectly given in the *Œuvres complètes* as $\pm rh/n$.
66. Some details are repeated on p.351 of the supplement.
67. In the corresponding passage in the *Théorie analytique des probabilités* the reference to Daniel Bernoulli, Euler and Gauss is replaced by the phrase "des géomètres célèbres".
68. Jaynes [1976, p.233] suggests that "an historical study would show that the reasons for the interest of both Laplace and Jeffreys in probability theory arose from the problem of extracting 'signals' (i.e., new systematic effects) from the 'noise' of imperfect observations, in astronomy and geophysics respectively."
69. Note also the brief discussion of Laplace's *Exposition du système du monde* in his *Essai philosophique sur les probabilités.*
70. For details see Fabry [1893-1895, p.5].
71. For comment on the difficulty of this memoir see Sheynin [1976, p.164].
72. Fabry [1893-1895, p.2] points out that "l'absence des orbites hyperboliques est une objection contre cette théorie".
73. The radius of this sphere is taken here to be 10^5 Astronomical Units.
74. Gauss [1874, p.582] gives Laplace's quaesitum as follows: "Laplace findet für das Verhältniss der Wahrscheinlichkeit einer solchen Hyperbel, wo die halbe grosse Axe 100 Halbmesser der Erdbahn nicht übersteigt, zu der Wahrscheinlichkeit der übrigen Fälle, nemlich einer Hyperbel von grösserer Axe, einer Parabel oder einer Ellipse."
75. See also Pollard [1966].
76. For reasons why attention need be restricted only to values of $\beta < \pi/2$ see Fabry [1893-1895, p.7].

77. For a proof see Fabry [1893-1895, pp.7,8].
78. Fabry [1893-1895, p.8] terms comets with perihelion distance less than d, "comètes visibles".
79. See Fabry [1893-1895, §4].
80. His definition runs as follows: "... la probabilité d'un événement est le rapport du nombre des cas favorables au nombre total des cas possibles" [p.3].
81. Fabry [1893-1895] considers something similar in his Articles 42 and 53.
82. Seeliger, in his paper of 1890, introduces a term $\varphi(D)$ at the outset: I have experienced some difficulty in following his argument.
83. Gauss [1874, p.582] finds this "einer sehr plausibeln Hypothese."
84. See Fabry [1893-1895, p.5] for details of these and other papers.
85. Seeliger's [1890] corrections, though at first sight different from those of Fabry, can in fact be shown to be in agreement with the latter's: see Fabry [1893-1895, pp.31-34].
86. See Schiaparelli [1874, p.80].
87. See Fabry [1893-1895, p.19].
88. Compare Schiaparelli [1874, p.80].
89. Fabry [1893-1895, p.20] comments on Laplace's procedure as follows: "Il est à remarquer que par la manière dont Laplace conduit son calcul, il fait U infinie seulement implicitement en supposant i infinie; c'est peut-être pour cela qu'il n'a pas réfléchi aux conséquences de cette supposition."
90. See Fabry [1893-1895, pp.25-26] for details.
91. The expression given in (40) differs from that given by Gauss and that given by Seeliger: see Fabry [1893-1895, p.34] for a discussion.
92. See Schiaparelli [1874, p.80].
93. Note Fabry [1893-1895, pp.31-43]. Further development of the matter discussed in this memoir may be found in the papers by Fabry, Schiaparelli and Seeliger.
94. Sheynin [1977, p.59], in writing of the first Supplement, says that it is "essentially compiled from two memoirs", but a cursory examination shows that the two early papers are not just reprinted in the Supplement.
95. The first edition of 1812 was dedicated to Napoleon: for a discussion of the suppression of this dedication in later editions see Pearson [1929]. The full text of the dedication is given in Todhunter [1865, art.931].
96. For a general discussion of this *Leçon* see Fagot [1980, pp.59-77].
97. A copy of this *Notice* is provided in Gillispie [1979, pp.265 et seqq.].

98. General comments on the *Essai* may be found in Maistrov [1974, §III.9] and Neyman & Le Cam [1965, pp.iv-ix]. More detailed studies are given in Pearson [1978, pp.651-703] and Todhunter [1865, arts 933-947]. See also Herschel [1857, p.393].
99. See Coolidge [1949, §XIII.2] and Zabell [1988a].
100. See also Pearson [1978, p.660] and Todhunter [1865, art.643].
101. For comment on this matter see Keynes [1921, chap. XVI, §§16-19] and Pearson [1978, pp.671-672, 682-683].
102. See Pearson [1978, p.674] for comment on a similar four-fold division elsewhere in Laplace's work. For a discussion of Laplace's work on testimony see Zabell [1988b].
103. In an anonymous review of 1837, de Morgan describes this work as "the Mont Blanc of mathematical analysis", but he qualifies this with the words "the mountain has this advantage over the book, that there are guides always ready near the former, whereas the student has been left to his own method of encountering the latter" [p.347]. Bertrand [1972, p.v] writes "Le Calcul des probabilités est une des branches les plus attrayantes des Sciences mathématiques et cependant l'une des plus négligées. Le beau livre de Laplace en est peut-être une des causes."
104. Some discussion may be found in Molina [1930] and, *in extenso*, in Todhunter [1865, arts 948-968].
105. Page numbers throughout this section are to the seventh volume of the *Œuvres complètes de Laplace* of 1886.
106. See Sheynin [1971b, p.237] for a discussion of the looseness of this definition of probability.
107. See Clero [1988] and Shafer [1982] for discussion of the importance of the consideration of time-dependence in conditional probability.
108. Compare the discussion of Principle VII.
109. For comment on Laplace's two methods of inversion of Bernoulli's Theorem see Dale [1988b, §6] and Monro [1874, pp.74, 77]. Note also de Morgan [1838b].
110. See Neyman & Le Cam [1965, p.vii] for comment on Laplace's use of the posterior distribution of a parameter given a large number of observations.
111. The data seem to be from baptismal records — see p.384 of the *Théorie analytique des probabilités.*
112. More accurately, Laplace considers *births* in London and *baptisms* in Paris.
113. Lidstone [1941] provides a discussion, based on a suggestion of G.F. Hardy, of a suitable choice of prior.

114. A similar problem had been earlier considered by Laplace in his *Sur les naissances...*: see §6.7.
115. For details of the earlier treatment see Todhunter [1865, art.1032].
116. See Todhunter [1865, art.1036] for references to earlier discussions of this problem.
117. No divorces!
118. Independence is implicit here.
119. It might well be argued whether the assumptions of what is now called a Bernoulli trial in fact hold here.
120. Laplace follows this with the following words: "On formera ainsi, d'année en année, une Table des valeurs de i. En faisant ensuite une somme de tous les nombres de cette Table, et en la divisant par n, on aura la durée moyenne des mariages faits à l'âge a pour les garçons et à l'âge a' pour les filles" [p.424].
121. The original has "infinis".
122. See Whittaker and Watson [1973].
123. Todhunter's [1865] analysis in his Article 1037 is slightly different. In his next article he derives as a natural consequence of this analysis an extension of Bernoulli's Theorem.
124. This case is described by Todhunter [1865, art.1040] as "a modification of the problem just considered, which may be of more practical importance."
125. Todhunter [1865, art.1042] in fact sees this entire chapter as "mainly a reproduction of the memoir by Daniel Bernoulli."
126. For a discussion of this chapter see Shafer [1978, pp.348-349].
127. See Walker [1929, p.21] for further comment.
128. There are some passages in the early parts of this Supplement which are not to be found in the memoirs.
129. Comment on the matter of this section may be found in Bertrand [1907, chap. XIII] and Poisson [1837, pp.2-7]. Todhunter [1865] does not discuss this section at all. Sheynin [1976] provides a brief summary, while Pearson [1978, pp.690-692] gives a more detailed investigation.
130. What Laplace means by an *equitable* ("juste") opinion of the tribunal is discussed earlier in this Section, where the following question is posed: "la preuve du délit de l'accusé-a-t-elle le haut degré de probabilité nécessaire pour que les citoyens aient moins à redouter les erreurs des tribunaux, s'il est innocent et condamné, que ses nouveaux attentats et ceux des malheureux qu'enhardirait l'example de son impunité, s'il était coupable et absous?" [p.521]. A little later on [p.522] he states that the decision of a tribunal is equitable if it conforms to the true ("vraie") solution of the question.

131. Laplace talks of the integral in the denominator below as a "somme", and, as Pearson [1978, p.692] has pointed out, no mention is made of the Euler-Maclaurin bridge.
132. Laplace in fact discussed incomplete beta-functions in considering the incomplete binomial summation. This was perhaps not realized by Pearson (see Molina [1930, p.376]).
133. See Pearson [1978, pp.691-692].
134. For details see the *Théorie analytique des probabilités*, p.535.
135. For a discussion of this assumption see p.536 of the *Théorie analytique des probabilités.*
136. See Lancaster [1966] for a discussion of Laplace's determination of the posterior distribution of h, this distribution being viewed there as a forerunner of the χ^2 distribution.
137. See p.549 of this Supplement for a discussion of what happens if this latter assumption is not met.
138. For some references to general remarks on the *Théorie analytique des probabilités* see Todhunter [1865, art.1052].
139. Laplace's work is discussed from the "inductive behaviour" versus the "inductive reasoning" point of view by Neyman [1957, pp.19-21].
140. The following discussion owes much to Karl Pearson — see Pearson [1978, pp.366-369].
141. A generalization of Bayes's theorem is given in Pearson [1924b].

Chapter 7

1. On the contribution actually made by Bernoulli see Pearson [1925]: Sheynin [1968] considers Pearson's judgement of Part 4 of the *Ars Conjectandi* as unsatisfactory to be "hardly fair".
2. On the history of the Poisson distribution see Dale [1989], Good [1986, p.166], Haight [1967] and Stigler [1982b].
3. The term $2/\sqrt{\pi}$ below is mistakenly given in the original [p.271] as $1/\sqrt{\pi}$.
4. This example is again considered in Poisson's *Recherches sur la probabilité des jugements*: see Sheynin [1978, p.272].
5. Poisson's reference is to p.383 of the *Théorie analytique des probabilités*: this is p.391 of the *Œuvres complètes* edition.
6. According to Haight [1967] "Certain authors give the date 1832 for Poisson's *Recherches*" [p.113]: none are cited by name, and I have found no evidence of a publication date preceding 1837. See also Maistrov [1974, p.158].

7. For further comment on Poisson's distinction between "chance" and "probability" see von Kries [1927, p.275]. Good [1986, pp.157-158] finds Poisson's concept of probability to be more that of logical probability, or credibility, than that of subjective, or personal, probability.
8. On the importance of time-order in connexion with such conditional probabilities see Shafer [1982].
9. There is no twelfth section.
10. On this paradox see Feller [1968, §X.4], Martin-Löf [1985] and Todhunter [1865, arts 389-393].
11. In this generalization it is supposed that the random variables, although still two-valued and independent, are in general differently distributed: i.e. $\Pr[X_i = 1] = p_i = 1 - \Pr[X_i = 0]$ for each i. Good [1986, p.160] regards Poisson's Law of Large Numbers as "perhaps [his] main direct contribution to the mathematical theory of probability and statistics."
12. Compare Poisson's Article 46 discussed earlier.
13. For a detailed discussion of this result, and Poisson's generalization of it, see Keynes [1921, chap. XXIX], where Bernoulli's theorem is viewed as one which "exhibits algebraical rather than logical insight" [p.341]. Keynes declares that the conditions under which the theorem is valid are usually not realized in practice.
14. Keynes [1921, chap. XXIX, §2] points out that the approximation in fact requires that μpq be large.
15. Keynes [1921, chap. XXIX, §2] considers the simpler approximation $(2/\sqrt{\pi})\int_0^u \exp(-t^2)\,dt$ satisfactory, in practice, in view of all the approximations involved in the derivation.
16. For comment on an oversight made by Poisson in the derivation of this result see §7.4 and de Morgan [1838b].
17. The factor "3" in the denominator is missing in Todhunter [1865, p.556].
18. The Bayesian nature of Poisson's approach here is commented on by Good [1986, p.161].
19. The reference is to his *Essai d'arithmétique morale.* Further details of the experiment may be found in Keynes [1921, chap. XXIX].
20. There is no "not proven" verdict.
21. Good [1986, pp.167-168] sees in this work of Poisson's a sequential use of Bayes factors. Note also the comment by Solomon [1986, pp.174-176] on the Poisson jury model.
22. Apart from those cases mentioned here, Poisson considers $\ell = 1/2$, $\ell' = 1$ and $\ell = 0$, $\ell' = 1/2$.
23. In the first supplement to his *Théorie analytique des probabilités.*

24. The copy of this tract in the Wishart Library of Cambridge University bears on the cover "De Morgan on Probability". Inside, an inked inscription reads "J.C. Adams Esq. from the Authors J.W. Lubbock & J.E.D. Bethune." In its biographical note on Lubbock the *Dictionary of National Biography* says of this work "A binder's blunder caused this work to be often attributed to De Morgan, despite his frequent disclaimers" [Vol. XII, p.227]. According to the ninth edition [1877] of the *Encyclopædia Britannica*, de Morgan found that this error "seriously annoyed his nice sense of bibliographical accuracy." The matter was only settled after fifteen years and a letter from de Morgan to the *Times*.
25. The generally accepted date of publication seems to be 1830.
26. See Kyburg and Smokler [1980, pp.13-15].
27. Kneale's [1949, pp.203-204] criticism of the rule of succession is based on ignorance of this extension.
28. For a short discussion of Bolzano's introduction of probability see Nový [1980, pp.30-31].
29. A biography of de Morgan may be found in Heath's edition of 1966 of de Morgan's *On the Syllogism and Other Logical Writings.*
30. MacFarlane [1916, p.19].
31. Heath [1966, p.vii] asserts that the defective eye was the right.
32. Smith [1982] lists 200-odd items, with certain deliberate omissions.
33. This excludes references to probability in logical works.
34. See Stigler [1975, p.507] and Smith [1982, p.142].
35. Commenting on this quotation Keynes [1921, chap. XVI, §14] writes "If this were true the principle of Inverse Probability would certainly be a most powerful weapon of proof, even equal, perhaps, to the heavy burdens which have been laid on it. But the proof given in Chapter XIV. makes plain the necessity in general of taking into account the *à priori* probabilities of the possible causes."
36. Particular mention is made on p.64 of the case in which $n = 0$.
37. The date is variously given as 1837 (Smith [1982] and the Edinburgh University Library Catalogue), 1845 (Encyclopædia Britannica, 14th edition, and the National Union Catalogue), and 1849 (Keynes [1921]). Stigler [1986a, p.378], citing Sophia de Morgan's *Memoir of Augustus De Morgan* of 1882, says that this article was written in 1836-1837.
38. For further discussion of this memoir and other works by Bienaymé see Heyde & Seneta [1977].
39. His birth and death dates are 12th September 1801 and 20th December 1861, a. St., or 24th September 1801 and 1st January 1862, n. St.

40. For a discussion of Ostrogradskiĭ's work see Gnedenko [1951]: I have not seen this paper. See also Maistrov [1974, pp.180-187].
41. I have been unable to see this paper: the discussion here follows Maistrov [1974, pp.182-184].
42. Ostrogradskiĭ gives his answers in terms of the Vandermonde symbol $[x]^n = x(x-1)\dots(x-n+1)$.
43. Todhunter [1865, p.558] remarks that Galloway's article may be viewed as "an abridgement of Poisson'[s] *Recherches ... sur la Prob.*" The article was in fact also published as a book.
44. Just as all roads lead to Rome — at least in two dimensions (see Feller [1968, §XIV.7]).
45. In a note to this theorem Catalan points out that "probability" is to be understood here as meaning what some mathematicians have called *probabilité subjective* (or *probabilité extrinsèque*, as he prefers to call it) in contrast to *probabilité intrinsèque.*
46. In a footnote Catalan points out that a similar result had been obtained by J.B.J. Liagre in his *Calcul des probabilités et théorie des erreurs, avec des applications aux sciences d'observation en général, et à la géodésie en particulier* of 1852.
47. Porter [1986, p.85] regards Friess ("competent if not original in mathematics") as the introducer in Germany of the frequentist viewpoint.
48. There is much discussion in Cournot's book of chance and probability in objective and subjective settings: in fact, Cournot makes the following distinction: "le terme de *possibilité* se prend dans un sens *objectif*, tandis que le terme de *probabilité* implique dans ses acceptions ordinaires un sens *subjectif* " [p.81]. For further comment on this point see Hacking [1971, p.343], Keynes [1921, chap. XXIV, §3] and Porter [1986, pp.84-85]. Zabell [1988c] in fact finds, in Cournot's work, three distinct categories of probability — objective, subjective and philosophical, "the last involving situations whose complexity precluded mathematical measurement" [p.178]. According to Sheynin [1986, p.308], Chuprov [1910, p.30] described Cournot as "one of the most original and profound thinkers of the 19th century, whom his contemporaries ... had failed to appreciate and who rates higher and higher in the eyes of posterity."
49. For details of these changes see Robson's edition, of 1974, of Mill's *Collected Works.*
50. In the first edition Mill in fact concluded that "the condition which Laplace omitted is not merely one of the requisites for the possibility of a calculation of chances; it is the only requisite" [§2]. Mill's change of heart was to a large extent brought about by criticism invited by

Mill from John Herschel. This criticism was incorporated into the *Logic* as early as 1846. For further details see Strong [1978, §3] and Porter [1986, p.83]. Porter (op. cit. p.82) regards Mill's comments in the first edition of his *Logic* as "one of the harshest denunciations of classical probability written in the nineteenth century."

51. Strong [1978, p.34] notes Mill's argument for a frequency theory of probability, some 25 years before John Venn.
52. *La Grande Encyclopédie* has no record of his death: it was, however, certainly after 1876. His christian names are given in varying orders.
53. In his paper "On the application of the theory of probabilities to the question of the combination of testimonies or judgements" of 1857, Boole writes of Ellis "There is no living mathematician for whose intellectual character I entertain a more sincere respect than I do for that of Mr. Ellis" [Boole 1952, p.350].
54. Salmon [1980a] does not regard Ellis, rather than Venn, as the first frequentist: he in fact concludes that "Ellis ... took us to the very threshold of a frequency theory of probability; Venn opened the door and led us in" [p.143]. Boldrini [1972, p.124] finds the relative frequency conception of probability "Formulated many years ago in Italy by G. Mortara". For some discussion see Boldrini, op. cit. pp.140-141.
55. For a general discussion of Donkin's work see Zabell [1988c, §6.1]. Zabell (op. cit. p.180) regards Donkin as representing "what may be the highwater mark in the defense of the Laplacean position".
56. Porter [1986, p.122] goes so far as to refer to "the subjectivist W.F. Donkin".
57. See Newcomb [1860e, §23].
58. If, in Donkin's theorem, we consider n probabilities $p_1, p_2, \ldots, p_n$, the first r of which are unchanged by the new information while the remainder are altered to $q_{r+1}, \ldots, q_n$ (say), we find, on setting $\beta = \sum_{r+1}^{n} q_i$, that $p_1 + \cdots + p_r = 1 - \beta$. On defining $p_i' = p_i/\beta$, for $i \in \{1, 2, \ldots, r\}$, we obtain $p_i' : p_j' :: p_i : p_j$, as asserted by Donkin. If the p_i's are initially the same ($= 1/n$), the entropy $H_n = -\sum_1^n p_i \log p_i$ is maximal, and this maximality is preserved, under the changes in probabilities mentioned above, by replacing each of the unchanged p_i by $p_i = (1-\beta)/r$.
59. As noted by Keynes [1921, chap. XVI, §13].
60. Notation altered: here all probabilities are supposed to be conditional on some fundamental state of knowledge.
61. For a full discussion of Boole's work in probability see Hailperin [1976].
62. See Boole [1952, p.261] for further details.
63. Details of the discussants are given in Boole [1952, p.271] and Keynes [1921, chap. XVII, §2].

64. Keynes [1921, chap. XVI, §6] finds Boole's error to arise from his adoption of two inconsistent definitions of "independence". For comment on Boole's confounding of "conditional probability" and "probability of a conditional" see Jaynes [1976, pp.241-242].
65. The correct solution was given by MacColl in 1897.
66. For further discussion of this problem see Keynes [1921, chap. XVII, §2].
67. See Jaynes [1976, p.241] for comment on this and other criticism.
68. In Chapter XVI, §7 Boole gives a summary of principles taken chiefly from Laplace: this summary, as Molina [1930, p.384] has noted, does not include the Laplacean generalization of Bayes's Theorem.
69. This chapter Stigler [1984] in his review of Smith [1982], views as "an early contribution to the theory of upper and lower probabilities and the combination of evidence."
70. See Boole [1854, p.362].
71. See Keynes [1921, chap. XXX, §14].
72. This problem also received attention in the nineteenth century from Hagen [1837].
73. Jaynes [1976, p.241] notes that Boole "did *not* reject it [Laplace's work] in the ground of the actual performance of Laplace's results in the case of the uniform prior because he, like Laplace's other critics, never bothered to examine the actual performance under these conditions". For further comment on Boole and the principle of insufficient reason see Zabell [1988c, §6.2]. Zabell [1989a, p.249] has noted that Boole's objection is to the principle of *insufficient* reason rather than to the principle of *cogent* reason. Boole himself amplified his thoughts on this point in a paper published in 1862 (see Boole [1952, p.390]).
74. See Edgeworth [1884a, p.208] and Keynes [1921, chap. IV, §9].
75. Jaynes [1976, p.242] in fact claims that "all of 'Boolean algebra' was contained already in the rules of probability theory given by Laplace".
76. This section is commented on by Keynes [1921, chap. XVI, §16].
77. Terrot, elected Bishop of Edinburgh and Pantonian professor in 1841, was chosen primus of Scotland in 1857, an office which he held until a stroke of paralysis forced his resignation in 1862. He was a Fellow and Vice-president of the Royal Society of Edinburgh, and contributed several papers to its journals.
78. On the assimilation of the problem as stated to the "bag and balls" case compare Boole's *An Investigation of the Laws of Thought*, chap. XX, §23.
79. The extra binomial coefficient required here if order is *not* considered will cancel out in the final analysis.

80. The solution given here is in fact that given by Keynes [1921, chap. XXX, §11].
81. Meyer's year of birth is sometimes given as 1803: the day was the 31st May, so the difference cannot be due to old versus new style.
82. Correct, that is, except for a few misprints.
83. The 31st chapter of Keynes [1921] is devoted to this result.
84. For comment on the contributions of Bernoulli and de Moivre see Pearson [1925].
85. See MacKenzie [1981, pp.236-237].
86. Venn in fact collaborated with Galton in the study of heredity — see Porter [1986, p.271].
87. Porter [1986, p.87] considers this work to be "The most influential nineteenth-century work on the philosophy of probability".
88. References throughout this section are to this edition. Venn made considerable changes in the second and third editions, but we have restricted our attention here to the last, which no doubt presented his considered views on probability and allied matters.
89. Venn's contention that the question of *time* is extraneous to probability considerations is in conflict with the views expressed by Shafer [1982].
90. The term is introduced, on p.190, with "A word of apology". For a discussion of this concept see Salmon [1980b, pp.131-132].
91. Salmon [1980b, p.133] finds "basic misunderstandings" in this chapter.
92. The reference given in this quotation is to Fisher's *Statistical Methods and Scientific Inference*, chapter II, §3. Fisher states that Venn's examples "seem to be little more than rhetorical sallies intended to overwhelm an opponent with ridicule. They scarcely attempt to conform with the conditions of Bayes' theorem, or of the rule of succession based upon it" [pp.25-26].
93. For names of others who rejected the rule of succession — and of those who accepted it — see Keynes [1921, chap. XXX, §14]. Edgeworth [1884b], while agreeing in the main with Venn's views, was led to conclude that "the particular species of inverse probability called the 'Rule of Succession' may not be so inane as Mr. Venn would have us believe" [p.234].

Chapter 8

1. Laurent usually used only the third of his christian names.
2. Given in Article 16 of Book II of the *Théorie analytique des probabilités*.

3. For later discussion of the inversion of Bernoulli's theorem see Castoldi [1959], Dale [1988b] and Jordan [1923], [1925], [1926a], [1926b] and [1933].
4. For biographies of Jevons see FitzPatrick [1960, pp.53-58] and Keynes [1936].
5. In his interpretation of probability Jevons was diametrically opposed to Venn: his views followed from those of de Morgan, Laplace and Poisson. See Porter [1986, pp.175-176] and Strong [1976, §6].
6. Further comment may be found in Keynes [1921, chap. IV, §4].
7. Jevons notes that "The probability that an event has a particular condition entirely depends upon the probability that if the condition existed the event would follow" [1877, p.240].
8. This extension, as we have already seen, seems to be due to Lubbock and Drinkwater-Bethune [c.1830, art.52].
9. For a more recent discussion of the value of further information see Horwich [1982, pp.122-129].
10. There is some discussion of Bing's work in A. Fisher [1926].
11. The translations in this section are by the Foreign Language Service of the South African Council for Scientific and Industrial Research: the original texts may be found in Appendix 8.2.
12. Bing finds in his paradox an analogy with one discussed by de Morgan in his *Essay on Probabilities*, but declares that the latter's explanation is "most unsatisfactory".
13. Translation: The only possible form of the function which does not give rise to contradictions has thus been proved to be unusable, and I would accordingly claim to have demonstrated that there simply is no such thing as *a posteriori* probability in problems in which no information is available about causes.
14. An even stronger statement has been made by de Finetti: see the preface to his [1974].
15. For further comment see Fisher [1926], Kroman [1908] and Whittaker [1920, §5].
16. The surname is incorrectly given as "McAlister" in this first contribution to the problem. Sir Donald MacAlister, Senior Wrangler at Cambridge in 1877, and later a medical man and principal of the University of Glasgow, produced, in response to a request from Galton, the log-normal distribution (see MacKenzie [1981, p.235]).
17. The problem was later discussed in Chapter VII, §17, of the third edition of 1888 of Venn's *The Logic of Chance.* Venn finds the assumption of equal *a priori* probabilities in this case less arbitrary than usual.

18. According to FitzPatrick [1960] Edgeworth's first names were originally in reverse order.
19. For a study of Edgeworth's work see Bowley [1928] and Stigler [1986a, chap. 9].
20. For a discussion of Gosset's work on the t-distribution see Welch [1958].
21. Bowley [1928] claims that this paper and that of 1922 (considered later in this section) provide the best insight into Edgeworth's original and final thoughts on his conception of probability. For a discussion of Edgeworth's compromise between subjectivism and frequentism see Porter [1986, p.259].
22. For comments on this formula see Sobel [1987, pp.170-171].
23. Edgeworth in fact realizes his example by considering the pattern of fragments of an exploding shell.
24. I trust that the reader will not attribute the lack of clarity between an estimator and an estimate evinced here to ignorance on the part of the author.
25. The article is signed merely with the initials M.W.C.
26. As to the meaning of "cause" Crofton states "The term 'cause' is not here used in its metaphysical sense, but as simply equivalent to 'antecedent state of things' " [p.773].
27. This problem, and the next, were also considered by Crofton in the chapter he contributed on mean value and probability to Williamson [1896].
28. See Porter [1986, p.86].
29. The formula is attributed by Hardy to Laplace, but its antecedents are certainly to be found in Price's appendix to Bayes's *Essay.*
30. Good [1965, p.17] has suggested that "It seems possible that G.F. Hardy was the first to suggest a 'continuum of inductive methods,' to use Carnap's phrase." For further comment on the choice of a beta prior see Good, op. cit., §§3.2, 4.1.
31. From China.
32. Le Cam, writing on the Central Limit Theorem, says "Bertrand and Poincaré wrote treatises on the calculus of probability, a subject neither of the two appeared to know. Except for some faint praise for Gauss' circular argument, Bertrand's book consists mainly of repeated claims that his predecessors made grievous logical mistakes" [1986, p.81].
33. The pagination is different.
34. A similar problem is discussed by Stabler [1892].

35. As Keynes [1921, chap. XXX, §11] has pointed out, the further assumption is needed that the number of balls is infinite: he also gives the correct solution that obtains when the urn contains only finitely many balls.
36. See his *An Investigation of the Laws of Thought* [1854, chap. XVI, §3].
37. This view was supported by Govan [1920, p.228] in his comments on Makeham (see §8.13).
38. For comment on Chrystal's three-ball problem see Zabell [1989a, p.252].
39. Described by Perks [1947, p.286] as an "unfortunate onslaught".
40. Makeham explicates Laplace's term by "p is a quantity such that the true value of p' in any particular urn is just as likely to be *above* as *below* it" [p.246].
41. In 1947 Perks proposed $p_x\, dx = dx/\left[\pi x^{1/2}(1-x)^{1/2}\right]$ as a new indifference rule, a rule which yields $(m+1/2)/(m+n+1)$ as the posterior probability of the next trial's being a success. Perks (op. cit. p.304) notes the conformity between this new result and the expression $(m+k)/(m+n+2k)$ as obtained by W.E. Johnson (according to H. Jeffreys), and also notes that it fits Makeham's empirical "general" formula. He finds, however, that "Makeham's work is marred by serious confusions of thought" [op. cit. p.304].
42. In Gini's paper of 1949 a yet more general formula is suggested: viz., if p is the observed frequency in n events, the probable value of its probability is to be taken as $(np+k)/(n+k+h)$, where k and h are determined by previous experience.
43. Makeham's own words are "If an observed event may be the result of one of n different causes; their probabilities are, respectively, as the probabilities of the event derived from their existence" [p.450]: it seems clear, however, that the formulation as given in the text is what is intended. A similar comment occurs in connexion with the fourth principle.
44. According to J.B.S. Haldane [1957] it was during Pearson's stay in Germany during the early 1880's that he began to spell his name with a K rather than a C.
45. Porter [1986, p.274] describes Pearson as "an astute historian of science".
46. MacKenzie [1981, p.73] mentions that Pearson's writings included "poetry, a 'passion play', art history, studies of the Reformation and mediaeval Germany, philosophy, biography and essays on politics, quite apart from his contributions — in the form of over four hundred articles — to mathematical physics, statistics and biology." Pearson was also a supporter of the feminist movement. For a partial listing of Pearson's work see E.S. Pearson [1938].

47. The second edition of this work was published in 1900, a third edition following in 1911. In his biography of his father E.S. Pearson [1938] wrote "It was because Pearson felt in later years that the task of bringing his Grammar up to date was beyond his powers, that he would not consent to its republication although all editions were out of print" [p.132].
48. In Chapter I, §5, Pearson avers "The man who classifies facts of any kind whatever, who sees their mutual relation and describes their sequences, is applying the scientific method and is a man of science."
49. Compare the postulates in the Appendix on Eduction in Johnson [1924].
50. See MacKenzie [1981, p.92].
51. See also Note 12 to Chapter 8 in MacKenzie [1981].
52. MacKenzie's approach provides a posterior distribution for the variance θ in sampling from $N(0, \theta)$; Welch's yields a posterior distribution for the correlation coefficient ρ.
53. Consideration is also given, in Pearson and Filon's paper, to multivariate Normal situations.
54. Porter [1986, p.306] suggests that Pearson's philosophy of probability was borrowed from Edgeworth.
55. This expression includes the term $B(r+1, s+1)$ omitted by Laplace.
56. On Pearson's evaluation of the beta-integrals, and the earlier work done by Laplace in this connexion, see Molina [1930, p.376].
57. See Dale [1988b, p.351].
58. Of this term Moroney [1951, p.114], citing an unnamed source, says "it is neither an error nor probable."

Appendix

Reading Note
L.J. Savage

4 March 1960

Thomas Bayes, "Essay towards solving a problem in the doctrine of chances," The Philosophical Transactions[1], 53 (1763), 370-418.

This famous paper and another from the same volume (269-271) of the transactions were reproduced photographically by the Department of Agriculture in 1941 (?)[2] with some commentary by W.E. Deming and E.C. Molina. The other essay shows that Stirling's series for $n!$ is asymptotic only; this is apparently the first notice ever taken of asymptotic series. Both papers were edited posthumously by Bayes's friend Richard Price, who made at least some contribution to them. Biometrika 45 (1958), 293-315, republished the essay with a biographical note by G.A. Barnard[3].

Though the essay is not long, it is rich, and I find need to prepare myself a special sort of abstract of it. My interest for this purpose is not in mathematical aspects of the paper such as validity of demonstrations but in certain ideas. In what way, or ways, does Bayes view probability? What propositions does he consider important? In what form is "Bayes's Theorem" among them?

The essence of the essay is stated to be this "Problem": "Given the number of times in which an unknown event has happened and failed: Required the chance that the probability of its happening in a single trial lies somewhere between any two degrees of probability that can be named." Bayes says elsewhere that chance and probability are synonymous for him and seems to stick by that. The problem is of the kind we now associate with Bayes's name, but it is confined from the outset to the special problem of drawing the Bayesian inference, not about an arbitrary sort of parameter, but about a "degree of probability" only.

Price, in a letter introducing the essay, says that Bayes saw clearly how to solve his problem if an initial distribution were given and that Bayes thought there was good but not perfect reason to postulate the uniform prior distribution. It is thus with good reason that the term "Bayes's Postulate" is sometimes used for this assumption. What Price says convinces me that Bayes was aware of Bayes's theorem[4] in full generality, except possibly that he confined himself to unknown "degrees of probability." Price declares the problem to be central to the philosophy of induction and therefore to the "argument taken from final causes for the existence of the Deity." De Moivre's theorem[5] he holds as nothing compared to this in importance; the converse problems must not be confused with each other.

After baldly stating his Problem, Bayes presents, as Section I, a whole short course on probability. Using modern terms freely, it may be paraphrased thus:

Definitions:

1. Inconsistent events = incompatible events.

2. Contrary events = a two-fold partition, a dichotomy.

3. "An event is said to fail, when it cannot happen; or, which comes to the same thing, when its contrary has happened."

4. "An event is said to be determined when it has either happened or failed."

5. "The probability of any event is the ratio between the value at which an expectation depending on the happening of the event ought to be computed, and the value of the thing expected upon it's [sic] happening."

6. "By chance I mean the same as probability."

7. "Events are independent when the happening of any one of them does neither increase nor abate the probability of the rest."

These definitions could provoke many remarks. For example, apparently Bayes thinks that pairwise independence is independence[6]. The definition of probability is of course most interesting. Apparently, an expectation was clearly understood by contemporaries as a payment contingent on an event, and such things must have sometimes been bought, sold, and used as collateral. What does "ought" mean?

Next some propositions and Corollaries are derived. I have not taken great pains to check the derivations, but in general they take Bayes's definition of probability seriously; they are beclouded by the idea that numbers are a little more shameful than ratios.

Prop. 1. Simple additivity.

Cor. The sum of probabilities over a partition and, in particular, over a dichotomy is 1.

Prop. 2. "If a person has an expectation depending on the happening of an event, the probability of the event is to the probability of its failure as his loss if it fails to his gain if it happens."

That is, the odds p/q for winning a simple fair lottery is the ratio of the prize minus the price of the ticket to the price of the ticket.

Prop. 3. $\Pr[A \text{ and } B] = \Pr[A]\Pr[B|A]$.

Cor. $\Pr[B|A] = \Pr[A \text{ and } B] / \Pr[A]$.

Prop. 4. "If there be two subsequent events to be determined every day, and each day the probability of the 2nd is b/N and the probability of both P/N, and I am to receive N if both the events happen the first day on which the 2nd does; I say, according to these conditions, the probability of my obtaining N is P/b."

That is, if A_i and B_i are independent from index to index, the probability that the first occurrence of a B_i will be accompanied by that of the corresponding A_i is what it should be.

Cor. "Suppose after the expectation given me in the foregoing proposition, and before it is at all known whether the 1st event has happened or not, I should find that the 2nd event has happened; from hence I can only infer that the event is determined on which my expectation depended, and have no reason to esteem the value of my expectation either greater or less than it was before. ... But the probability that an event has happened is the same as the probability I have to guess right if I guess it has happened. Wherefore the following proposition is evident."

Prop. 5. "If there be two subsequent events, the probability of the 2nd b/N and the probability of both together P/N, and it being first discovered that the 2nd event has happened, from hence I guess that the 1st event has also happened, the probability I am in the right is P/b."[7]

Prop. 6. Product rule for "several independent events."

Cor. 1. The probability of prescribed sequences of successes and failures of independent events.

Cor. 2. Ditto when all these events have probability α.

Definition. In effect, defines Bernoulli trials. Here Bayes deliberately introduces an ambiguity that helps and hinders us to this day: "And hence it is manifest that the happening or failing of the same event in so many diffe[rent] trials, is in reality the happening or failing of so many distinct independent events exactly similar to each other."

Prop. 7. Derives the binomial distribution for Bernoulli trials.

This concludes Section I, the short course on first principles. It is admirable and shows good insight into conditional probability, but there is no trace of what we think of as characteristic of Bayes, a theorem about the probability of causes. The germ of that, but the germ only, is to come in the next and final section.

Section II

Straining over the rigoritis of his own time, but showing perfectly modern insight into the thing itself, Bayes describes a schematic Monte Carlo procedure based on a levelled table and two balls. Throwing the first ball once selects an α uniformly between 0 and 1. Then, throwing the second ball n times yields n trials, that, given α, are independent with probability α.

Prop. 8. Calculates by means of a beta integral the probability that α will fall in a preassigned interval and p of the n trials will be successful.

Cor. Gives the probability of just p successes indirectly in terms of the ratio of a beta integral to a binomial coefficient. Bayes knows that this is the same, namely $1/(n+1)$, for all p, but he is too formal to mention it here. A little later he adduces this uniformity in p as a particularly telling justification for Bayes's postulate as a description of the blank mind.

Prop. 9. Gives the probability of any interval in α given p (couched in the language of guessing).

Cor. Gives the cumulative distribution of α given p.

Scholium. A deliberately extra-mathematical argument in defense of Bayes's postulate, already mentioned by me in connection with the corollary to Proposition 8.

Prop. 10. Restates Prop. 9 in terms of the newly gained knowledge of the value of the complete beta integral.

The essay concludes with practical rules for computing the incomplete beta integral, which is outside the province of this abstract.

Price has an appendix of numerical examples and philosophy. Below is an instance that Laplace has later made famous. Such a discussion necessarily seems old-fashioned today, but the second paragraph is far from naïve and the third seems important against those who take universals seriously.

"Let us imagine to ourselves the case of a person just brought forth into this world, and left to collect from his observation of the order and course of events what powers and causes take place in it. The Sun would, probably, be the first object that would engage his attention; but after losing it the first night he would be entirely ignorant whether he should ever see it again. He would therefore be in the condition of a person making a first experiment about an event entirely unknown to him. But let him see a second appearance or one return of the Sun, and an expectation would be raised in him of a second return, and he might know that there was an odds of 3 to 1 for some probability of this. This odds would increase, as before represented, with the number of returns to which he was witness. But no finite number of returns would be sufficient to produce absolute or physical certainty. For let it be supposed that he has seen it return at regular and stated intervals a million of times. The conclusions this would warrant would be such as follow. There would be the odds of the millionth power of 2, to one, that it was likely that it would return again at the end of the usual interval. There would be the probability expressed by 0.5352, that the odds for this was not greater than 1,600,000 to 1; and the probability expressed by 0.5105, that it was not less than 1,400,000 to 1.

"It should be carefully remembered that these deductions suppose a previous total ignorance of nature. After having observed for some time the course of events it would be found that the operations of nature are in general regular, and that the powers and laws which prevail in it are stable and permanent. The consideration of this will cause one or a few experiments often to produce a much stronger expectation of success in further experiments than would otherwise have been reasonable; just as the frequent observation that things of a sort are disposed together in any place would lead us to conclude, upon discovering there any object of a particular sort, that there are laid up with it many others of the same sort. It is obvious that this, so far from contradicting the foregoing deductions, is only one particular case to which they are to be applied.

"What has been said seems sufficient to shew us what conclusions to draw from uniform experience. It demonstrates, particularly, that instead of proving that events will always happen agreeably to it, there will be always reason against this conclusion. In other words, where the course of nature has been the most constant, we can have only reason to reckon upon a recurrency of events proportioned to the degree of this constancy; but we can have no reason for thinking that there are no causes in nature which will ever interfere with the operations of the causes from which this constancy is derived, or no circumstances of the world in which it will fail. And if this true, supposing our only data derived from experience, we shall find additional reason for thinking thus if we apply other principles, or have recourse to such considerations as reason, independently of experience, can suggest."

Notes

1. Often referenced as Phil. Trans. Roy. Soc. London.

2. Actually 1940. *Facsimiles of two papers by Bayes.* (ed. W.E. Deming.) The Graduate School, United States Department of Agriculture, Washington.

3. A further publication is in *Studies in the History of Statistics and Probability* (ed. E.S. Pearson and M.G. Kendall.) London: Griffin (1970).

4. The suggestion has been made that Bayes is not the originator of the theorem that is now named after him. See S.M. Stigler (1983), *American Statistician* 37, 290-296.

5. It is not quite clear what is meant here; presumably the binomial distribution and its normal approximation. See A.W.F. Edwards (1986), *American Statistician* 40, 109-110.

6. In a private communication D.V. Lindley has suggested that a more charitable view than that taken here would show that the usual definition of independence (for three events) follows from Bayes's definition. (Note by A.I.D.)

7. A careful analysis of this has been given by G. Shafer (1982), *American Statistician* 10, 1075-1089.

EPIPHONEMA

Non est necesse haec omnia ad felicitatem observare; sed tamen qui haec omnia observaverit felix erit.
Longe autem facilius est haec scire quam exequi.

Girolamo Cardano,
"Præceptorum ad filios libellus".

Bibliography

Many authors entertain, not only a foolish, but a really dishonest objection to acknowledge the sources from whence they derive much valuable information. We have no such feeling.

Charles Dickens, Pickwick Papers.

Ackland, T.G. & Hardy, G.F. 1889. *Graduated Exercises and Examples for the Use of Students of the Institute of Actuaries Textbook.* London: C. & E. Layton.

Anderson, J.G. 1941. The Reverend Thomas Bayes. *The Mathematical Gazette* 25: 160-162.

Anon. (a) 1923. Early Presbyterianism at Tunbridge Wells. pp.222-224 in *The Journal of the Presbyterian Historical Society of England*, 2 (parts 1-4, 1920-1923), ed. Rev. J. Hay Coolidge. (This article is from issue № 4, May 1923.)

Anon. (b). *A view of the Dissenting Intrest in London of the Presbyterian & Independent Denominations from the year 1695 to the 25 of December 1731. With a postscript of the present state of the Baptists.* Housed in Dr Williams's Library.

Arbuthnott, J. 1710. An Argument for Divine Providence, taken from the constant Regularity observ'd in the Births of both Sexes. *Philosophical Transactions* 27: 186-190. Reprinted in Kendall & Plackett [1977], pp.30-34.

—— 1770. *Miscellaneous Works of the late Dr. Arbuthnot. With an Account of the Author's Life.* 2 vols. London.

Archibald, R.C. 1926. A rare pamphlet of Moivre and some of his discoveries. *Isis* 8: 671-676 + 7.

Baker, K.M. 1975. *Condorcet: from natural philosophy to social mathematics.* Chicago: Chicago University Press.

Balguy, J. 1730. *Divine Rectitude, or a Brief Inquiry concerning the Moral Perfections of the Deity; Particularly in respect of Creation and Providence.* London.

Barnard, G.A. 1958. Thomas Bayes — a biographical note. *Biometrika* 45: 293-295. (Bayes's essay is reprinted as pp.296-315.)

—— 1964. Article on Thomas Bayes. *Encyclopædia Britannica*, 14th edition.

—— 1987. R.A. Fisher — a true Bayesian? *International Statistical Review* 55: 183-189.

—— 1988. The future of statistics: teaching and research. pp.17-24 in Bernardo et al. [1988].

Bayes, T. Notebook.

—— 1731. *Divine Benevolence, or an attempt to prove that the Principal End of the Divine Providence and Government is the Happiness of his Creatures. Being an answer to a Pamphlet entitled: 'Divine Rectitude: or an Inquiry concerning the Moral Perfections of the Deity'. With a Regulation of the Notions therein advanced concerning Beauty and Order, the Reason of Punishment, and the Necessity of a State of Trial antecedent to perfect Happiness.* London: John Noon.

—— 1736. *An Introduction to the Doctrine of Fluxions, and Defence of the Mathematicians against the Objections of the Author of the Analyst, so far as they are designed to affect their general Methods of Reasoning.* London: John Noon.

—— 1763a. (published 1764.) An Essay towards solving a Problem in the Doctrine of Chances. By the late Rev. Mr. Bayes, F.R.S. communicated by Mr. Price, in a letter to John Canton, A.M. F.R.S. *Philosophical Transactions* 53: 370-418.

—— 1763b. (published 1764.) A letter from the late Reverend Mr. Thomas Bayes, F.R.S. to John Canton, M.A. and F.R.S. *Philosophical Transactions* 53: 269-271.

——— 1764. (published 1765.) A Demonstration of the Second Rule in the Essay towards the Solution of a Problem in the Doctrine of Chances, published in the Philosophical Transactions, Vol. LIII. Communicated by the Rev. Mr. Price, in a Letter to Mr. John Canton, M.A. F.R.S. *Philosophical Transactions* 54: 296-325.

Beguelin, N. de. 1765. (published 1767.) Sur les suites ou séquences dans la lotterie de Genes. *Histoire de l'Academie royale des Sciences et Belles-Lettres, Berlin* : 231-280.

——— 1767. (published 1769.) Sur l'usage du principe de la raison suffisante dans le calcul des probabilités. *Histoire de l'Academie royale des Sciences et Belles-Lettres (Classe de Philosophie spéculative), Berlin* 23: 382-412.

Bellhouse, D.R. 1988a. The Reverend Thomas Bayes F.R.S. — c.1701-1761. *The Institute of Mathematical Statistics Bulletin* 17: 276-278.

——— 1988b. The Reverend Thomas Bayes, F.R.S. — c.1701-1761. *The Institute of Mathematical Statistics Bulletin* 17: 483.

Berkeley, G. 1734. *The Analyst; or, a Discourse addressed to an Infidel Mathematician. Wherein it is examined whether the object, principles, and inferences of the modern analysis are more distinctly conceived, or more evidently deduced, than religious Mysteries and points of Faith. By the author of The Minute Philosopher.* London.

——— 1735. *A Defence of Free-thinking in Mathematics. In answer to a pamphlet of Philalethes Cantabrigiensis intituled Geometry no friend to Infidelity, or a defence of Sir Isaac Newton and the British Mathematicians. Also an appendix concerning Mr. Walton's Vindication of the principles of fluxions against the objections contained in the Analyst; wherein it is attempted to put this controversy in such a light as that every reader may be able to judge thereof. By the author of The Minute Philosopher.* London.

Bernardo, J.M., DeGroot, M.H., Lindley, D.V. & Smith, A.F.M. (Eds) 1988. *Bayesian Statistics 3: Proceedings of the Third Valencia International Meeting, June 1-5, 1987.* Oxford: Clarendon Press.

Bernoulli, D. 1735. De inclinatione mutua orbitarum planetarum. Recherches physiques et astronomiques sur le problème proposé pour la seconde fois par l'Academie royale des Sciences de Paris. Quelle est la cause physique de l'inclinaison des plans des orbites des planètes par rapport au plan de l'equateur de la révolution du soleil autour de son axe, et d'où vient que les inclinaisons de ces orbites sont différentes entre elle. *Pieces qui ont remporté le prix de l'Academie royale des Sciences en 1734, Paris*, 3: 95-122. (The original Latin version occupies pp.125-144.)

Bernoulli, J. 1713. *Ars Conjectandi.* Basileæ: Thurnisiorum.

Bertrand, J.L.F. 1889. *Calcul des Probabilités.* Paris. 2nd edition, 1907. Reprinted as 3rd edition, 1972, New York: Chelsea.

Bienaymé, I.J. 1838. Mémoire sur la probabilité des résultats moyens des observations; démonstration directe de la règle de Laplace. *Mémoires présentés par divers savants a l'académie royale des sciences de l'institut de France* 5: 513-558.

Bing, F. 1879a. Om aposteriorisk Sandsynlighed. *Tidsskrift for Mathematik, Series 4*, 3: 1-22.

—— 1879b. Svar til Professor L. Lorenz. *Tidsskrift for Mathematik, Series 4*, 3: 66-70.

—— 1879c. Nogle Bemærkninger i Anledning af „Gjensvaret" fra Professor L. Lorenz. *Tidsskrift for Mathematik, Series 4*, 3: 122-131.

Blake, F. 1741. *An Explanation of Fluxions, in a short essay on the theory.* London: W. Innys.

Bloom, D.M. 1973. A birthday problem. *American Mathematical Monthly* 80: 1141-1142.

Bogue, D. & Bennett, J. 1808-1812. *History of Dissenters, from the Revolution, in 1688, to the year 1808.* 4 vols (1808, 1809, 1810, 1812). London: printed for the authors.

Boldrini, M. 1972. *Scientific Truth and Statistical Method.* London: Charles Griffin & Company. (Translation of the 2nd Italian edition of 1965.)

Bolzano, B. 1837. *Wissenschaftslehre. Versuch einer ausführlichen und größtenheils neuen Darstellung der LOGIK mit steter Rücksicht auf deren bisherige Bearbeiter.* Translated as *Theory of Science. Attempt at a Detailed and in the main Novel Exposition of LOGIC. With Constant Attention to Earlier Authors.* Edited and translated by Rolf George, 1972. Berkeley and Los Angeles: University of California Press.

Boole, G. 1851a. On the theory of probabilities, and in particular on Mitchell's problem of the distribution of fixed stars. *The Philosophical Magazine, Series 4*, 1: 521-530. Reprinted in Boole [1952], pp.247-259.

—— 1851b. Further observations on the theory of probabilities. *The Philosophical Magazine, Series 4*, 2: 96-101. Reprinted in Boole [1952], pp.260-267.

—— 1851c. Proposed question in the theory of probabilities. *The Cambridge and Dublin Mathematical Journal 6*: 286. Reprinted in Boole [1952], pp.268-269.

—— Sketch of a Theory and Method of Probabilities founded upon the Calculus of Logic. First published in Boole [1952], pp.141-166.

—— 1854a. *An Investigation of the Laws of Thought, on which are founded the mathematical theories of logic and probabilities.* London: Macmillan. Reprinted in 1958; New York: Dover.

—— 1854b. Solution of a question in the theory of probabilities. *The Philosophical Magazine, Series 4*, 7: 29-32. Reprinted in Boole [1952], pp.270-273.

—— 1854c. Reply to some observations by Mr. Wilbraham on the theory of chances. *The Philosophical Magazine, Series 4*, 8: 87-91. Reprinted in Boole [1952], pp.274-279.

—— 1854d. Further observations relating to the theory of probabilities in reply to Mr. Wilbraham. *The Philosophical Magazine, Series 4*, 8: 175-176. Reprinted in Boole [1952], pp.289-290.

—— 1857. On the application of the theory of probabilities to the question of testimonies or judgements. *Transactions of the Royal Society of Edinburgh 21*: 597-652. Reprinted in Boole [1952], pp.308-385.

—— 1952. *Collected Logical Works, vol. 1. Studies in Logic and Probability.* La Salle, Illinois: Open Court.

Bowley, A.L. 1928. *F.Y. Edgeworth's Contributions to Mathematical Statistics.* London: Royal Statistical Society.

Boyer, C.B. 1968. *A History of Mathematics.* New York: John Wiley & Sons.

Brewer, E.C. 1978. *The Dictionary of Phrase and Fable*, classic edition. New York: Avnel Books.

Bru, B. & Clero, J.P. 1988. Essai en vue de resoudre un probleme de la doctrine des chances. *Cahiers d'Histoire et de Philosophie des Sciences (Paris)*, 18. (A translation of Bayes's Essay.)

Bru, B. & Crepel, P. 1989. Le calcul des probabilités et la mathématique sociale chez Condorcet. *Actes du colloque Condorcet*, Paris: Minerve.

Buffon, G.L.L. Comte de. 1777. Essai d'Arithmetique Morale. pp.67-216 in *Œuvres Complètes de M. le Cte. de Buffon, vol. X. Histoire Naturelle, générale et particulière, servant de suite à l'Histoire Naturelle de l'Homme, 1778.* Paris.

Bühler, W.K. 1981. *Gauss: a biographical study.* Berlin, Heidelberg, New York: Springer-Verlag.

Burnside, W. 1924. On Bayes' formula. *Biometrika* 16: 189.

—— 1928. *Theory of Probability.* Cambridge: Cambridge University Press. Reprinted in 1959, New York: Dover.

Burr, T.B. 1766. *The History of Tunbridge Wells.* London.

Cajori, F. 1893. *A History of Mathematics.* London: Macmillan. Reprinted, second edition, 1919a; London: Macmillan.

—— 1919b. *A History of the Conceptions of Limits and Fluxions in Great Britain from Newton to Woodhouse.* Chicago & London: Open Court.

Calamy, E. 1830. *An Historical Account of My Own Life, with some reflections on the times I have lived in (1671-1731).* 2 vols. Second edition. London: Henry Colburn & Richard Bentley.

Cantor, M. B. (Ed.) 1908. *Vorlesungen über Geschichte der Mathematik. Viertelband von Jahre 1759 bis zum Jahre 1799.* Reprinted 1965; New York: Johnson Reprint Corporation.

Cardano, G. c.1564. *Liber de Ludo Aleae.* First published in *Opera Omnia* 1663. Reprinted 1967; New York and London: Johnson Reprint Corporation.

Carnap, R. 1952. *The Continuum of Inductive Methods.* Chicago: Chicago University Press.

Castoldi, L. 1959. Inversione delle formule di Bayes per la determinazione di probabilità condizionali dirette. *Rendiconti del Seminario della Facolti di Scienze della Università di Cagliori, Fasc. 3-4*, 29: 26-31.

Catalan, E.C. 1841. Deux problèmes de probabilités. *Journal de Liouville* 6: 75-80.

—— 1877. Un nouveau principe de probabilités. *Académie royale des Sciences, des Lettres et des Beaux-Arts de Belgique. Bulletin.* 44, pt. 2: 463-468.

—— 1886. Problèmes et théorèmes de probabilités. *Académie royale des Sciences, des Lettres et des Beaux Arts de Belgique* 46: 1-16.

—— 1888. Sur une application du théorème de Bayes, faite par Laplace. *Mémoires de la Société royale des Sciences de Liège, Deuxième Série, Tome XV. Mélanges Mathématiques* 299: 255-258.

Catalogue of the printed books in the library of the University of Edinburgh, 1918-1923. Edinburgh: Edinburgh University Press, T. & A. Constable.

Chang, W.-C. 1976. Statistical theories and sampling practice. pp.297-315 in Owen [1976].

Chrystal, G. 1891. On some fundamental principles in the theory of probability. *Transactions of the Actuarial Society of Edinburgh* 11, part 13: 421-439.

Chuprov, A.A. 1910. *Essays on the Theory of Statistics.* [In Russian.] Moscow.

Ciesielski, K. 1987. Lost legends of Lvov 1: The Scottish Café. *The Mathematical Intelligencer* 9, no. 4: 36-37.

Clay, J.W. (Ed.) 1895. *Familiae Minorum Gentium.* vol. III. Volume 39 of *The Publications of the Harleian Society*, London.

Clero, J.P. 1988. Temps, cause et loi d'apres l'essai de Thomas Bayes. *Cahiers d'Histoire et de Philosophie des Sciences (Paris)* 18: 129-159. (Postscript to Bru & Clero [1988].)

Coles, E. 1674. *The Newest, Plainest, and the Shortest Shorthand.* London.

Condorcet, M.J.A.N. Caritat, le Marquis de. 1772(?), 1774. MSS 883, ff.216-221, and 875, ff.84-99 (copy 100-109). Bibliothèque de l'Institut de France.

—— *Mémoire sur le calcul des probabilités* in six parts.
1781. Réflexions sur la règle générale qui prescrit de prendre pour valeur d'un évènement certain, la probabilité de cet évènement, multipliée par la valeur de l'évènement en lui-même. *Histoire de l'Académie royale des Sciences*: 707-728.
1781. Application de l'analyse à cette question: Déterminer la probabilité qu'un arrangement régulier est l'effet d'une intention de la produire. *Histoire de l'Académie royale des Sciences*: with Part 1.
1782. Sur l'évaluation des droits éventuels. *Histoire de l'Académie royale des Sciences*: 674-691.
1783. Réflexions sur la méthode de déterminer la probabilité des évènemens futurs, d'après l'observation des évènemens passés. *Histoire de l'Académie royale des Sciences*: 539-553.
1783. Sur la probabilité des faits extraordinaires. *Histoire de l'Académie royale des Sciences*: 553-559.
1784. Application des principes de l'article précédent à quelques questions de critique. *Histoire de l'Académie royale des Sciences*: 454-468.

—— 1785a. *Probabilité*, from *Encyclopédie Méthodique, ou par ordre de matières: par une société de gens de lettres, de savans et d'artistes; Précéde d'un Vocabulaire universel, servant de Table pour tout l'Ouvrage, ornére des Portraits de MM. Diderot & D'Alembert, premiers Editeurs de l'Encylcopédie.* vol. 2, pp.640-663.

—— 1785b. *Essai sur l'application de l'analyse à la probabilité des décisions rendues à la pluralité des voix.* Paris: de l'imprimerie royale. Reprinted 1972; New York: Chelsea.

—— 1787. Discours sur l'astronomie et le calcul des probabilités. Lu au Lycée en 1787. Reprinted in volume 1 of Condorcet-O'Connor & Arago [1847-1849], pp.482-503.

—— 1805. *Eléméns du calcul des probabilités et son application aux jeux de hasard, a la loterie, et aux jugemens des hommes. Avec un discours sur les avantages des mathématiques sociales. Et une notice sur M. de Condorcet.* Paris: Royez.

Condorcet-O'Connor, A. & Arago, F. 1847-1849. *Œuvres de Condorcet.* 12 vols. Paris: Firmin Didot Frères.

Coolidge, J.L. 1949. *The Mathematics of Great Amateurs.* Oxford: Clarendon Press.

Cournot, A.A. 1843. *Exposition de la Théorie des Chances et des Probabilités.* Paris: L. Hachette.

Crepel, P. 1987. Le premier manuscrit de Condorcet sur le calcul des probabilités (1772). *Historia Mathematica* 14: 282-284.

——— 1988a. Condorcet, la theorie des probabilités et les calculus financiers. pp.267-325 in Rashed et al. [1988].

——— 1988b. Condorcet et l'estimation statistique. *Journal de Societe de Statistique de Paris* 129: 43-64.

——— 1989a. A quoi Condorcet a-t-il applique le calcul des probabilités? *Actes du colloque Condorcet*, Paris: Minerve.

——— 1989b. Condorcet, un mathématicien du social. *La Recherche* 20: 248-249.

——— 1989c. De Condorcet à Arago: l'enseignement des probabilités en France de 1786 à 1830. *Bulletin de la Société des Amis de la Bibliothéque de l'Ecole Polytechnique* 4: 29-55.

Crofton, M.W. 1885. Article on Probability. *Encyclopædia Britannica*, 9th edition, 19: 768-788.

Dale, A.I. 1982. Bayes or Laplace? An examination of the origin and early applications of Bayes' theorem. *Archive for History of Exact Sciences* 27: 23-47.

——— 1986. A newly-discovered result of Thomas Bayes. *Archive for History of Exact Sciences* 35: 101-113.

——— 1988. The Reverend Thomas Bayes, F.R.S. — c.1701-1761. *The Institute of Mathematical Statistics Bulletin* 17: 278.

——— 1988b. On Bayes' theorem and the inverse Bernoulli theorem. *Historia Mathematica* 15: 348-360.

——— 1989. An early occurrence of the Poisson distribution. *Statistics & Probability Letters* 7: 21-22.

——— 1990. Thomas Bayes: some clues to his education. *Statistics & Probability Letters* 9: 289-290.

——— 1991. Thomas Bayes's work on infinite series. *Historia Mathematica* 18: 312-327.

Dale, R.W. 1907. *A History of English Congregationalism*, ed. A.W.W. Dale. London: Hodder & Stoughton.

Dalzel, A. 1862. *History of the University of Edinburgh from its Foundation.* 2 vols. Edinburgh: Edmonston & Douglas.

Darling, J. 1852-1854. *Cyclopaedia Bibliographica: a library manual of theological and general literature, and guide for authors, preachers, students, and literary men. Analytical, bibliographical, and biographical.* London.

Daston, L.J. 1979. D'Alembert's critique of probability theory. *Historia Mathematica* 6: 259-279.

——— 1988. *Classical Probability in the Enlightenment.* New Jersey: Princeton University Press.

David, F.N. 1962. *Games, Gods and Gambling. The origins and history of probability and statistical ideas from the earliest times to the Newtonian era.* London: Charles Griffin & Co.

——— 1965. Some notes on Laplace. pp.30-44 in Neyman and Le Cam [1965].

Davis, C.H. 1857. See Gauss [1809].

de Finetti, B. 1932. Funzione caratteristica di un fenomeno aleatorio. *Atti del Congresso Internazionale dei Matematici Bologna, 3-10 settembre 1928* 6: 179-190.

——— 1937. La prévision: ses lois logiques, ses sources subjectives. *Annales de l'Institut Henri Poincaré (Paris)* 7: 1-68. Reprinted in Kyburg and Smokler [1964].

——— 1972. *Probability, Induction and Statistics: the art of guessing.* London, New York, Sydney, Toronto: John Wiley & Sons.

——— 1974. *Theory of Probability. A critical introductory treatment.* 2 vols. London, New York, Sydney, Toronto: John Wiley & Sons.

de Moivre, A. 1733. *Approximatio ad Summam Terminorum Binomii* $(a+b)^n$ *in Seriem expansi.* Reprinted in Archibald [1926] and in de Moivre [1756], pp.243-250.

—— 1756. *The Doctrine of Chances: or, a Method of Calculating the Probabilities of Events in Play*, 3rd edition. London: A. Millar (1st edition 1718). Reprinted 1967; New York: Chelsea.

de Morgan, A. 1837. Review of Laplace's Théorie Analytique des Probabilités (3rd edition). *Dublin Review* 2: 338-354 & 3: 237-248.

—— 1838a. *An Essay on Probabilities and their Application to Life Contingencies and Insurance Offices.* London: Longman, Orme, Brown, Green, & Longmans.

—— 1838b. On a question in the theory of probabilities. *Transactions of the Cambridge Philosophical Society* 6: 423-430.

—— 1843. Article on Theory of Probabilities in the *Encyclopaedia Metropolitana, 1st Division, Pure Sciences.* vol. 2: 393-490. (ed. H.J. Rose.)

—— 1847. *Formal Logic.* London: Taylor & Walton.

—— 1860. Rev. Thomas Bayes, etc. *Notes and Queries*, 2nd Series, 9: 9-10.

Dempster, A.P. 1966. New methods for reasoning towards posterior distributions based on sample data. *Annals of Mathematical Statistics* 37: 355-374.

Derham, W. 1798. *Physico-Theology; or a Demonstration of the Being and Attributes of God, from his Works of Creation: being the Substance of 16 Sermons preached at the Honble Mr. Boyle's Lectures, with large Notes and many curious Observations.* London. (1st edition 1713.)

Dickens, C.J.H. 1852-1854. *A Child's History of England.* 3 vols. London: Bradbury and Evans.

Dinges, H. 1983. Gedanken zum Aufsatz der Rev. Th. Bayes (1702-1761). *Mathematische Semesterberichte* 30 Heft 1: 61-105.

Donkin, W.F. 1851. On certain questions relating to the theory of probabilities: in three parts. *The London, Edinburgh, and Dublin Philosophical Magazine and Journal of Science, Series 4*, 1: 353-368; 1: 458-466; 2: 55-60.

Drysdale, A.H. 1889. *History of the Presbyterians in England: their rise, decline and revival.* London: Publication Committee of the Presbyterian Church of England.

Dutka, J. 1981. The incomplete beta function — a historical profile. *Archive for History of Exact Sciences* 24: 11-29.

Edgeworth, F.Y. 1883. The method of least squares. *The London, Edinburgh, and Dublin Philosophical Magazine and Journal of Science, Series 5*, 16: 360-375.

—— 1884a. À priori probabilities. *The London, Edinburgh, and Dublin Philosophical Magazine and Journal of Science, Series 5*, 18: 204-210.

—— 1884b. The philosophy of chance. *Mind* 9: 223-235.

—— 1911. Article on Probability. *Encyclopædia Britannica*, 11th edition, 22: 376-403.

—— 1921. Molecular statistics. *Journal of the Royal Statistical Society* 84: 71-89.

—— 1922. The philosophy of chance. *Mind* 31: 257-283.

Edwards, A.W.F. 1974. A problem in the doctrine of chances. *Proceedings of Conference on Foundational Questions in Statistical Inference, Aarhus, May 7-12, 1973.* (Eds O. Barndorff-Nielsen, P. Blaesild, G. Schou.) Department of Theoretical Statistics, Institute of Mathematics, University of Aarhus. Memoirs No. 1: 41-60.

—— 1978. Commentary on the arguments of Thomas Bayes. *Scandinavian Journal of Statistics* 5: 116-118.

Ellis, R.L. 1844. On the foundations of the theory of probabilities. *Transactions of the Cambridge Philosophical Society* 8: 1-6.

Emerson, W. 1776. *Miscellanies, or a Miscellaneous Treatise; containing Several Mathematical Subjects.* London: J. Nourse.

—— 1793. *Tracts: A New Edition, to which is prefixed, Some Account of the Life and Writings of the Author, by the Rev. W. Bowe.* London: F. Wingrove.

Encyclopædia Britannica. 9th edition, 1875-1889; Edinburgh: Black. 11th edition, 1910-1911; Cambridge: Cambridge University Press. 14th edition, 1929-1973; Chicago: Encyclopædia Britannica Inc. 15th edition, 1974; Chicago: Encyclopædia Britannica Inc.

Epstein, R.A. 1967. *The Theory of Gambling and Statistical Logic.* New York, San Francisco, London: Academic Press.

Evans, J. *Dissenting Congregations in England and Wales. The list compiled between 1715 and 1729.* Housed in Dr Williams's Library.

Fabry, L. 1893-1895. Étude sur la probabilité des comètes hyperboliques et l'origine des comètes. *Annales de la Faculté des Sciences de Marseilles* 3-4: 1-214.

Fagot, A.M. 1980. Probabilities and causes: on life tables, causes of death, and etiological diagnoses. pp.41-104 in Hintikka et al. [1980].

Feller, W. 1957. *An Introduction to Probability Theory and Its Applications.* vol. 1. 2nd edition. New York, London, Sydney: John Wiley & Sons. (1st edition 1950; 3rd edition 1968.)

—— 1966. *An Introduction to Probability Theory and Its Applications.* vol. 2. New York: John Wiley & Sons.

Ferguson, T.S. 1976. Development of the decision model. pp.333-346 in Owen [1976].

Fine, T.A. 1973. *Theories of Probability: an examination of foundations.* New York, San Francisco, London: Academic Press.

Fisher, A. 1926. *The Mathematical Theory of Probability and its Application to Frequency Curves and Statistical Methods.* (Translated from the Danish by C. Dickson.) New York: Macmillan. (1st edition 1915.)

Fisher, R.A. 1922. On the mathematical foundations of theoretical statistics. *Philosophical Transactions of the Royal Society of London, A* 222: 309-368.

—— 1956. *Statistical Methods and Scientific Inference.* Edinburgh: Oliver & Boyd. (2nd edition 1959.) Re-issued as Fisher [1990].

—— 1970. *Statistical Methods for Research Workers.* Edinburgh: Oliver & Boyd. (1st edition 1925.) Re-issued as Fisher [1990].

—— 1990. *Statistical Methods, Experimental Design, and Scientific Inference.* A Re-issue of *Statistical Methods for Research Workers, The Design of Experiments*, and *Statistical Methods and Scientific Inference*, ed. J.H. Bennett. Oxford: Oxford University Press.

Fisher, T. 1865. Article on Thomas Bayes in Waller [1865].

FitzPatrick, P.J. 1960. Leading British statisticians of the nineteenth century. *Journal of the American Statistical Association* 55: 38-70. Reprinted in Kendall & Plackett [1977], pp.180-212.

Forbes, J.D. 1849. On the alleged evidence for a physical connexion between stars forming binary or multiple groups, arising from their proximity alone. *The London, Edinburgh, and Dublin Philosophical Magazine and Journal of Science, Series 3*, 35: 132-133.

—— 1850. On the alleged evidence for a physical connexion between stars forming binary or multiple groups, deduced from the doctrine of chances. *The London, Edinburgh, and Dublin Philosophical Magazine and Journal of Science, Series 3*, 37: 401-427.

Forbush, W.B. (Ed.) 1926. *Fox's Book of Martyrs. A History of the Lives, Sufferings and Triumphant Deaths of the Early Christian and the Protestant Martyrs.* Grand Rapids 6, Michigan: Zondervan.

Friess, J.F. 1842. *Versuch einer Kritik der Principien der Wahrscheinlichkeitsrechnung.* Braunschweig: Friedr. Vieweg und Sohn.

Fujisawa, R. 1891. An elementary demonstration of a theorem in probability. *Tokyo sugaka butsurigaku kwai kiji* 4: 351-352.

Galloway, T. 1839. *A Treatise on Probability: forming the article under that head in the seventh edition of the Encyclopædia Britannica.* Edinburgh: Adam and Charles Black.

Gardner, M. 1957. *Fads and Fallacies in the Name of Science.* New York: Dover.

Gauss, C.F. 1809. *Theoria Motus Corporum Coelestium.* Hamburg. English translation by C.H. Davis, 1963; New York: Dover.

—— 1815. Compte rendu du Mémoire de Laplace. *Göttingische gelehrte Anzeigen* 40: 385-396. Reprinted in *Carl Friedrich Gauss Werke* 6: 581-586; Königlichen Gesellschaft der Wissenschaften zu Göttingen, 1874.

Geisser, S. 1988. The future of statistics in retrospect. pp.147-158 in Bernardo et al. [1988].

The Gentleman's Magazine and Historical Chronicle 31 (1761): 188; 59 (1789): 961.

Gillies, D.A. 1987. Was Bayes a Bayesian? *Historia Mathematica* 14: 325-346.

Gillispie, C.C. 1970. *Dictionary of Scientific Biography.* vol. I. New York: Charles Scribner's Sons.

——— 1972. Probability and politics: Laplace, Condorcet and Turgot. *Proceedings of the American Philosophical Society* 116: 1-20.

——— 1979. Mémoires inédits ou anonymes de Laplace sur la théorie des erreurs, les polynômes de Legendre, et la philosophie des probabilités. *Revue d'Histoire des Sciences* 32: 223-279.

Gini, C. 1949. Concept et mesure de la probabilité. *Dialectica* 3: 36-54.

Gnedenko, B.V. 1951. On the works of M.V. Ostrogradskiĭ on probability theory. [In Russian.] *Istoriko-inst. Issledovanija* 4: 99-123.

Good, I.J. 1965. *The Estimation of Probabilities: An Essay on Modern Bayesian Methods.* Cambridge, Mass.: M.I.T. Press.

——— 1986. Some statistical applications of Poisson's work. *Statistical Science* 1: 157-180.

——— 1988. Bayes's red billiard ball is also a herring, and why Bayes withheld publication. *Journal of Statistical Computing and Simulation* 29: 335-340.

Gosiewski, W. 1886. Ein leichter Beweis der Umkehrung des Bernoulli'schen Satzes. *Sitzungsberichte der mathematisch-naturwissenschaftlichen. Section der Krakauer Akademie (Krakau)* 13: 153-159. [In Polish.]

Gouraud, C. 1848. *Histoire du Calcul des Probabilités, depuis ses origines jusqu'à nos jours.* Paris: D'Auguste Durand.

Govan, J. 1920. The theory of inverse probability, with special reference to Professor Chrystal's paper "On some fundamental principles in the theory of probability". *Transactions of the Faculty of Actuaries* 8, part 6: 207-230.

Gower, B. 1982. Astronomy and probability: Forbes versus Michell on the distribution of the stars. *Annals of Science* 39: 145-160.

Grant, A. 1884. *The Story of the University of Edinburgh during its first three hundred years.* 2 vols. London: Longmans, Green, and Co.

Grosart, A.B. 1885. Bayes, Joshua (1671-1746). pp.439-440 in Stephen [1885].

Grove, H. 1734. *Wisdom, the first Spring of Action in the Deity; a discourse, in which, Among other Things, the Absurdity of God's being actuated by Natural Inclinations, and of an unbounded Liberty, is shewn. The Moral attributes of God are explained. The Origin of Evil is considered. The Fundamental Duties of Natural Religion are shewn to be reasonable; and several things advanced by some late authors, relating to these subjects, are freely examined.* London.

Hacking, I. 1965. *Logic of Statistical Inference.* Cambridge: Cambridge University Press.

—— 1970. Bayes, Thomas. pp.531-532 in Gillispie [1970].

—— 1971. Equipossibility theories of probability. *British Journal for the Philosophy of Science* 22: 339-355.

—— 1975. *The Emergence of Probability: a philosophical study of early ideas about probability, induction and statistical inference.* Cambridge: Cambridge University Press.

Hagen, G. 1837. *Grundzüge der Wahrscheinlichkeits-Rechnung*, 3rd edition, 1882. Berlin: Ernst & Korn.

Haight, F.A. 1967. *Handbook of the Poisson Distribution.* New York: John Wiley & Sons.

Hailperin, T. 1976. *Boole's Logic and Probability.* Amsterdam: North-Holland.

Hald, A. 1990. *A History of Probability and Statistics and Their Applications before 1750.* New York: John Wiley & Sons.

Haldane, J.B.S. 1957. Karl Pearson, 1857 (1957). A centenary lecture delivered at University College London. *Biometrika* 44: 303-313.

Halkett, S. & Laing, J. 1926. *Dictionary of Anonymous and Pseudonymous English Literature*, ed. J. Kennedy, W.A. Smith and A.F. Johnson. Edinburgh: Oliver and Boyd.

Hardy, G.F. 1889. Correspondence in the *Insurance Record.* Summarized in Whittaker [1920], pp.174-182.

Harper, W.L. & Hooker, C.A. 1976. *Foundations of Probability Theory, Statistical Inference, and Statistical Theories of Science. vol. II: Foundations and Philosophy of Statistical Inference.* Dordrecht: D. Reidel.

Harter, H.L. 1974. The method of least squares and some alternatives. Part 1. *International Statistical Review* 42: 147-174.

Hartigan, J.A. 1983. *Bayes Theory.* New York, Berlin, Heidelberg, Tokyo: Springer-Verlag.

Hastings, J. (Ed.) 1967. *Encyclopaedia of Religion and Ethics*, 6th impression. Edinburgh: T. & T. Clark.

Heath, P. (Ed.) 1966. *On the Syllogism and Other Logical Writings, by Augustus de Morgan.* London: Routledge & Kegan Paul.

Herschel, J.F.W. 1849. *Outlines of Astronomy.* New edition, 1873; London: Longmans, Green & Co.

—— 1850. Quetelet on probabilities. *The Edinburgh Review* 92: 1-57. (Reprinted in Herschel [1857].)

—— 1857. *Essays from the Edinburgh and Quarterly Reviews, with addresses and other pieces.* London: Longman, Brown, Green, Longmans and Roberts.

Heyde, C.C. & Seneta, E. 1977. *I.J. Bienaymé: Statistical Theory Anticipated.* New York: Springer-Verlag.

Hicks, H. 1887. *History of the Bunhill Fields Burial Ground, with some of the Principal Inscriptions.* London: Charles Skipper and East, Printers.

Hinkley, D. 1979. Predictive likelihood. *Annals of Statistics* 7: 718-728.

Hintikka, J., Gruender, D. & Agazzi, E. (Eds) 1980. *Pisa Conference Proceedings. vol. II.* Dordrecht: D. Reidel.

Holland, J.D. 1962. The Reverend Thomas Bayes, F.R.S. (1702-1761). *Journal of the Royal Statistical Society, Series A*, 125: 451-461.

—— 1968. An eighteenth-century pioneer: Richard Price, D.D., F.R.S. (1723-1791). *Notes and Records of the Royal Society of London* 23: 42-64.

Home, R.W. 1974-1975. Some manuscripts on electrical and other subjects attributed to Thomas Bayes, F.R.S. *Notes and Records of the Royal Society of London* 29: 81-90.

Horwich, P. 1982. *Probability and Evidence.* Cambridge: Cambridge University Press.

Hume, D. 1886. *A Treatise of Human Nature, being an attempt to introduce the experimental method of reasoning into moral subjects, and dialogues concerning natural religion.* 2 vols, ed. T.H. Green & T.H. Grose. London: Longmans, Green. Reprinted 1964, Scientia Verlag Aalen. (Originally printed in 1739.)

—— 1894. *Essays Literary, Moral and Political.* [Sir John Lubbock's Hundred Books]. London & New York: George Routledge & Sons.

Huygens, C. 1657. *De Ratiociniis in Ludo Aleae.* pp.517-534 in F. van Schootens' *Exercitationum Mathematicarum* Leiden: Johannes Elsevirii. Reprinted, with commentary, in Bernoulli [1713]. English translation in Arbuthnott [1770]. vol. II, pp.257-294.

James, T.S. 1867. *The History of the Litigation & Legislation respecting Presbyterian Chapels & Charities in England and Ireland, between 1816 and 1849.* London: Hamilton Adams & Co.

Jaynes, E.T. 1976. Confidence intervals vs Bayesian intervals. pp.175-257 (with discussion) in Harper & Hooker [1976].

Jevons, W.S. 1874. *The Principles of Science: a treatise on logic and scientific method.* 2 vols. (Reprinted in one volume in 1877.) London: Macmillan.

Johnson, W.E. 1924. *Logic, Part III. The Logical Foundations of Science.* Cambridge: Cambridge University Press.

—— 1932. Probability; the deductive and inductive problems, with an appendix (ed. R.B. Braithwaite) *Mind* 41: 409-421, 421-423.

Jones, J.A. (Ed.) 1849. *Bunhill Memorials. Sacred Reminiscences of Three Hundred Ministers and other persons of note, who are buried in Bunhill Fields, of every denomination. With the inscriptions on their tombs and gravestones, and other historical information respecting them, from authentic sources.* London: James Paul.

Jordan, C. 1923. On the inversion of Bernoulli's theorem. *The London, Edinburgh, and Dublin Philosophical Magazine and Journal of Science, Series 6*, 45: 732-735.

—— 1925. On probability. *Proceedings of the Physico-Mathematical Society of Japan, Series 3*, 7: 96-109.

—— 1926a. Sur l'inversion du théorème de Bernoulli. *Comptes Rendus de l'Academie des Sciences, Paris* 182: 431-432.

—— 1926b. Sur la probabilité des épreuves répétés, le théorème de Bernoulli et son inversion. *Bulletin de la Societé Mathématique de France* 54: 101-137.

—— 1933. Inversione della formula di Bernoulli relativa al problema della prove ripetute a più variabili. *Giorn. Ist. Ital. Attuari* 4: 505-513.

Jurin, J. 1734. *Geometry no friend to infidelity; or, a defence of Sir Isaac Newton and the British mathematicians, in a letter to the author of the Analyst ... By Philalethes Cantabrigiensis.* London.

—— 1735. *The minute mathematician; or, the free-thinker no just-thinker: set forth in a second letter to the author of the Analyst; containing a defence of Sir Isaac Newton and the British mathematicians, against a late pamphlet, entituled, A defence of free-thinking in mathematicks. By Philalethes Cantabrigiensis.* London.

Kac, M. 1985. *Enigmas of Chance: an autobiography.* New York: Harper and Row.

Kendall, M.G. 1963. Isaac Todhunter's History of the Mathematical Theory of Probability. *Biometrika* 50: 204-205. Reprinted in Pearson & Kendall [1970], pp.253-254.

Kendall, M.G. & Plackett, R.L. 1977. *Studies in the History of Statistics and Probability.* vol. II. London: Charles Griffin & Co.

Keynes, J.M. 1921. *A Treatise on Probability.* London: Macmillan. Reprinted, 1973; London: Macmillan.

—— 1936. William Stanley Jevons, 1835-1882. *Journal of the Royal Statistical Society* 99: 516-555.

Kleiber, J. 1887. On "random scattering" of points on a surface. *The London, Edinburgh, and Dublin Philosophical Magazine and Journal of Science, Series 5*, 24: 439-445.

—— 1888. Michell's problem. *Nature* 38: 342.

Kneale, W. 1949. *Probability and Induction.* Oxford: Clarendon Press.

Kroman, K. 1908. Den aposterioriske Sandsynlighed. *Kjöbenhavn Oversigt* 3: 133-166.

Kruskal, W. 1988. Miracles and statistics: the casual assumption of independence. *Journal of the American Statistical Association* 83: 929-940.

Kyburg, H.E. & Smokler, H.E. 1964. *Studies in Subjective Probability.* New York, London, Sydney: John Wiley & Sons. Reprinted 1980; Huntington, New York: Robert E. Krieger.

Lacroix, S.F. 1816. *Traité Élémentaire du Calcul des Probabilités.* Paris: Courcier.

Lagrange, J.L. 1770-1773. Mémoire sur l'utilité de la méthode de prendre le milieu entre les résultats de plusiers observations, dans lequel on examine les avantages de cette méthode par le calcul des probabilités, et où l'on résout différénts problèmes relatifs a cette matière. *Miscellanea Taurinensia* 5: 167-232. Reprinted in Serret [1868].

Lancaster, H.O. 1966. Forerunners of the Pearson χ^2. *Australian Journal of Statistics* 8: 117-126.

—— 1968. *Bibliography of Statistical Bibliographies.* Edinburgh: Oliver and Boyd.

Laplace, P.S. 1773. (published 1776.) Recherches sur l'intégration des équations différentielles aux différences finies et sur leur usage dans la théorie des hasards. *Mémoires de l'Académie royale des Sciences de Paris (Savants étrangers).* 7 (the probability section occupies pp.113-163). Reprinted in *Œuvres complètes de Laplace* 8: 69-197.

—— 1774a. Mémoire sur les suites récurro-récurrentes et sur leurs usages dans la théorie des hasards. *Mémoires de l'Académie royale des Sciences de Paris (Savants étrangers)* 6: 353-371. Reprinted in *Œuvres complètes de Laplace* 8: 5-24.

—— 1774b. Mémoire sur la probabilité des causes par les événements. *Mémoires de l'Académie royale des Sciences de Paris (Savants étrangers)* 6: 621-656. Reprinted in *Œuvres complètes de Laplace* 8: 27-65.

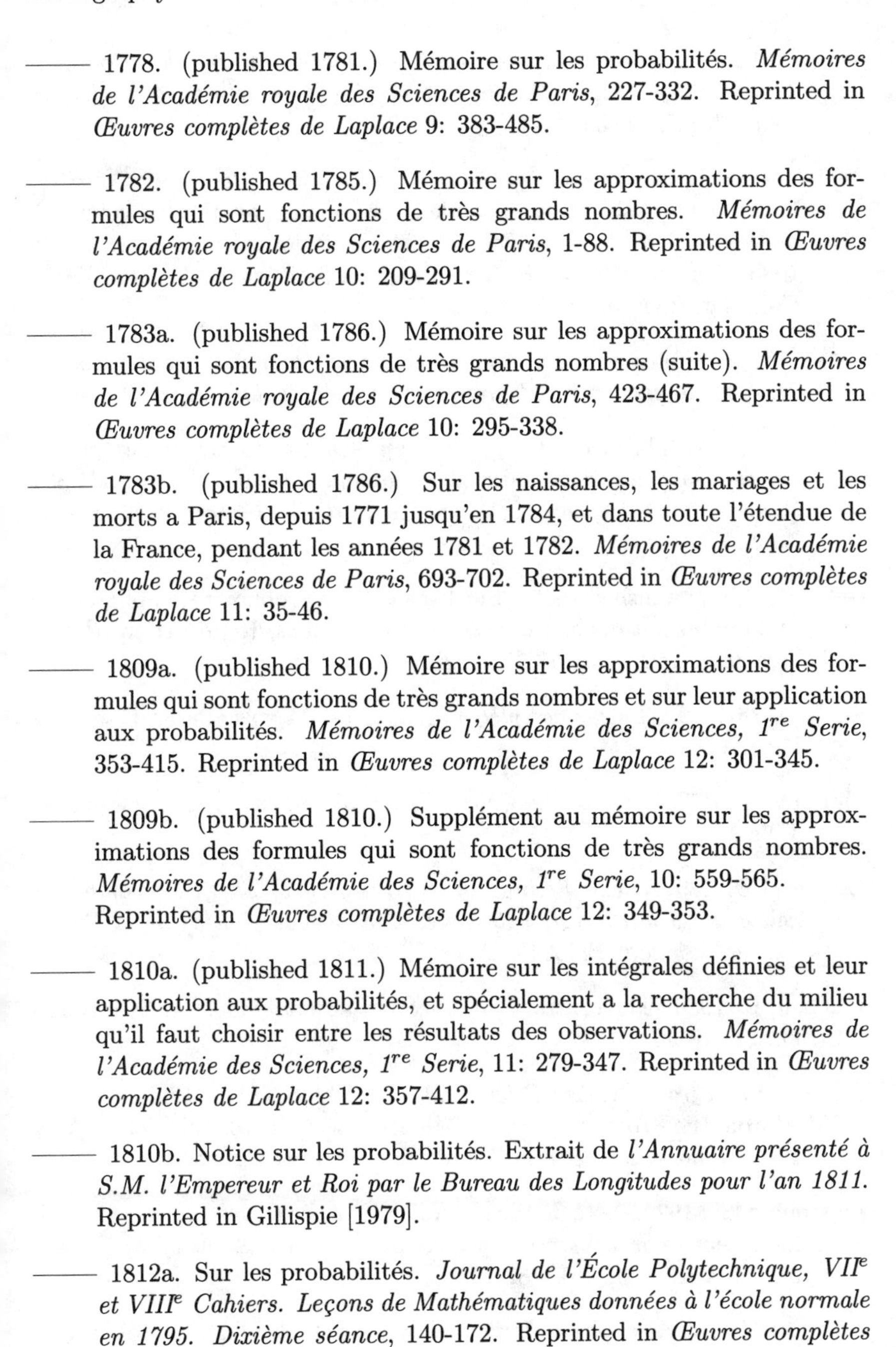

——— 1778. (published 1781.) Mémoire sur les probabilités. *Mémoires de l'Académie royale des Sciences de Paris*, 227-332. Reprinted in *Œuvres complètes de Laplace* 9: 383-485.

——— 1782. (published 1785.) Mémoire sur les approximations des formules qui sont fonctions de très grands nombres. *Mémoires de l'Académie royale des Sciences de Paris*, 1-88. Reprinted in *Œuvres complètes de Laplace* 10: 209-291.

——— 1783a. (published 1786.) Mémoire sur les approximations des formules qui sont fonctions de très grands nombres (suite). *Mémoires de l'Académie royale des Sciences de Paris*, 423-467. Reprinted in *Œuvres complètes de Laplace* 10: 295-338.

——— 1783b. (published 1786.) Sur les naissances, les mariages et les morts a Paris, depuis 1771 jusqu'en 1784, et dans toute l'étendue de la France, pendant les années 1781 et 1782. *Mémoires de l'Académie royale des Sciences de Paris*, 693-702. Reprinted in *Œuvres complètes de Laplace* 11: 35-46.

——— 1809a. (published 1810.) Mémoire sur les approximations des formules qui sont fonctions de très grands nombres et sur leur application aux probabilités. *Mémoires de l'Académie des Sciences, 1re Serie*, 353-415. Reprinted in *Œuvres complètes de Laplace* 12: 301-345.

——— 1809b. (published 1810.) Supplément au mémoire sur les approximations des formules qui sont fonctions de très grands nombres. *Mémoires de l'Académie des Sciences, 1re Serie*, 10: 559-565. Reprinted in *Œuvres complètes de Laplace* 12: 349-353.

——— 1810a. (published 1811.) Mémoire sur les intégrales définies et leur application aux probabilités, et spécialement a la recherche du milieu qu'il faut choisir entre les résultats des observations. *Mémoires de l'Académie des Sciences, 1re Serie*, 11: 279-347. Reprinted in *Œuvres complètes de Laplace* 12: 357-412.

——— 1810b. Notice sur les probabilités. Extrait de *l'Annuaire présenté à S.M. l'Empereur et Roi par le Bureau des Longitudes pour l'an 1811.* Reprinted in Gillispie [1979].

——— 1812a. Sur les probabilités. *Journal de l'École Polytechnique, VIIe et VIIIe Cahiers. Leçons de Mathématiques données à l'école normale en 1795. Dixième séance*, 140-172. Reprinted in *Œuvres complètes de Laplace* 14: 146-177.

—— 1812b. *Théorie analytique des probabilités.* Paris: Ve Courcier. Third edition, with supplements, reprinted as volume 7 in *Œuvres complètes de Laplace.*

—— 1814. *Essai philosophique sur les probabilités.* Paris: Ve Courcier. Reprinted, considerably expanded, as the Introduction to volume 7 in *Œuvres complètes de Laplace.* Fifth edition of 1825 reprinted in 1986 with a preface by R. Thom and a postscript by B. Bru. Paris: Christian Bourgois.

—— 1816. (published 1813.) Sur les comètes. *Connaissance des Temps,* 213-220. Reprinted in *Œuvres complètes de Laplace* 13: 88-97.

—— 1818a. (published 1815.) Sur l'application du calcul des probabilités a la philosophie naturelle. *Connaissance des Temps,* 361-381 (together with Laplace [1818b]). Reprinted in *Œuvres complètes de Laplace* 13: 98-116.

—— 1818b. (published 1815.) Sur le calcul des probabilités appliqué a la philosophie naturelle. *Connaissance des Temps.* Reprinted in *Œuvres complètes de Laplace* 13: 117-120.

—— 1878-1912. *Œuvres complètes de Laplace.* 14 vols. Paris: Gauthier-Villars.

Laurent, H. 1873. *Traité du Calcul des Probabilités.* Paris: Gauthier-Villars.

Leader, J.D. 1897. *The Records of the Burgery of Sheffield, commonly called the Town Trust, with notes and introduction by John Daniel Leader.* London: Elliot Stock.

Le Cam, L. 1986. The central limit theorem around 1935. *Statistical Science* 1: 78-96.

Letters to and from Richard Price D.D. F.R.S. 1767-1790. 1903. Reprinted from the Proceedings of the Massachusetts Historical Society, May 1903. John Wilson & Son, Cambridge University Press.

Lidstone, G.J. 1920. Note on the general case of the Bayes-Laplace formula for inductive or *a posteriori* probabilities. *Transactions of the Faculty of Actuaries* 8, part 6: 182-192.

—— 1941. Review of *Facsimiles of two papers by Bayes* (Molina and Deming, 1940). *The Mathematical Gazette* 25: 177-180.

Liebermeister, C. 1877. *Ueber Wahrscheinlichkeitsrechnung in Anwendung auf Therapeutische Statistik.* Sammlung Klinischer Vorträge, ed. R. Volkmann, No. 110. (Innere Medicin No. 39), pp.935-962.

The London Magazine, or Gentleman's Monthly Intelligencer 30 (1761): 220.

Lorenz, L. 1879. Bemærkninger til Hr. Bings Afhandelung „Om aposteriorisk Sandsynlighed". *Tidsskrift for Mathematik, Series 4*, 3: 57-66.

—— 1879. Gjensvar til Hr. Direktør F. Bing. *Tidsskrift for Mathematik, Series 4*, 3: 118-122.

Lotze, R.H. 1874. *Logik; drei Bücher, vom Denken, vom Untersuchen und vom Erkennen.* Leipzig: S. Hirzel. 2nd edition, 1880. 3rd edition, 1912. (Translated by B. Bosanquet, 1884, as *Logic, in three books, of thought, of investigation, and of knowledge.* Oxford: Clarendon Press.)

Lubbock, J.W. 1830a. On the calculation of annuities, and on some questions in the theory of chances. *Transactions of the Cambridge Philosophical Society* 3: 141-154, + 8 pages of tables.

—— 1830b. On the comparison of various tables of annuities. *Transactions of the Cambridge Philosophical Society* 3: 321-341, + 7 pages of tables.

Lubbock, J.W. & Drinkwater-Bethune, J.E. c.1830. *On Probability.* London: Baldwin & Craddock.

Lupton, S. 1888a. Michell's problem. *Nature* 38: 272-274.

—— 1888b. Michell's problem. *Nature* 38: 414.

MacAlister, D. 1881. Problem on Listerism. *The Educational Times*, problem 6929 (posed in December). Comments by (i) D. MacFarlane & the proposer (February 1882), (ii) the editor, the proposer, H. MacColl & E. Blackwood (March 1882), (iii) W.A. Whitworth, MacColl & Blackwood (April 1882), (iv) Blackwood (May 1882).

MacFarlane, A. 1916. *Lectures on Ten British Mathematicians of the Nineteenth Century.* New York: John Wiley & Sons.

MacKenzie, D.A. 1981. *Statistics in Britain 1865-1930. The Social Construction of Scientific Knowledge.* Edinburgh: Edinburgh University Press.

McLachlan, H. 1931. *English Education under the Test Acts: being the history of the non-conformist academies, 1662-1820.* Manchester: Manchester University Press.

Maistrov, L.E. 1974. *Probability Theory: a historical sketch.* New York and London: Academic Press. Originally published in Russian in 1967.

Makeham, W.M. 1891a. (published 1892.) On the theory of inverse probabilities. *Journal of the Institute of Actuaries* 29: 242-251; 444-459.

—— 1891b. (published 1892.) On a problem in probabilities. *Journal of the Institute of Actuaries* 29: 475-476.

Martin-Löf, A. 1985. A limit theorem which clarifies the "Petersburg Paradox". *Journal of Applied Probability* 22: 634-643.

Matthews, A.G. 1934. *Calamy Revised. Being a revision of Edmund Calamy's Account of the ministers ejected and silenced, 1660-1662.* Oxford: Clarendon Press.

Mendelssohn, M. 1777. *Philosophische Schriften.* Berlin: C. Friederich.

Meyer, A. 1856. Note sur le théorème inverse de Bernoulli. *Bulletin de l'Académie royale des Sciences, des Lettres et des Beaux-Arts de Belgique* 23, no. 1: 148-155.

—— 1857. *Essai sur une Exposition nouvelle de la Théorie analytique des Probabilités a posteriori.* Liege: H. Dessain.

Michell, J. 1767. An Inquiry into the probable Parallax, and Magnitude of the fixed Stars, from the Quantity of Light which they afford us, and the particular Circumstances of their Situation. *Philosophical Transactions* 57: 234-264.

Miles, R.E. & Serra, J. (Eds) 1978. *Geometrical Probability and Biological Structures: Buffon's 200th Anniversary.* New York: Springer-Verlag.

Mill, J.S. 1843. *A System of Logic, Ratiocinative and Inductive: Being a Connected View of the Principles of Evidence and the Methods of Scientific Investigation.* 8th edition 1872; London: Longmans, Green, Reader, and Dyer. Reprinted 1961, Longmans. Also published as volumes 7 & 8 in *Collected Works of John Stuart Mill*, ed. J.M. Robson, 1974; Toronto: University of Toronto Press, and London: Routledge and Kegan Paul.

Minute Books of the Body of Protestant Dissenting Ministers of the Three Denominations in and about the cities of London and Westminster. 3 vols. (11 July 1727-7 July 1761; 6 Oct. 1761-11 April 1797; 3 April 1798-3 April 1827). Housed in Dr Williams's Library.

Molina, E.C. 1930. The theory of probability: some comments on Laplace's Théorie Analytique. *Bulletin of the American Mathematical Society* 36: 369-392.

——— 1931. Bayes' theorem: an expository exposition. *Annals of Mathematical Statistics* 2: 23-37. Also printed in 1931 in *The Bell System Technical Journal* 10: 273-283.

Molina, E.C. & Deming, W.E. 1940. *Facsimiles of two papers by Bayes.* The Graduate School, The Department of Agriculture, Washington.

Monro, C.J. 1874. Note on the inversion of Bernoulli's theorem in probabilities. *Proceedings of the London Mathematical Society* 5: 74-78. (Errata pp.145-146.)

Morgan, W. 1815. *Memoirs of the Life of The Rev. Richard Price, D.D. F.R.S.* London: R. Hunter.

Moroney, M.J. 1951. *Facts from Figures.* Harmondsworth, Middlesex: Penguin.

Munford, A.G. 1977. A note on the uniformity assumption in the birthday problem. *The American Statistician* 31: 119.

Murray, F.H. 1930. Note on a scholium of Bayes. *Bulletin of the American Mathematical Society* 36: 129-132.

Nekrassof, P.A. c.1890. Die Umkehrung des Bernoulli'schen Theorem. *Mosk. Physik. Abt.* III: 45-47. [In Russian.]

Netto, E. 1908. Kombinatorik, Wahrscheinlichkeitsrechnung, Reihen-imaginaïres. §XXI in Cantor [1908].

Neveu, J. 1965. *Mathematical Foundations of the Calculus of Probability.* San Francisco, London, Amsterdam: Holden-Day. Originally published in French in 1964.

Newcomb, S. 1860a. Notes on the theory of probabilities (d). *Mathematical Monthly* 2, no. 4: 134-140.

—— 1860b. Notes on the theory of probabilities (e). *Mathematical Monthly* 2, no. 8: 272-275.

Neyman, J. 1957. "Inductive behavior" as a basic concept of philosophy of science. *Revue Internationale de Statistique* 25: 7-22.

Neyman, J. & Le Cam, L.M. 1965. *Bernoulli 1713, Bayes 1763, Laplace 1813.* New York: Springer-Verlag.

Nicholson, F. & Axon, E. 1915. *The Older Nonconformity in Kendal. A history of the Unitarian Chapel in the Market Place with transcripts of the registers and Notices of the Nonconformist Academies of Richard Frankland, M.A., and Caleb Rotherham, D.D.* Kendal: Titus Wilson.

Nový, L. 1980. Some remarks on the calculus of probability in the eighteenth century. pp.25-32 in *Pisa Conference Proceedings.* vol. II, ed. J. Hintikka et al. Dordrecht: Reidel.

O'Donnell, T. 1936. *History of Life Assurance in Its Formative Years. Compiled from approved sources by Terence O'Donnell.* Chicago: American Conservation Company.

O'Hagan, A. 1988. The Reverend Thomas Bayes, F.R.S. — c.1701-1761. *The Institute of Mathematical Statistics Bulletin* 17: 482.

Ore, O. 1953. *Cardano, the Gambling Scholar. With a translation from the Latin of Cardano's 'Book on Games of Chance', by Sydney Henry Gould.* Princeton: Princeton University Press.

Ostrogradskiĭ, M.V. 1838. Extrait d'un mémoire sur la probabilités des erreurs des tribunaux. *Bulletin Scientifique No. 3, Mémoires de l'Académie Impériale des Sciences de Saint-Petersbourg. VIe series. Sciences Mathématiques, Physiques et Naturelles, Tome III. Première Partie, Sciences Mathématique et Physique, Tome I*: xiv-xxv.

—— 1846. (published 1848.) Sur une question des probabilités. *Mémoires de l'Académie Impériale des Sciences de Saint Petersbourg* VI: 321-346.

Owen, D. (Ed.) 1976. *On the History of Statistics and Probability.* New York & Basel: Marcel Dekker.

—— 1987. Hume *versus* Price on miracles and prior probabilities: testimony and the Bayesian calculation. *The Philosophical Quarterly* 37: 187-202.

Parker, I. 1914. *Dissenting Academies in England. Their Rise and Progress and their Place among the Educational Systems of the Country.* Cambridge: Cambridge University Press.

Patel, J.K. & Read, C.B. 1982. *Handbook of the Normal Distribution.* New York and Basel: Marcel Dekker.

Pearson, E.S. 1938. *Karl Pearson: An Appreciation of Some Aspects of His Life and Work.* Cambridge: Cambridge University Press.

——— 1967. Some reflexions on continuity in the development of mathematical statistics, 1885-1920. *Biometrika* 54: 341-355.

Pearson, E.S. & Kendall. M. (Eds) 1970. *Studies in the History of Statistics and Probability.* vol. I. London: Charles Griffin & Co.

Pearson, K. 1892. *The Grammar of Science.* London: Walter Scott.

——— 1900. On the criterion that a given system of deviations from the probable in the case of a correlated system of variables is such that it can be reasonably supposed to have arisen from random sampling. *The London, Edinburgh, and Dublin Philosophical Magazine and Journal of Science, Series 5*, 50: 157-175.

——— 1907. On the influence of past experience on future expectation. *The London, Edinburgh, and Dublin Philosophical Magazine and Journal of Science, Series 6*, 13: 365-378.

——— 1920a. The fundamental problem of practical statistics. *Biometrika* 13: 1-16.

——— 1920b. Note on the "Fundamental problem of practical statistics." *Biometrika* 13: 300-301.

——— 1924a. Historical note on the origin of the normal curve of errors. *Biometrika* 16: 402-404.

——— 1924b. Note on Bayes' theorem. *Biometrika* 16: 190-193.

——— 1925. James Bernoulli's theorem. *Biometrika* 17: 201-210.

——— 1928. On a method of ascertaining limits to the actual number of marked members in a population of given size from a sample. *Biometrika* 20: 149-174.

——— 1929. Laplace, being extracts from lectures delivered by Karl Pearson. *Biometrika* 21: 202-216.

—— 1978. *The History of Statistics in the 17th and 18th Centuries, against the changing background of intellectual, scientific and religious thought. Lectures given at University College London during the academic sessions 1921-1933*, ed. E.S. Pearson. London: Charles Griffin & Co.

Pearson, K. & Filon, L.N.G. 1898. Mathematical contributions to the theory of evolution. —IV. On the probable errors of frequency constants and on the influence of random selection on variation and correlation. *Philosophical Transactions of the Royal Society of London*, A 191: 229-311.

Perks, W. 1947. Some observations on inverse probability including a new indifference rule. *Journal of the Institute of Actuaries* 73: 285-334.

Phippen, J. 1840. *Colbran's New Guide for Tunbridge Wells.* London: Bailey & Co.

Plackett, R.L. 1972. The discovery of the method of least squares. *Biometrika* 59: 239-251.

Poisson, S-D. 1830. Mémoire sur la proportion des naissances des filles et des garçons. *Mémoires de l'Académie des Sciences, Paris* 9: 239-308.

—— 1837. *Recherches sur la probabilité des jugements en matière criminelle et en matière civile, précédées des règles générales du calcul des probabilités.* Paris: Bachelier.

Pollard, H. 1966. *Mathematical Introduction to Celestial Mechanics.* Englewood Cliffs, New Jersey: Prentice-Hall.

Porter, T.M. 1986. *The Rise of Statistical Thinking 1820-1900.* Princeton: Princeton University Press.

Press, S.J. 1989. *Bayesian Statistics: Principles, Models, and Applications.* New York: John Wiley & Sons.

Prevost, P. and Lhuilier, S.A.J. 1796a. (published 1799.) Sur les probabilités. *Mémoires de l'Académie royale des Sciences et Belles-Lettres, Classe de Mathématique, Berlin*: 117-142.

—— 1796b. (published 1799.) Mémoire sur l'art d'estimer la probabilité des causes par les effets. *Mémoires de l'Académie royale des Sciences et Belles-Lettres, Classe de Philosophie spéculative, Berlin*: 3-25.

—— 1796c. (published 1799.) Remarques sur l'utilité & l'étendue du principe par lequel on estime la probabilité des causes. *Mémoires de l'Académie royale des Sciences et Belles-Lettres, Classe de Philosophie spéculative, Berlin*: 25-41.

Price, R. 1768. *Four Dissertations.* 2nd edition with additions. London: A. Millar & T. Cadell. (1st edition 1767.)

—— 1787. *A Review of the Principal Questions in Morals. Particularly Those respecting the Origin of our Ideas of Virtue, its Nature, Relation to the Deity, Obligation, Subject-Matter, and Sanctions. The third edition corrected, and Enlarged by an Appendix, containing Additional Notes, and a Dissertation on the Being and Attributes of the Deity.* London: T. Cadell.

Proctor, R.A. 1872. *Essays on Astronomy: A series of papers on planets and meteors, the sun and sun-surrounding space, stars and star cloudlets; and a dissertation of the approaching transits of Venus. Preceded by a sketch of the life and work of Sir John Herschel.* London: Longman, Green, and Co.

Rashed, R. (Ed.) 1988. *Sciences a l'èpoque de la revolution française.* Paris: Albert Blanchard.

The Record of the Royal Society of London , 3rd edition, 1912. London: Oxford University Press.

Robinson, A. 1966. *Non-standard Analysis.* Amsterdam: North-Holland.

Roger, J. 1978. Buffon and mathematics. pp.29-35 in Miles and Serra [1978].

Rose, H.J. (Ed.) 1817-1845. *The Encyclopaedia Metropolitana*, issued in 59 parts. London: B. Fellowes.

—— 1848. *New General Biographical Dictionary.* vol. 3. London: B. Fellowes.

Salmon, W.C. 1980a. Robert Leslie Ellis and the frequency theory. pp.139-143 in *Pisa Conference Proceedings.* vol. II, ed. J. Hintikka et al. Dordrecht: Reidel.

—— 1980b. John Venn's *Logic of Chance.* pp.125-138 in *Pisa Conference Proceedings.* vol. II, ed. J. Hintikka et al. Dordrecht: Reidel.

Savage, L.J. 1960. Reading note on Bayes' theorem. Printed as the Appendix to the present work.

Schiaparelli, G.V. 1874. Sul calcolo di Laplace intorno alla probabilità delle orbite cometarie iperboliche. *Rendiconti del Reale Instituto Lombardo di Science e Lettere, Ser. II,* 7: 77-80.

Schneider, I. 1968. Der Mathematiker Abraham de Moivre (1667-1754). *Archive for History of Exact Sciences* 5: 177-317.

Seal, H.L. 1949. The historical development of the use of generating functions in probability theory. *Bulletin de l'Association des Actuaires suisses* 49: 209-228.

Seeliger, H. 1890. Ueber die Wahrscheinlichkeit des Vorkommens von hyperbolischen Cometenbahnen. *Astronomische Nachrichten* 124 (no. 2968): 257-262.

Serret, J.A. (Ed.) 1868. *Œuvres de Lagrange.* vol. 2: 173-234. Paris: Gauthier-Villars.

'sGravesande, G.J. (or W.J.S.) 1774. *Œuvres Philosophiques et Mathematiques*, ed. J.S.N. Allemand. Amsterdam: Marc Michel Ray.

Shafer, G. 1976. *A Mathematical Theory of Evidence.* Princeton, New Jersey: Princeton University Press.

——— 1978. Non-additive probabilities in the work of Bernoulli and Lambert. *Archive for History of Exact Sciences* 19: 309-370.

——— 1982. Bayes's two arguments for the rule of conditioning. *Annals of Statistics* 10: 1075-1089.

——— 1985. Conditional probability. *International Statistical Review* 53: 261-277.

Shairp, J.C., Tait, P.G. & Adams-Reilly, A. 1873. *Life and Letters of James David Forbes, F.R.S.* London: Macmillan and Co.

Shapiro, B.J. 1983. *Probability and Certainty in Seventeenth-Century England: a study of the relationships between natural science, religion, history, law, and literature.* Princeton: Princeton University Press.

Shelton, T. 1641. *Tachygraphy the most exact and compendius methode of short and swift writing that hath ever yet beene published by any.* Cambridge. Republished as *Tachygraphia: sive, Exactissima & compendiosissima breviter scribendi methodus.* London: Tho. Creake.

—— 1654. *Zeiglographia: or, a new art of short-writing never before published. More easie, exact, short and speedie than any heretofore.* London.

Sheynin, O.B. 1968. On the early history of the law of large numbers. *Biometrika* 55: 459-467. Reprinted in Pearson & Kendall [1970], pp.231-239.

—— 1969. On the work of Thomas Bayes in probability theory. [In Russian]. *The Academy of Sciences of the USSR, the Institute of History of Natural Sciences and Technology. Proceedings of the 12th Conference of Doctorands and Junior Research Workers. Section of History of Mathematics and Mechanics.* Moscow.

—— 1971a. On the history of some statistical laws of distribution. *Biometrika* 58: 234-236. Reprinted in Kendall & Plackett [1977], pp.328-330.

—— 1971b. Newton and the classical theory of probability. *Archive for History of Exact Sciences* 7: 217-243.

—— 1976. P.S. Laplace's work on probability. *Archive for History of Exact Sciences* 16: 137-187.

—— 1977. Laplace's theory of errors. *Archive for History of Exact Sciences* 17: 1-61.

—— 1978. S.D. Poisson's work in probability. *Archive for History of Exact Sciences* 18: 245-300.

—— 1986. A. Quetelet as a statistician. *Archive for History of Exact Sciences* 36: 281-325.

Signatures in the first Journal-Book and the Charter-Book of the Royal Society: being a facsimile of the signatures of the founders, patrons and fellows of the society from the year 1660 down to the present time, 1912. London: printed for the Royal Society, Oxford University Press.

Smith, G.C. 1980. Thomas Bayes and fluxions. *Historia Mathematica* 7: 379-388.

—— 1982. *The Boole-de Morgan Correspondence 1842-1864.* Oxford: Clarendon Press.

Sobel, J.H. 1987. On the evidence of testimony for miracles: a Bayesian interpretation of David Hume's analysis. *The Philosophical Quarterly* 37: 166-186.

Solomon, H. 1986. Comment on Good [1986]. *Statistical Science* 1: 174-176.

Sorel, G. 1887. Le calcul des probabilités et l'expérience. *Revue Philosophique de la France et de l'Étranger* 23: 50-66.

Sprott, D.A. 1978. Gauss's contributions to statistics. *Historia Mathematica* 5: 183-203.

Stabler, E.L. 1892. On Mr. Makeham's theory of inverse probabilities. *Journal of the Institute of Actuaries* 30: 239-244. (Errata, p.580).

Stephen, L. (Ed.) 1885. *The Dictionary of National Biography.* vol. 3. London: Smith, Elder & Co.

Stigler, S.M. 1975. Napoleonic statistics: the work of Laplace. *Biometrika* 62: 503-517.

——— 1978. Laplace's early work: chronology and citations. *Isis* 69: 234-254.

——— 1980a. *Springer Statistics Calendar 1981.* New York: Springer-Verlag.

——— 1980b. Stigler's law of eponymy. *Transactions of the New York Academy of Sciences, Series II*, 39: 147-158.

——— 1982a. Thomas Bayes's Bayesian inference. *Journal of the Royal Statistical Society, A* 145: 250-258.

——— 1982b. Poisson on the Poisson distribution. *Statistics and Probability Letters* 1: 33-35.

——— 1984. Review of Smith [1982]. *Journal of the American Statistical Association* 79: 947-948.

——— 1986a. *The History of Statistics. The Measurement of Uncertainty before 1900.* Cambridge, Mass., and London, England: the Belknap Press of Harvard University.

——— 1986b. Laplace's 1774 memoir on inverse probability. *Statistical Science* 1: 359-378.

——— 1986c. John Craig and the probability of history: from the death of Christ to the birth of Laplace. *Journal of the American Statistical Association* 81: 879-887.

—— 1988. The Reverend Thomas Bayes, F.R.S. — c.1701-1761. *The Institute of Mathematical Statistics Bulletin* 17: 278.

Stout, Q.F. & Warren, B. 1984. Tree algorithms for unbiased coin tossing with a biased coin. *Annals of Probability* 12: 212-222.

Strange, C.H. 1949. *Nonconformity in Tunbridge Wells.* Tunbridge Wells: the Courier Printing and Publishing Co., Ltd.

Strong, J.V. 1976. The infinite ballot box of nature: De Morgan, Boole, and Jevons on probability and the logic of induction. pp.197-211 in Suppe & Asquith [1976].

—— 1978. John Stuart Mill, John Herschel, and the "Probability of Causes". *Proceedings of the 1978 biennial meeting of the Philosophy of Science Association.* vol. I, ed. P.D. Asquith and I. Hacking. East Lansing, Michigan: Philosophy of Science Association.

Struve, F.G.W. 1827. *Catalogus novus stellarum duplicium et multiplicium. Maxima ex parte in specula Universitatis Caesareae Dorpatensis par magnum telescopium achromaticum Fraunhoferi detectarum.* Dorpati, pp. xxxvii-xlviii.

Suppe, F. & Asquith, P.D. (Eds) 1976. *Proceedings of the 1976 biennial meeting of the Philosophy of Science Association.* vol. I. East Lansing, Michigan: Philosophy of Science Association.

Terrot, C.H. 1853. Summation of a compound series, and its application to a problem in probabilities. *Transactions of the Royal Society of Edinburgh* 20: 541-545.

Thomas, R. 1924. *Richard Price: Philosopher and Apostle of Liberty.* Oxford: Oxford University Press.

Timerding, H.E. (Ed.) 1908. *Versuch der Lösung eines Problems der Wahrscheinlichkeitsrechnung von Thomas Bayes.* Ostwalds Klassiker der Exakten Wissenschaften, Nr.169. Leipzig: Wilhelm Engelmann.

Timpson, T. 1859. *Church History of Kent: from the earliest period to MDCCCLVIII.* London: Ward & Co.

Todhunter, I. 1865. *A History of the Mathematical Theory of Probability: from the time of Pascal to that of Laplace.* Cambridge.

Trembley, Io. 1795-1798. De probabilitate causarum ab effectibus oriunda: disquisitio mathematica. *Commentationes Societatis Regiae Scientiarum Gottingensis, Commentationes Mathematica* 13: 64-119.

Turner, G.L. (Ed.) 1911. *Original Records of Early Nonconformity under Persecution and Indulgence.* London & Leipsic: T. Fisher Unwin.

Tweedie, C. 1922. *James Stirling: a sketch of his life and works along with his scientific correspondence.* Oxford: Clarendon Press.

van Dantzig, D. 1955. Laplace probabiliste et statisticien et ses précurseurs. *Archives Internationales d'Histoire des Sciences* 34: 27-37.

van Rooijen, J.P. 1942. De waarschijnlijkheidsrekening et het theorema van Bayes. *Euclides (Groningen)* 18: 177-199.

Venn, J. 1866. *The Logic of Chance.* London: Macmillan. Reprinted as the 4th edition 1962 (unaltered reprint of the 3rd edition of 1888). New York: Chelsea.

von Kries, J. 1927. *Die Principien der Wahrscheinlichkeits-Rechnung.* (2nd edition). Tübingen: J.C.B. Mohr (Paul Siebeck). (1st edition 1886.)

Walker, H.M. 1929. *Studies in the History of Statistical Method.* Baltimore: The Williams & Wilkins Company.

—— 1934. Abraham de Moivre. *Scripta Mathematica* 2: 316-333. Reprinted in de Moivre [1967].

Waller, J.F. (Ed.) 1865. *The Imperial Dictionary of Universal Biography.* London, Glasgow & Edinburgh: Wm. MacKenzie.

Walton, J.A. 1735. *Vindication of Sir Is. Newton's Fluxions.* London.

Ward, J. 1720. Letter to Thomas Bayes. (MS.6224) *Index of Manuscripts in the British Library.* Cambridge: Chadwyck-Healey.

—— 1740. *The Lives of the Professors of Gresham College: To which is prefixed the Life of the Founder Sir Thomas Gresham.* London. Reprinted 1967; New York and London: Johnson Reprint Corporation.

Welch, B.L. 1958. "Student" and small sample theory. *Journal of the American Statistical Association* 53: 777-788.

Westergaard, H. 1968. *Contributions to the History of Statistics.* New York: Agathon Press.

Whiston, W. 1749. *Memoirs of the Life and Writings of Mr. William Whiston. Containing Memoirs of Several of his Friends also.* 2nd edition, corrected, 1753. London: J. Whiston & B. White.

Whittaker, E.T. 1920. On some disputed questions of probability (with notes and discussion). *Transactions of the Faculty of Actuaries* 8, part 6: 163-206.

—— 1949. Laplace. *American Mathematical Monthly* 56: 369-372.

—— 1951. *A History of the Theories of Aether and Electricity * The Classical Theories.* London: Thomas Nelson and Sons. (1st edition 1910.)

Whittaker, E.T. & Watson, G.N. 1973. *A Course of Modern Analysis: an introduction to the general theory of infinite processes and of analytic functions; with an account of the principal transcendental functions.* 4th edition. Cambridge: Cambridge University Press. (1st edition 1902.)

Whitworth, W.A. 1878. *Choice and Chance, An Elementary Treatise on Permutations, Combinations, and Probability, with 300 Exercises.* 3rd edition. Cambridge: Deighton, Bell. (1st edition 1867.)

Wilbraham, H. 1854. On the theory of chances developed in Professor Boole's "Laws of Thought". *The Philosophical Magazine, Series 4, supplement to vol. VII.* Reprinted as Appendix B in Boole [1952], pp.473-486.

Wild, A. 1862. *Die Grundsätze der Wahrscheinlichkeits-Rechnung und ihre Anwendung.* München: E.A. Fleischmann.

Williamson, B. 1896. *An Elementary Treatise on the Integral Calculus, containing applications to plane curves and surfaces, and also a chapter on the calculus of variations, with numerous examples.* 7th edition. London: Longman's, Green & Co.

Wilson, E.B. 1922-1923. First and second laws of error. *Journal of the American Statistical Association* 18: 841-851.

—— 1934. Boole's challenge problem. *Journal of the American Statistical Association* 29: 301-304.

Wilson, W. 1814. *The History and Antiquities of Dissenting Churches and Meeting Houses, in London, Westminister, and Southwark; including the lives of their ministers, from the rise of Nonconformity to the Present time. With an appendix on the origin, progress, and present state of Christianity in Britain.* 4 vols. London: W. Button.

Winkler Prins Encyclopaedie. 1948. 6th impression. Amsterdam & Brussels: Elsevier.

Winsor, C.P. 1948. Probability and Listerism. *Human Biology* 20: 161-169.

Wishart, J. 1927. On the approximate quadrature of certain skew curves, with an account of the researches of Thomas Bayes. *Biometrika* 19: 1-38.

Zabell, S.L. 1982. W.E. Johnson's "sufficientness" postulate. *Annals of Statistics* 10: 1091-1099.

——— 1988a. Buffon, Price, and Laplace: scientific attribution in the 18th century. *Archive for History of Exact Sciences* 39: 173-181.

——— 1988b. The probabilistic analysis of testimony. *Journal of Statistical Planning and Inference* 20: 327-354.

——— 1988c. Symmetry and its discontents. pp.155-190 in *Causation, Chance, and Credence.* vol. I, ed. B. Skyrms & W.L. Harper. Norwell, Mass.: Kluwer Academic Publishers.

——— 1989a. R.A. Fisher on the history of inverse probability. *Statistical Science* 4: 247-263.

——— 1989b. The rule of succession. *Erkenntnis* 31: 283-321.

Index

References of primary importance are indicated by bold-faced numerals.